Ex-BR DIESELS IN INDUSTRY

**Full details of those British Rail diesel locomotives
of under 1,000hp which were sold for industrial service
and preservation - past and present - at home and abroad**

by
Adrian Booth

HANDBOOK 9BRD
INDUSTRIAL RAILWAY SOCIETY 2024

9BRD

ISBN 978 1 912995 18 9

Ninth Edition
© Industrial Railway Society
26 Great Bank Road, Herringthorpe,
Rotherham, South Yorkshire, S65 3BT

www.irsociety.co.uk

Front cover photograph:
ZM32 at the Steeple Grange Light Railway
on 8th September 2022.   (Adrian Booth)

Printed in the UK by Biddles Books Ltd

This book is copyright under the Berne Convention. All rights reserved. No portion of this publication may be reproduced, stored in a retrieval system, or transmitted in any form or by any means without prior permission in writing from the Industrial Railway Society. Within the UK, exceptions are allowed in respect of any fair dealing for the purpose of private research or private study or criticism or review, as permitted under the Copyright, Designs & Patents Act 1988.

CONTENTS:

INTRODUCTION:

Fifty-one years ago, in May 1973 to be precise, I attended a quarterly meeting of the IRS Committee in Birmingham. At the conclusion of the formal business we were all having a cup of tea when I was approached by Eric Tonks, the Society President, who asked if he could have a private word. He asked me if I would be prepared to take over from him as the Society's records officer for ex-BR diesels in industry (BRD). He explained what the job involved and said that I was his choice because I was already working as a member of the Society's team of records officers, plus I was known to be interested in diesels. I agreed to take on this additional job. By a 'truly remarkable coincidence' Eric just happened to have his hand-written BRD records in his briefcase, and these were promptly handed over to me together with a copy of the first (and very thin) edition of the BRD Handbook which the Society had published in 1972. Thereafter I soon made my own set of hand-written BRD records, which were later typed on my behalf. Much later I moved nervously into the modern age when I obtained my first PC – a giant step forward – and I duly computerised the BRD records. Over the years I compiled 2BRD in 1981, 3BRD in 1987, 4BRD in 1991, 5BRD in 2000, 6BRD in 2007, 7BRD in 2011 and 8BRD in 2019. Now here I am in 2024, fifty-one years after that chat with Eric Tonks, still maintaining the BRD records and now presenting 9BRD...which is a landmark issue: 9BRD will be the last to be published in book form, because the IRS Committee feels the way forward is for 10BRD onwards to be digital downloads.

Readers who are familiar with the earlier editions of BRD will now observe an expansion of the contents. This ninth edition contains a tremendous amount of additional information that I have inserted since publication of 8BRD. I have trawled through all my personal notebooks since 1966, and have incorporated tit-bits of information, dated sightings, plus details of locomotives' liveries, numbers and names. Any such personally-acquired information is prefixed 'seen' throughout this book's main tables and appendices. I have also looked through countless other publications and internet sites, searching for similar items of information, plus have inserted 'new' and 'withdrawn' dates for every locomotive. I have also endeavoured to standardise the text as regards locomotive owners' names and have eliminated the majority of abbreviations. The IRS archive holds all letters which have been submitted over the years to the Society's team of records officers; these letters for years 1969 to 1976 were looked through by John Cartwright and those containing ex-BR locomotive reports were scanned and forwarded to me. I have subsequently trawled through all the letters from 1967 to 2005. Brian Cuttell and Rob Cooke have also undertaken extensive research on the internet and in old magazines, sourcing photographs which have been used for dating and livery purposes. Sightings sourced in this way, and all second-hand information, are prefixed 'noted' throughout the main tables and appendices.

I would like to thank: Alex Betteney, John Cartwright, Rob Cooke, Brian Cuttell, and Peter Hall (for help and information); Peter Hall (for permission to refer to the RCTS website); Robert Pritchard (for photographs); Anthony Sayer (for Paxman information); Mark Jones, Peter Hall and Ivor Thomas (email forum); the BR database website, https://brdatabase.info (for information); various IRS members (for visit reports); John Wade (for 03129 information, and for being the leading light of the Heritage Shunters Trust); Martin Shill and Huw Williams (proofreading); and Andrew Smith (proofreading and production). The locomotive listings incorporate all information received by me up to 13th August 2024.

Adrian Booth, Rotherham, 14th August 2024   email: brlocos@irsociety.co.uk

FRONTISPIECE:

0-6-0 shunter number 11125 was delivered new to 32A Norwich Depot on 17th August 1955 and was later renumbered D2219. It was transferred from Norwich to 86C Hereford Depot during week-ending 15th August 1964, being one of five Class 04s transferred to Hereford towards the end of steam. All five were reallocated to 85A Worcester Depot during week-ending 29th November 1964, and then D2219 moved to 8H Birkenhead Depot during week-ending 12th November 1966. It was withdrawn from Birkenhead during week-ending 6th April 1968 and was sold to the Barnsley District Coking Co Ltd, arriving at the company's Barrow Coking Plant, near Barnsley, in October 1968. It is seen working at Barrow Coking Plant on 28th June 1969.

(photograph by Adrian Booth)

EXPLANATORY NOTES:

Layout

The book is divided into 28 sections, each devoted to one class, and brief details appropriate to each are given. The engines, etc, apply to locomotives as built, and not necessarily as when sold out of service. All are standard gauge unless otherwise stated. Sections 1 to 28 are followed by Appendices A to E. Each locomotive has eight columns of information, thus:

BR number	Builder	Works number	Year built	Last depot	Date wdn	P/F	Title

These will be simple to understand, but the following notes are pertinent:-

BR number

The 1957 numbering scheme (with the 'D' prefix) has been used to dictate the order of the classes in the book, although the later TOPS computer numbers (where allocated) are also given.

Builder

Many locomotives were built at British Railways own workshops at Ashford, Crewe, Darlington, Derby, Doncaster, Horwich and Swindon, and these are shown where appropriate. Locomotives which were constructed at private companies' workshops are indicated by standard IRS abbreviations, as under. In the case of Drewry locomotives two numbers are given: this company was basically a sales organisation, which did not erect locomotives but rather sub-contracted the construction work, in practice to RSH or VF. Both organisations' works numbers are given in these cases.

AB	Andrew Barclay, Sons & Co Ltd, Kilmarnock
CE	Clayton Equipment Co Ltd, Hatton, Derby
DC	Drewry Car Co Ltd, London
EE	English Electric Co Ltd, London
HC	Hudswell, Clarke & Co Ltd, Leeds
HE	The Hunslet Engine Co Ltd, Leeds
NB	North British Locomotive Co Ltd, Glasgow
RH	Ruston & Hornsby Ltd, Lincoln
RSHD	*Robert Stephenson & Hawthorns Ltd, Darlington
RSHN	*Robert Stephenson & Hawthorns Ltd, Newcastle upon Tyne
VF	*Vulcan Foundry Ltd, Newton-le-Willows, Lancashire
YE	Yorkshire Engine Co Ltd, Sheffield

*(latterly owned by English Electric)

Works number

This only applies to locomotives built by private companies. Those locomotives constructed at British Railways workshops were not allocated individual works numbers.

Year built

The year quoted is that in which the locomotive was officially added to British Railways stock, or the worksplate date if this is known to be different.

Last depot

During their 'main line' life, all the locomotives listed in this book were allocated to various British Railways motive power depots. Up to 6th May 1973 these depots were identified by an official code that comprised a number and letter; after this date a new system of two-letter codes was adopted. The list below shows all the depots and workshops from which locomotives listed in this book were withdrawn. The original and new codes for all of these depots are shown below although, in most cases, the older style code is used in the listings.

1A	WN	Willesden (London)	41J	SB	Shirebrook West
1E	BY	Bletchley	50B	--	Dairycoates (Hull)
2E	SY	Saltley (Birmingham)	50C	BG	Botanic Gardens (Hull)
2F	BS	Bescot (Walsall)	50D	GO	Goole
5A	CD	Crewe Diesel	51A	DN	Darlington
6A	CH	Chester	51L	TE	Thornaby (Middlesbrough)
6G	LJ	Llandudno Junction	52A	GD	Gateshead
8C	--	Speke Junction (Liverpool)	52B	HT	Heaton (Newcastle)
8F	SP	Springs Branch (Wigan)	--	TY	Tyne Yard (Gateshead)
8H	BC	Mollington Street	55A	HO	Holbeck (Leeds)
		(Birkenhead)	55B	YK	York
	BD	Birkenhead North	--	YC	York Clifton (from Oct. 1983)
8J	AN	Allerton (Liverpool)	55C	HM	Healey Mills (Wakefield)
9A	LO	Longsight (Manchester)	55F	HS	Hammerton Street
9D	NH	Newton Heath (Manchester)			(Bradford)
10D	LH	Lostock Hall (Preston)	55G	KY	Knottingley
12A	KD	Kingmoor Diesel (Carlisle)	55H	NL	Neville Hill (Leeds)
12B	CL	Upperby (Carlisle)	60A	IS	Inverness
12C	BW	Barrow	62A	TJ	Thornton Junction
15A	LR	Leicester Midland			(Kirkcaldy)
16A	TO	Toton (Nottingham)	62C	DT	Townhill (Dunfermline)
16B	--	Colwick (Nottingham)	64B	HA	Haymarket (Edinburgh)
16C	DY	Derby	64H	--	Leith Central (Edinburgh)
16F	BU	Burton upon Trent	--	EC	Craigentinny (Edinburgh)
30A	SF	Stratford (London)	65A	ED	Eastfield (Glasgow)
	TM	Temple Mills (sub of 30A)	66A	PO	Polmadie (Glasgow)
30E	CR	Colchester	66B	ML	Motherwell
31A	CA	Cambridge	67C	AY	Ayr
31B	MR	March	70D	EH	Eastleigh
32A	NR	Norwich Thorpe	70F	BM	Bournemouth West
	NC	Norwich Crown Point	70H	RY	Ryde (Isle of Wight)
34E	PB	New England	73C	HG	Hither Green (London)
		(Peterborough)	73F	AF	Chart Leacon (Ashford)
36A	DR	Doncaster	75A	BI	Brighton
36C	FH	Frodingham (Scunthorpe)	75C	SU	Selhurst (London)
40A	LN	Lincoln	81A	OC	Old Oak Common (London)
40B	IM	Immingham	81D	RG	Reading
41A	TI	Tinsley (Sheffield)	81F	OX	Oxford

82A	BR	Bath Road (Bristol)
82B	PM	St Philip's Marsh (Bristol)
82C	SW	Swindon
82D	WY	Westbury
83B	--	Taunton
84A	LA	Laira (Plymouth)
85A	WS	Worcester
85B	GL	Gloucester
86A	CF	Canton (Cardiff)
86E	ST	Severn Tunnel Junction
87B	MG	Margam
87E	LE	Landore (Swansea)
	ZC	Crewe Works
	ZF	Doncaster Works
	ZG	Eastleigh Works
	ZH	St Rollox (Glasgow) Works
	ZJ	Horwich Works
	ZI	Ilford Works
	ZL	Swindon Works
	ZN	Wolverton Works

The following are codes that have been adopted for this book only:-

BSD	Beeston Sleeper Depot, near Nottingham
CJ	Chesterton Junction Permanent Way Materials Depot, Cambridge
HHC	Hall Hills Creosoting Depot, Boston
LSD	Lowestoft Sleeper Depot, Suffolk
MQ	Meldon Quarry, near Okehampton, Devon
RSD	Reading Signal Depot

Date withdrawn

This is the official date of withdrawal from British Railways stock. Official BR information often quoted a locomotive being withdrawn during a period, as for example "7-12-86 to 3-1-87". In such cases the last date has been taken, and the example quoted would be shown as 1/87 in the following tables. As regards post-BR days, policy changed so that locomotives tended not to be allocated to specific depots, whilst the TOCs (Train Operating Companies) tended not to release official information regarding withdrawals. It has thus proved difficult to find information as regards withdrawal depots and dates, resulting in many locomotives not having this information shown.

P/F and Title

These columns indicate whether the stated number and/or name is the Present (P) or Former (F) one carried by the locomotive. NPT indicates 'No Present Title', meaning the locomotive presently carries neither name nor number. Former titles given for scrapped locomotives are those carried by the locomotive at the time it was cut up. NFT indicates 'No former title', meaning the locomotive carried neither name nor number at the time it was scrapped.

Notes

These follow on from the basic locomotive data. The first industrial user (or preservation society) is stated, together with the date the locomotive arrived on site. Locomotives were probably moved from their last main line depot unless otherwise stated, and in some cases despatch details are shown where positive information has been found. All known subsequent movements and ultimate disposals are given, with the dates when known. Certain abbreviations are used:

APCM	Associated Portland Cement Manufacturers Ltd	BREL	British Rail Engineering Ltd
		BSC	British Steel Corporation
BHES	Barrow Hill Engine Shed	BTC	British Transport Commission
BR	British Railways	CCD	Coal Concentration Depot

CEGB	Central Electricity Generating Board	NCB	National Coal Board
DBS	Deutsch Bahn Schenker	NCBOE	National Coal Board Opencast Executive
ECC	English China Clay	NSF	National Smokeless Fuels Ltd
EWS	English Welsh & Scottish Railway Ltd	OOU	Out of use
		PLC	Public Limited Company
GBRf	GB Railfreight Ltd	RAF	Royal Air Force
GNER	Great North Eastern Railway Ltd	RTC	Railway Technical Centre
HNRC	Harry Needle Railroad Company	SYRPS	South Yorkshire Railway Preservation Society
ICI	Imperial Chemical Industries Ltd		
L&NWR	London & North Western Railway		
MoD	Ministry of Defence		

The class 08, a design which can trace its origins back to the mid-1930s, has come to dominate the ex-BR diesels industrial scene, with a number still in service today. 08650 is seen at Foster Yeoman Quarries Ltd, Isle of Grain Stone Terminal, on 8th March 1992.

(Adrian Booth)

LOCOMOTIVE LISTINGS:

BR number	Builder	Works number	Year built	Last depot	Date wdn	P/F	Title

SECTION 1:

British Railways 0-6-0 diesel mechanical locomotives, numbered D2000-D2199, and D2370-D2399, built at BR's Doncaster and Swindon Works and introduced 1957. Used on every region of British Railways except the Scottish Region. Fitted with a Gardner 8L3 engine developing 204bhp at 1200rpm, a five speed gearbox, and driving wheels of 3ft 7in diameter. Later classified TOPS Class 03.

D2010   Swindon   1958   51L   11/74 F   03010
03010   new 12th February 1958; withdrawn, 10th November 1974; sold to Shipbreaking (Queenborough) Ltd; noted at Thornaby Depot, 22nd April 1976; despatched by rail from Tyne Yard to Stranraer, 18th May 1976; moved by road to Shipbreaking (Queenborough) Ltd, Cairnryan Port works; exported, May 1976 (see Appendix C).

D2012   Swindon   1958   31B   12/75 F   03012 / F135L
03012   new 12th February 1958; withdrawn, 15th December 1975; sold to A. King & Sons Ltd, Snailwell, Cambridgeshire, and moved 28th July 1976; noted working at Snailwell, 26th October 1976 and 20th July 1979; disused by February 1987; supplied parts for rebuild of 03020, January 1988; noted disused, Snailwell, 18th September 1988; remains scrapped, 28/29th January 1991.

D2018   Swindon   1958   32A   11/75 P   D2018
03018   new 14th April 1958; withdrawn, 9th November 1975; sold to George Cohen, Sons & Co Ltd, Cransley, Northamptonshire, and moved 29th April 1976; noted in BR blue livery with number painted over, Cransley, 2nd March 1980; to 600 Fragmentisers Ltd, Willesden, London, October 1980; noted in blue livery with number 600/2, Willesden, 9th July 1983; to Mayer Parry Ltd, Snailwell, Cambridgeshire, 17th July 1995; re-sold to Harry Needle Railroad Company; to South Yorkshire Railway Preservation Society, Meadowhall, Sheffield, 11th November 1998; to Lavender Line, Isfield, for storage, June 2001; re-sold to Mangapps Railway Museum, Burnham-on-Crouch, and moved 1st March 2004; noted in green livery with number D2018, Mangapps, 25th October 2015.

D2019   Swindon   1958   32A   7/71   F   1
new 6th May 1958; withdrawn, 25th July 1971; noted at Norwich Depot, 1st July 1972; sold to Shipbreaking (Queenborough) Ltd, Kent, and moved July 1972; noted at Queenborough, 31st July 1972 and 9th August 1972; exported from Sheerness Docks, August 1972 (see Appendix C).

D2020   Swindon   1958   32A   12/75 P   03020 / F134L
03020   new 6th May 1958; withdrawn, 3rd December 1975; sold to A. King & Sons Ltd, Snailwell, Cambridgeshire, and moved July 1976; noted disused, 26th October 1976; disused to 1987; noted in BR blue livery with numbers 03020 and F134L, Snailwell, 31st May 1987; rebuilt with parts from 03012, 1987; returned to use, November 1987; seen in faded BR blue livery with number 03020, Snailwell, 9th March 1992; re-sold to Harry Needle Railroad Company, about October 1995; to South Yorkshire Railway Preservation Society, Meadowhall, Sheffield, 9th November 1995; noted in blue livery with numbers 03020 and

F134L, Meadowhall, 27th April 1996; to Lavender Line, Isfield, for storage, 27th July 2001; to Sonic Rail Services Ltd, East Newlands, St Lawrence, Southminster, Essex, for restoration, 13th December 2010.

D2022 Swindon   1958   52A   11/82 P   03022
03022   new 27th May 1958; noted at Gateshead Depot, 11th February 1982; withdrawn, 7th November 1982; despatched from Gateshead, 7th January 1983; noted in BR blue livery with number 03022, Swindon Works, 2nd April 1983; sold to Swindon & Cricklade Railway; despatched from BREL Swindon Works, by road, 18th November 1983; to Swindon & Cricklade Railway; repainted and renumbered D2022; noted at Swindon & Cricklade Railway, 16th October 1985; to Coopers (Metals) Ltd, Gipsy Lane Works, Swindon, on hire (but serviced by Coopers), early October 1989; noted in green livery with cast 2022 numberplate, Coopers, 8th July 1993; returned to Swindon & Cricklade Railway, summer 1995; noted in green livery with number 2022, Swindon & Cricklade Railway, October 2009; noted in blue livery with number 03022, Swindon & Cricklade Railway, October 2019; to Somerset & Dorset Railway Heritage Trust, Midsomer Norton, on loan, 12th July 2022; to Swindon & Cricklade Railway, 25th October 2022.

D2023 Swindon   1958   40A   7/71   P   D2023
new 12th August 1958; withdrawn, 4th July 1971; sold to Tees & Hartlepool Port Authority, Middlesbrough Docks, and moved July 1972; seen in yellow livery, with T&HPA logo and number 5, with no BR number, Middlesbrough Docks, 15th June 1979; to T&HPA, Grangetown Docks, for storage, 15th September 1980; seen in yellow livery, Grangetown Docks, 15th May 1982 and 3rd August 1983; re-sold to Kent & East Sussex Railway, Tenterden, and moved 14th August 1983; noted with number 46 and named FAITH in 1988; noted in green livery, Kent & East Sussex Railway, Tenterden, 7th July 1994; noted in black livery with number D2023, Kent & East Sussex Railway, July 2018.

D2024 Swindon   1958   40A   7/71   P   D2024
new 19th August 1958; withdrawn, 4th July 1971; sold to Tees & Hartlepool Port Authority, Middlesbrough Docks, and moved July 1972; seen repainted yellow, with T&HPA logo and number 4, with no BR number, Middlesbrough Docks, 15th June 1979; to T&HPA, Grangetown Docks, 15th September 1980; seen in yellow livery, Grangetown Docks, 15th May 1982 and 3rd August 1983; re-sold to Kent & East Sussex Railway, Tenterden, and moved 4th September 1983; noted with number 7 in 1989; noted in yellow livery, Kent & East Sussex Railway, Tenterden, 7th July 1994; noted in green livery with number D2024, Kent & East Sussex Railway, 24th July 2018.

D2027 Swindon   1958   30E   1/76   P   NPT
03027   new 16th September 1958; withdrawn, 30th January 1976; sold to Shipbreaking (Queenborough) Ltd; despatched from BR Colchester Depot; noted in Hither Green Yard, 7th and 16th July 1976; moved in a four-locomotive convoy to Shipbreaking (Queenborough) Ltd, Kent, about 20th July 1976; noted at Queenborough, 30th October 1976; seen still in original BR blue livery with number 03027, two-way arrow emblem and CR shed-code sticker, Queenborough, 13th September 1979; noted in blue livery with no number, Queenborough, December 1983 and 28th February 1986; re-sold to South Yorkshire Railway Preservation Society, Meadowhall, Sheffield, and moved 9th February 1991; noted in blue livery with number 18, Meadowhall, 4th May 1992; to Knights of Old Ltd, Old, Northamptonshire, 28th September 1993; re-sold to Amber Valley Loco Group and moved to Peak Rail, Darley Dale, 2nd January 1997; re-sold to Heritage Shunters Trust, Rowsley, about 2001; seen with number D2027, Heritage Shunters Trust, Rowsley,

12th December 2003; noted in blue livery with no number, Heritage Shunters Trust, 14th August 2022.

D2032   Swindon   1958   32A   7/71   F   2
new 25th November 1958; withdrawn, 25th July 1971; sold to Shipbreaking (Queenborough) Ltd: noted at Norwich Depot, 1st July 1972; to Shipbreaking (Queenborough) Ltd, Kent, July 1972; noted at Queenborough, 31st July 1972 and 9th August 1972; exported from Sheerness Docks, August 1972 (see Appendix C).

D2033   Swindon   1958   32A   12/71   F   PROFILATINAVE 2
new 31st December 1958; withdrawn, 18th December 1971; sold to Shipbreaking (Queenborough) Ltd; noted at Norwich Depot, 1st July 1972; noted at Sittingbourne Goods Yard, 1st August 1972; to Shipbreaking (Queenborough) Ltd, Kent, August 1972; exported from Sheerness Docks, August 1972 (see Appendix C).

D2036   Swindon   1959   32A   12/71   F   PROFILATINAVE 1
new 9th February 1959; withdrawn, 18th December 1971; sold to Shipbreaking (Queenborough) Ltd; noted at Norwich Depot, 1st July 1972; to Shipbreaking (Queenborough) Ltd, Kent, July 1972; noted at Queenborough, 31st July 1972 and 9th August 1972; exported from Sheerness Docks, August 1972 (see Appendix C).

D2037   Swindon   1959   32A   9/76   P   NPT
03037   new 9th February 1959; withdrawn, 29th September 1976; sold to NCBOE; despatched from 32A Crown Point Depot, Norwich, October 1977; to Hargreaves Industrial Services Ltd, NCBOE British Oak Disposal Point, Crigglestone; seen in original BR blue livery, with 03037 number and two-way arrow logo, Crigglestone, 16th April 1979 and 11th June 1983; to NCBOE West Hallam Disposal Point, Mapperley, November 1983; noted at West Hallam Disposal Point, Mapperley, 7th December 1983; to NCBOE British Oak Disposal Point, Crigglestone, July 1984; seen at Crigglestone, 19th September 1984; seen with no wheels, being overhauled, Crigglestone, 12th March 1986; noted at Crigglestone, 7th February 1987; noted on low-loader, westbound on M62, 17th March 1987; noted back at Crigglestone, 28th March 1987; to NCBOE Oxcroft Disposal Point, Clowne, November 1988; noted at Oxcroft, 6th January 1989; re-sold to Harry Needle Railroad Company; to South Yorkshire Railway Preservation Society, Meadowhall, Sheffield, 18th July 1995; to Lavender Line, Isfield, for storage, about June 2001; to Heritage Shunters Trust, Rowsley, 18th May 2004; seen in black livery with no number, Heritage Shunters Trust, Rowsley, 9th March 2007; re-sold to private owner, near Burnham, Somerset, early 2012; to Foxfield Railway, Staffordshire, about 6th July 2012; to Royal Deeside Railway, Banchory, 16th October 2012; noted at Banchory, 16th September 2022.

D2041   Swindon   1959   75C   2/70   P   D2041
new 13th April 1959; withdrawn, 1st February 1970; sold to CEGB Richborough Power Station, and moved February 1970; to CEGB Rye House Power Station, Hoddesdon, Hertfordshire, March 1971; noted in yellow livery with no number and red lion emblem on cab-side, Rye House Power Station, April 1971; later allocated number 1; to CEGB Barking Power Station, about May 1971; to CEGB Rye House Power Station, Hoddesdon, August 1974; re-sold to Colne Valley Railway, Castle Hedingham, Essex, and moved 15th January 1981; noted with number 1 at Colne Valley Railway, 12th May 1981; noted in black livery with number D2041, Colne Valley Railway, April 2009; noted in green livery with number D2041, Colne Valley Railway, May 2015.

D2046   Doncaster   1958   51L   10/71   P   NPT
new 13th December 1958; rebuilt by The Hunslet Engine Co Ltd of Leeds (6644 of 1967) with flameproof equipment for long-term use at BP & Shell's Teesport Refinery, from late 1967; replaced by BP & Shell's locomotives, October 1971; withdrawn, 3rd October 1971; sold to Gulf Oil Co Ltd, Waterston, Milford Haven, and moved May 1972; given Gulf number 2; to BR Canton Depot, Cardiff, for tyre turning, 16th May 1975; returned to Gulf Oil Co Ltd, Waterston; noted at Waterston, 7th July 1980; to BR Canton Depot, Cardiff, for tyre turning, 5th November 1980; returned to Gulf Oil Co Ltd, Waterston; noted in blue livery, with no number, Waterston, 16th June 1994; to Petro-Plus, Waterston, Milford Haven, with site, 1998; re-sold to Beavor Power Ltd, Dowlais, Merthyr Tydfil, and moved January 2005; to Moveright International, Wishaw, about March 2006; noted in yellow livery with no number, Wishaw, 17th April 2007; re-sold to Plym Valley Railway, Marsh Mills, Plymouth, and moved on 28th May 2008; noted in yellow livery with no number, Marsh Mills, 18th October 2009; noted with no engine, under restoration, Marsh Mills, 23rd June 2013; noted at Marsh Mills, 4th September 2022; put on sale, October 2022; re-sold to Colne Valley Railway, and moved on 6th March 2023.

D2049   Doncaster   1958   50D   8/71   F   D2049
new 24th December 1958; withdrawn, 17th August 1971; sold to NCBOE; despatched from 50D Goole Depot, early May 1974; to Hargreaves Industrial Services Ltd, NCBOE Bowers Row Disposal Point, Astley, West Yorkshire, early May 1974; noted at Bowers Row, 12th May 1974; to NCBOE British Oak Disposal Point, Crigglestone, by 6th January 1975; seen with no number, Crigglestone, 31st July 1977; to NCBOE West Hallam Disposal Point, Mapperley, Derbyshire, March 1978; noted at Mapperley, 16th June 1979; seen in blue livery with no number, and with odometer reading of 53,574 miles, Mapperley, 25th October 1981; to NCBOE British Oak Disposal Point, Crigglestone, 28th January 1985; scrapped on site by Wath Skip Hire Ltd of Rotherham, November 1985.

D2051   Doncaster   1959   30E   12/72   P   D2051
new 27th January 1959; withdrawn, 14th December 1972; sold to Ford Motor Co Ltd; despatched from 30E Colchester Depot, June 1973; to Ford Motor Co Ltd, Dagenham; to BREL Swindon Works, for rebuild, 3rd October 1977; to Ford Motor Co Ltd, Dagenham, 23rd February 1978; renumbered 4, February 1978; noted with number 4, Ford, Dagenham, 19th August 1978; noted in blue livery with number 4, Ford, Dagenham, October 1988; re-sold to Rother Valley Railway, Robertsbridge, East Sussex, and moved 6th November 1997; re-sold to North Norfolk Railway, Sheringham, by 28th September 2000; noted in blue livery with number D2051, North Norfolk Railway, March 2014; re-sold to Telford Steam Railway, and moved 2nd November 2023.

D2054   Doncaster   1959   55B   11/72   F   CENTA
new 18th April 1959; withdrawn, 19th November 1972; noted at Sinfin Industrial Estate, Derby, 25th August 1973, 8th September 1973 and 24th September 1973; sold to Chair Centre Ltd, Derby, and moved October 1973; noted in St Mary's Yard, Derby, 10th July 1979; re-sold to British Industrial Sand Ltd, and moved to Middleton Towers, Norfolk, 20th July 1979; noted in white livery with no number, with camel logo on cab-side, Middleton Towers, 9th May 1981 and 24th August 1982; locomotive use ceased when shunting taken over by BR, late August 1982; to C.F. Booth Ltd, Rotherham, September 1982; scrapped, September 1982.

D2057   Doncaster   1959   51L   10/71   F   No.1
new 15th May 1959; rebuilt by The Hunslet Engine Co Ltd of Leeds (6645 of 1967) with flameproof equipment for long-term use on-hire at BP & Shell's Teesport Refinery, from late 1967; replaced by BP & Shell's own locomotives, October 1971; withdrawn, 3rd October 1971; sold to NCB; moved to Grimethorpe Colliery, Barnsley, 19th September 1972; noted at Grimethorpe Colliery, 10th January 1973; noted with number No.1, Grimethorpe Colliery, 24th March 1973; seen at Grimethorpe Colliery, 13th August 1978; noted in yellow livery with number No.1, Grimethorpe Colliery, 6th May 1979 and 17th September 1985; to C.F. Booth Ltd, Rotherham, 25th March 1986; seen in yellow livery, with number No.1, C.F. Booth Ltd, 27th March 1986 and 25th April 1986; scrapped, late April 1986.

D2059   Doncaster   1959   NC   7/87   P   NPT
03059   new 29th May 1959; withdrawn, 5th July 1987; despatched from BR Colchester Depot, 3rd November 1988; to Isle of Wight Steam Railway, Havenstreet, 4th November 1988; used as a shunter and on engineering trains; known to railway staff as 'Edward' but name not carried; seen in green livery with number D2059, Isle of Wight Steam Railway, 17th September 1992; to Island Line, on hire, 16th March 2002; returned to Isle of Wight Steam Railway, Havenstreet, 19th March 2002; noted in black livery with number D2059, Isle of Wight Steam Railway, October 2016; noted in black livery with no name or number, Isle of Wight Steam Railway, 23rd April 2024.

D2062   Doncaster   1959   32A   12/80   P   D.2062
03062   new 3rd July 1959; withdrawn, 12th December 1980; to BREL Swindon Works, 24th February 1981; despatched from BREL Swindon Works, 30th September 1982; to Dean Forest Railway, Lydney; noted in blue livery, Lydney, 18th July 1987; re-sold to East Lancashire Railway, Bury, and moved 13th October 1997; ran in BR blue with number 2062 for several years; to Metrolink, Manchester, on hire, (upgrade contract), 26th May 2007; returned to East Lancashire Railway, Bury, about 12th September 2007; noted in BR green livery with wasp stripes and number D2062, East Lancashire Railway, Bury, July 2011 and 6th May 2013; repainted in all-over green livery, 2015; noted in green livery with number D.2062, 16th April 2016.

D2063   Doncaster   1959   52A   11/87   P   D2063
03063   new 14th July 1959; withdrawn, 26th November 1987; put out to tender, August 1988; sold to Colne Valley Railway; despatched from BR Gateshead Depot; to Colne Valley Railway, Castle Hedingham, by road, 11th November 1988; to East Anglian Railway Museum, Wakes Colne, Essex, March 1995; re-sold to a member of North Norfolk Railway, Sheringham, and moved 23rd February 2000; noted in BR blue livery with number D2063, North Norfolk Railway, September 2007; to Mid-Norfolk Railway, Dereham, 7th November 2023; named PAUL A. MOBBS in a ceremony on 25th May 2024; locomotive bequeathed to Mid-Norfolk Railway upon Mobbs' death in 2024.

D2066   Doncaster   1959   52A   1/88   P   03066
03066   new 17th August 1959; withdrawn, 8th January 1988; despatched from BR Gateshead Depot, 28th June 1988; arrived at Horwich Foundry, Horwich, 30th June 1988; the locomotive passed to Horwich Foundry Ltd later in 1988; locomotive sold at auction; moved to South Yorkshire Railway Preservation Society, Meadowhall, Sheffield, 6th April 1991; re-sold to Barrow Hill Engine Shed Society, Staveley, 23rd March 1999; used as shed pilot; seen in blue livery with number 03066, Barrow Hill Engine Shed, 12th March 2022.

D2069   Doncaster   1959   52A   11/83 P D2069
03069   new 19th September 1959; stored, 27th November 1983; withdrawn, 18th December 1983; sold to The Vic Berry Company; despatched from BR Gateshead Depot; to The Vic Berry Company, Leicester, 4th January 1984; used as yard shunter at the famous scrapyard; noted at The Vic Berry Company, 9th February 1984; noted in blue livery with number 03069, The Vic Berry Company, February 1990; Berry closed after a yard fire in March 1991; 03 escaped the fire and was auctioned by Walker Walter Hanson of Nottingham, 1st August 1991; sold for £9,500; to the Gloucestershire Warwickshire Railway, Toddington, 2nd August 1991; to Wabtec, Doncaster, for open day, July 2003; returned to GWR Toddington, May 2004; noted in BR green livery with number D2069, GWR, October 2004; to Crewe Works for open day on 9th and 10th September 2005; returned to GWR Toddington, September 2005; noted on low-loader at Strensham North, M5 motorway (destination not known), 6th January 2006; returned to GWR Toddington by 28th January 2006; to Vale of Berkeley Railway, Sharpness, 7th December 2015; to Gloucestershire Steam & Vintage Rally, South Cerney, for display, early August 2018; noted in green livery with no number, South Cerney, 4th August 2018; returned to Vale of Berkeley Railway, Sharpness; to Gloucestershire Steam & Vintage Rally, South Cerney, for display, 2nd August 2019; noted in green livery with number D2069, South Cerney, 3rd August 2019; returned to Sharpness; to Dean Forest Railway, Lydney, for gala, 6th August 2019; returned to Sharpness; to Gwili Railway, 2020; to Dean Forest Railway, Lydney, 17th December 2020; to Chinnor & Princess Risborough Railway, 19th May 2022.

D2070   Doncaster   1959   51L   11/71 F D2070
new 2nd October 1959; withdrawn, 28th November 1971; sold to Shipbreaking (Queenborough) Ltd; despatched from BR Thornaby Depot; to Shipbreaking (Queenborough) Ltd, Kent, June 1972; noted at Queenborough on 11th July 1972; noted working, in green livery with no number, Queenborough, 14th February 1977; seen repainted in light blue livery with no number, Queenborough, 20th May 1978; noted at Queenborough, 4th May 1984; re-sold to South Yorkshire Railway Preservation Society, Meadowhall, Sheffield, and moved 5th October 1990; to Churnet Valley Railway, Cheddleton, 11th September 1993; re-sold to Cotswold Rail, RAF Quedgeley, near Gloucester, and moved about March 2001; used as source of spares; engine removed and exported to Egypt; frames went to Meyer Parry, Snailwell; remains noted at RAF Quedgeley, 2nd July 2001; remains scrapped, about July 2001.

D2072   Doncaster   1959   51A   3/81   P NPT
03072   new 31st October 1959; withdrawn, 8th March 1981; sold to Lakeside & Haverthwaite Railway; despatched from 51A Darlington Depot, August 1981; to Lakeside & Haverthwaite Railway, Cumbria; noted in blue livery with number 03072, Lakeside & Haverthwaite Railway, 1987; seen in green livery with number D2072, Lakeside & Haverthwaite Railway, 24th May 1996; noted in black livery with number D2072, Lakeside & Haverthwaite Railway, November 2008; noted in black livery with no number, Lakeside & Haverthwaite Railway, 15th July 2023.

D2073   Doncaster   1959   BD   5/89   P 03073
03073   new 5th November 1959; noted in BR blue livery with headboard 'Farewell to 03's's', Cavendish Sidings, Birkenhead (last time a BR shunter was used on Mersey Docks & Harbour Board tracks), 8th March 1989; to BR Chester for storage, 29th March 1989; withdrawn, 30th May 1989; sold to Heritage Centre, Crewe; despatched from BR Chester, 17th January 1991; stored at Basford Hall yard, Crewe, 17th January 1991; moved to Heritage Centre (later known as Railway Age), Crewe, 18th February 1991; displayed at

open day, Crewe Diesel Depot, 12th October 1991; noted at Heritage Centre, 2nd August 1992; to East Lancashire Railway, Bury, 1st October 1992; returned to Heritage Centre, Crewe, 8th October 1992; displayed at Crewe Railfair, Basford Hall Yard, 21st August 1994; noted in blue livery with number 03073, Crewe, March 1995; to Greater Manchester Metro Ltd, on hire, 19th August 1995; returned to Railway Age, Crewe, 16th September 1995; noted in blue livery, Heritage Centre, 30th March 1996; to Greater Manchester Metro Ltd, on hire, 26th September 1996; returned to Railway Age, Crewe, 8th October 1996; displayed at open day, Crewe Electric Depot, 3rd May 1997; to Greater Manchester Metro Ltd, on hire, 5th September 1997; returned to Railway Age, Crewe, 22nd September 1997; noted in blue livery with number 03073, Railway Age, March 1998; to Greater Manchester Metro Ltd, on hire, 12th March 1999; returned to Railway Age, Crewe, 30th March 1999; seen at Railway Age, 4th March 2000; to Greater Manchester Metro Ltd, on hire, 9th January 2001; returned to Railway Age, Crewe, by 3rd March 2001; to Crewe Works, (repainted in black livery), for open day, 3rd September 2005; displayed at open day on 9th and 10th September 2005; returned to Railway Age, Crewe, early 2006.

D2078   Doncaster   1959   52A   1/88   P   03078
03078   new 24th December 1959; withdrawn, 8th January 1988; sold to Stephenson Railway Museum, Chirton, near Newcastle upon Tyne, and moved 11th May 1988; seen in black livery with number D2078, Chirton, 12th April 2011; noted in black livery with number D2078 and 'North Tyneside Railway' on cab-side, Chirton, 1st June 2017; noted in blue livery with number 03078, Chirton, September 2019; to West Coast Railway Company, Carnforth, for tyre turning, 17th February 2021; returned to Chirton; to Embsay & Bolton Abbey Railway, on loan, 17th May 2022; to North Tyneside Railway, 21st November 2023.

D2079   Doncaster   1960   70H   6/96   P   03079
03079   new 7th January 1960; to Isle of Wight (as replacement for 05001) on road low-loader, 8th April 1984; rebuilt with cut-down cab, to enable it to pass through a low tunnel in Ryde; transferred to departmental stock and officially renumbered 97805 (although this number was never carried), 13th August 1984; withdrawn, 3rd June 1996; sold by Island Line, Isle of Wight (BR's successor) about May 1998 and left IoW, by road, 4th June 1998; arrived at Derwent Valley Light Railway Society, Murton, York, 6th June 1998; noted in BR blue livery with number 03079, Derwent Valley Light Railway, July 2012; renumbered D2054 for 1913-2013 centenary celebrations, 20th and 21st July 2013; later restored to number 03079; to North Tyneside Railway, for gala, 26th September 2019; stayed for repairs and repaint; returned to Derwent Valley Light Railway; to North Tyneside Railway, for gala, September 2020; returned to Derwent Valley Light Railway, 22nd September 2020; noted at DVLR, 20th April 2024.

D2081   Doncaster   1960   31B   10/80   P   03081 / LUCIE
03081   new 5th February 1960; withdrawn, 12th October 1980; moved to Swindon Works, 9th January 1981; sold to Sobermai; despatched by road from BREL Swindon Works, 13th November 1981; exported, November 1981 (see Appendix C); later returned to England; to Mangapps Railway Museum, Burnham-on-Crouch, 8th March 2004; noted in blue livery with number 03081, Mangapps, 30th April 2016.

D2084   Doncaster   1959   NC   7/87   P   D2084
03084   new 19th March 1959; withdrawn, 5th July 1987; moved to March Depot, for storage; privately purchased for preservation; despatched from 31B March Depot, 26th January 1992; to Knights of Old Ltd, Old, Northamptonshire, for storage; re-sold and moved to Peak Rail, Darley Dale, 4th January 1997; received name HELEN LOUISE; seen with

number 03084, Darley Dale, 12th December 2003; re-sold to Ecclesbourne Valley Railway, Wirksworth, 16th March 2005; to LH Group, Barton under Needwood, for repairs, 7th July 2006; noted in blue livery with number 03084 and name HELEN LOUISE, LH Group, March 2007; returned to Ecclesbourne Valley Railway, Wirksworth, 21st March 2007; to Lincolnshire Wolds Railway, Ludborough, 15th June 2009; re-sold to West Coast Railway Company, Carnforth, and moved 11th January 2011; to East Lancashire Railway, Bury, 3rd March 2016; stayed at Bury for gala held in mid-April 2016; noted in BR green livery with number D2084, East Lancashire Railway, Bury, 19th April 2016; to West Coast Railway Company, Carnforth, about May 2016; noted in BR green livery with number D2084, Carnforth, 21st January 2023; to Embsay & Bolton Abbey Railway, 23rd April 2024; put to work on engineering trains, early May 2024.

D2089   Doncaster   1960   NC   11/87 P   03089
03089   new 5th May 1960; withdrawn, 26th November 1987; sold to Nene Valley Railway; despatched from 31B March Depot, 2nd September 1988; to British Sugar Corporation Ltd, Woodston Factory, Peterborough, for storage; to Nene Valley Railway, Wansford, 31st August 1991; re-sold to Mangapps Railway Museum, Burnham-on-Crouch, and moved 3rd October 1991; noted in blue livery with number 03089, Mangapps, 30th April 2016; to Epping Ongar Railway, Essex, for gala held on 28/29th April 2018; returned to Mangapps Railway Museum, Burnham-on-Crouch.

D2090   Doncaster   1960   55B   7/76   P   D2090 / VIN
03090   new 5th May 1960; withdrawn, 18th July 1976; sold to National Railway Museum, York, and moved early August 1976; seen at NRM York, 7th August 1976; to National Railway Museum, Shildon, 10th June 2004; noted in blue livery with number 03090, NRM Shildon, June 2004; used as yard shunter; noted in green livery with number D2090, NRM Shildon, 14th October 2012; to Midland Road Depot, Leeds, for tyre turning, 13th November 2015; to NRM Shildon, 23rd November 2015; seen in green livery with number D2090 and 51F shedplate, NRM Shildon, 11th November 2023.

D2093   Doncaster   1960   51L   10/71 F   No.2
new 28th May 1960; rebuilt by The Hunslet Engine Co Ltd (6643 of 1967) with flameproof equipment for long-term use on-hire at BP & Shell's Teesport Refinery, from late 1967; replaced by BP & Shell's own locomotives, October 1971; withdrawn, 3rd October 1971; sold to NCB; moved to NCB Grimethorpe Colliery, Barnsley, 19th September 1972; noted at Grimethorpe Colliery, 10th January 1973; noted with number No.2, Grimethorpe Colliery, 24th March 1973; noted in yellow livery with number No.2, Grimethorpe Colliery, 6th May 1979; to C.F. Booth Ltd, Rotherham, for scrap, 26th March 1986; seen in yellow livery, with number No.2, C.F. Booth Ltd, 27th March 1986 and 11th April 1986; scrapped on 21st April 1986.

D2094   Doncaster   1960   52A   1/88   P   D2094
03094   new 28th June 1960; withdrawn, 8th January 1988; despatched from BR Gateshead Depot, 28th June 1988; arrived at Horwich Foundry, Horwich, 30th June 1988; the locomotive passed to Horwich Foundry Ltd later in 1988; sold by auction; to South Yorkshire Railway Preservation Society, Meadowhall, Sheffield, 6th April 1991; to Barrow Hill Engine Shed, Staveley, 23rd March 1999; noted in blue livery with number 03094, Barrow Hill Engine Shed, July 2002; re-sold to Cambrian Railway Trust, Llynclys, Oswestry, and moved 2nd June 2005; re-sold to Royal Deeside Railway, Banchory, and moved 13th May 2010; noted in green livery with number D2094, Royal Deeside Railway, Banchory, 30th September 2015 and 25th October 2020.

D2098   Doncaster   1960   51A   11/75   F   03098
03098   new 2nd July 1960; withdrawn, 8th November 1975; sold to Shipbreaking (Queenborough) Ltd; despatched by rail from Tyne Yard to Stranraer, 18th May 1976; moved by road to Shipbreaking (Queenborough) Ltd, Cairnryan Port yard; exported, May 1976 (see Appendix C).

D2099   Doncaster   1960   51L   2/76   P   03099
03099   new 15th July 1960; withdrawn, 10th February 1976; sold to NSF; noted at Thornaby Depot, 19th April 1976; to National Smokeless Fuels Ltd, Fishburn Coking Plant, County Durham, July 1976; noted in BR blue livery at Fishburn, 2nd August 1976; to Monkton Coking Plant, Hebburn, 26th February 1981; noted at Monkton Coking Plant, 14th March 1981; seen in yellow livery with no number, Monkton Coking Plant, 20th April 1987; seen in black livery with no number, Monkton Coking Plant, 16th April 1989; Monkton Coking Plant closed about December 1990; noted at Monkton Coking Plant, 15th February 1992; re-sold to South Yorkshire Railway Preservation Society, Meadowhall, Sheffield, and moved 23rd April 1992; to Exhibition of Steam and Speed, Doncaster, for display, July 2000; returned to South Yorkshire Railway Preservation Society, 17th July 2000; to Heritage Shunters Trust, Rowsley, March 2002; seen with number 03099, Rowsley, 12th December 2003.

D2112   Doncaster   1960   NC   7/87   P   D2112
03112   new 9th December 1960; withdrawn, 5th July 1987; sold to Nene Valley Railway; despatched from 31B March Depot, 2nd September 1988; to British Sugar Corporation Ltd, Woodston Factory, Peterborough, for storage; to Nene Valley Railway, Wansford, 31st August 1991; to Port of Boston, on hire, 20th January 1992; returned to Nene Valley Railway, 29th January 1992; seen in green livery with number D2112, Nene Valley Railway, 25th February 1993; to Port of Boston, on hire, 11th September 1998; to Nene Valley Railway, Wansford, 5th October 2006, to Port of Boston, on hire, 8th October 2006; re-sold to Harry Needle Railroad Company, June 2011; to Kent & East Sussex Railway, Tenterden, 9th November 2011; to Rother Valley Railway, Robertsbridge, East Sussex, 13th July 2012; noted in green livery with number D2112, Rother Valley Railway, July 2018 and 27th November 2022.

D2113   Doncaster   1960   55B   8/75   P   03113
03113   new 23rd December 1960; withdrawn, 3rd August 1975; sold to Gulf Oil Co Ltd; sent for repairs to BREL Doncaster Works, 30th April 1976; despatched from BREL Doncaster Works; to Gulf Oil Co Ltd, Waterston, Milford Haven, 17th May 1976; noted at Waterston, 7th July 1980; noted in use, July 1985; disused for several years until donated to Maritime & Heritage Museum, Milford Haven, and moved 17th October 1991; placed on static display on the dockside; noted in blue livery, Milford Haven, 16th June 1994; re-sold to Heritage Shunters Trust, Rowsley, and moved 16th November 2002; seen in blue livery with no number, with 'Milford Haven Oil Refinery' on engine panels, Heritage Shunters Trust, Rowsley, 12th December 2003 and 13th March 2004; restoration began in 2005; returned to working order, in BR blue livery and numbered 03113, 3rd September 2011; noted in blue livery with number 03113, 26th May 2019.

D2114   Swindon   1959   82C   5/68   F   D2114
new 2nd July 1959; withdrawn, 11th May 1968; sold to Bird's; despatched from BR Swindon Depot, August 1968; to Bird's Commercial Motors Ltd, Long Marston, Worcestershire; noted at Long Marston, 12th September 1969; scrapped, January 1975.

D2117   Swindon   1959   8F   10/71   P   NPT
new 4th September 1959; withdrawn, 9th October 1971; sold to Lakeside & Haverthwaite Railway; despatched from Springs Branch Depot, Wigan, and ran under its own power to Ulverston, 14th April 1972; completed journey by road to Lakeside & Haverthwaite Railway, Cumbria, 24th April 1972; noted in lined maroon livery, Lakeside & Haverthwaite Railway, 25th February 1973; seen at Lakeside & Haverthwaite Railway, 5th July 1975; seen in maroon livery, lettered L H R and No.8, Lakeside & Haverthwaite Railway, 18th May 1987; noted in red livery with no number, Lakeside & Haverthwaite Railway, November 2016; noted with no number, 15th July 2023.

D2118   Swindon   1959   12C   6/72   P   03118
new 17th September 1959; withdrawn, 18th June 1972; sold to Anglian Building Products Ltd, Atlas Works, Lenwade, Norfolk, and moved August 1973; noted in green livery with no number, Lenwade, 2nd October 1976 and 9th May 1981; BR rail connection lifted, November 1981; loco continued to work internally; site to Dowmac Concrete Ltd, circa March 1987; site to Costain Dowmac Ltd, circa July 1988; seen in blue livery with number D2118, Lenwade, 15th April 1993; to Costain Dowmac Ltd, Tallington Works, Lincolnshire, 7th October 1993; re-sold to Harry Needle Railroad Company; to South Yorkshire Railway Preservation Society, Meadowhall, Sheffield, 25th March 1996; re-sold to Rutland Railway Museum, Cottesmore, and moved to their site on 20th March 2001; to Peak Rail, Rowsley, 31st March 2005; noted in light blue livery with number D2118, Rowsley, 18th September 2005; seen with number D2118, Rowsley, 9th March 2007; re-sold to Great Central Railway (Nottingham), Ruddington, Nottingham, and moved 28th January 2011; noted in blue livery with number 03118 (which it never carried in BR service), Ruddington, July 2015.

D2119   Swindon   1959   87E   2/86   P   03119
03119 new 25th September 1959; rebuilt with cut down cab for working Burry Port & Gwendraeth Valley line; withdrawn, 23rd February 1986; noted at BR Landore Depot, 31st May 1986; sold to A.E. Knill Ltd; despatched from Landore Depot; to A.E. Knill Ltd, Barry, by 18th November 1986; noted at A.E. Knill Ltd, 4th December 1986; re-sold to Dean Forest Railway, Lydney, and moved by road on 19th December 1986; noted in blue livery, Dean Forest Railway, 18th July 1987; seen in green livery, with number 'D.2119' and name LINDA, Dean Forest Railway, 29th May 1990; to Rail & Maritime Engineering, Thingley Junction, Chippenham, for storage, February 1995; re-sold to West Somerset Railway, Minehead, and moved by road on 28th March 1996; re-sold to Epping Ongar Railway, Essex, and moved 3rd December 2011; noted in blue livery with number 03119, Epping Ongar Railway, 4th May 2024.

D2120   Swindon   1959   87E   2/86   P   D2120
03120 new 8th October 1959; rebuilt with cut down cab for working Burry Port & Gwendraeth Valley line; withdrawn, 21st February 1986; sold to Sir William McAlpine, The Fawley Hill Railway, near Henley-on-Thames, Buckinghamshire, and moved 18th December 1986; noted in green livery with number D2120, Fawley Hill, May 2012; to Didcot Railway Centre, Oxfordshire, for gala, 20th May 2015; returned to The Fawley Hill Railway, 28th May 2015; to Didcot Railway Centre, Oxfordshire, for gala, May 2016; returned to The Fawley Hill Railway, June 2016; noted in green livery with number D2120, The Fawley Hill Railway, 23rd June 2019 and 28th August 2023.

D2122   Swindon   1959   82A   11/72   F   D2122
new 30th October 1959; withdrawn, 26th November 1972; seen at BR Landore Depot, 22nd June 1973; despatched from BR Landore Depot; to Briton Ferry Steel Co Ltd, Glamorgan

(sold per R.E. Trem Ltd, Finningley, Doncaster), 13th August 1974; used for spares; noted (cut up into several sections) at Briton Ferry Steel Co Ltd, 24th August 1975 and 1st October 1976; these remains subsequently disposed of, date unknown.

D2123   Swindon   1959   82C   12/68   F   NFT

new 12th November 1959; withdrawn, 28th December 1968; noted at Swindon Depot, 1st June 1969; sold to Bird Group; despatched from 82C Swindon Depot, June 1969; to Bird's Commercial Motors Ltd, Long Marston, Worcestershire; noted at Long Marston, 28th June 1969; noted in BR green livery with number D2123, Long Marston, 30th March 1970; to Bird's (Swansea) Ltd, 40 Acre Site, Cardiff, April 1970; converted to a stationary generator with no wheels and re-painted yellow; noted in yellow livery with no number, Cardiff, 20th August 1970; to Derby Power Station, Full Street, Derby, by 8th November 1970, where used as a stationary generator on a contract to demolish the older part of the power station; to Bird's, Stapleton Road, Bristol, for storage, early June 1971; seen in yellow livery at Bird's, Stapleton Road, Bristol, 29th June 1971; used to power a baler; noted at Stapleton Road, Bristol, 18th August 1976; scrapped on site, November 1978.

D2125   Swindon   1959   82C   12/68   F   NFT

new 27th November 1959; noted stored at Swindon Depot, 1st December 1968; withdrawn, 28th December 1968; sold to Bird's; despatched from 82C Swindon Depot, June 1969; to Bird's Commercial Motors Ltd, Long Marston, Worcestershire; to Bird's (Swansea) Ltd, 40 Acre Site, Cardiff, late June 1969; noted at 40 Acre Site on several dates between 30th June 1969 and 1st August 1974; repainted yellow while at 40 Acre Site; scrapped, June 1976; 40 Acre Site closed in 1977.

D2128   Swindon   1960   82A   7/76   P   D2128

03128   new 7th January 1960; withdrawn, 4th July 1976; sold to Bird's Commercial Motors Ltd, Long Marston, and moved October 1976; re-sold for export, minus its engine; moved to Harwich Docks, 22nd December 1976; shipped shortly after from Harwich Docks (see Appendix C); returned to England and moved to Peak Rail, Darley Dale, 29th June 1993; repainted green, 1993; noted at Darley Dale, 17th October 1993; seen in green livery with no number, Darley Dale, 24th December 1994; seen in green livery with number D2128, Darley Dale, 7th December 1995; re-sold to Nottingham Sleeper Co Ltd, Elkesley, Retford, and moved 24th April 1996; stored on a short length of track; noted at Elkesley, 15th January 2000; fitted with a replacement Deutz engine, 2000; re-sold to Cotswold Rail, Fire Service College, Moreton in Marsh, and moved by road on 29th September 2000; noted at Moreton in Marsh, 4th January 2001; re-sold to Dean Forest Railway, Lydney Junction, about July 2002; noted at Lydney Junction, 20th August 2003; noted in light green livery with no number, 16th August 2007; re-sold to Andrew Briddon, about June 2008; to Blast Furnace Sidings, Corus, Scunthorpe, for storage, 8th July 2008; to Appleby Frodingham Railway Society, Scunthorpe, August 2008; noted in green livery with number D2128, Scunthorpe, August 2011; rebuilt as an 0-6-0 diesel-hydraulic with Cummins 350hp engine, and repainted black, September 2011; noted in black livery with no number, Scunthorpe, October 2014; renumbered 03901 by February 2015; to Andrew Briddon, Darley Dale, 8th September 2016; to Fox Productions Ltd, Longcross, Surrey, to assist in filming of 'Murder on the Orient Express', 16th December 2016; to Andrew Briddon, Darley Dale, 1st March 2017; to Somerset & Dorset Railway Heritage Trust, Midsomer Norton, initially for gala, 28th August 2019; noted in black livery with number 03901, Midsomer Norton, September 2019; noted in black livery with number D2128, Midsomer Norton, 31st October 2020; repainted in BR green livery with number D2128, Midsomer Norton, October 2021; to Andrew Briddon, Darley Dale, 13th July 2022.

D2129   Swindon   1960   41E   12/81   F   03129

03129   new 18th January 1960; withdrawn, 20th December 1981; coincidentally its last two BR shed allocations were Barrow and Barrow Hill; noted in BR blue livery, Barrow Hill Depot, 17th January 1982; despite being withdrawn, it worked coach rides for visitors at the depot open day, 3rd October 1982; sold to C.F. Booth Ltd, Rotherham, and moved 24th February 1983; it was earmarked to become yard shunter; initially placed in the non-ferrous sorting shed; used as yard shunter; quickly found unsuitable to negotiate the tight curves in the yard; decision reversed within a fortnight and thus it is possibly the shortest-lived ex-BR shunter to have worked in industry; seen in BR blue livery with number 03129, C.F. Booth Ltd, 8th, 12th and 15th March 1983; scrapped on 16th March 1983.

03129 at C.F.Booth Ltd on 12th March 1983.   *(photograph by Adrian Booth)*

D2132   Swindon   1960   82C   6/69   F   D2132 / LESLEY

new 12th February 1960; withdrawn, 18th May 1969; sold to NCB, May 1970; to NCB Bestwood Colliery, Nottinghamshire, October 1970; noted at Bestwood Colliery, 27th March 1971; to New Hucknall Colliery, Huthwaite, July 1971; named LESLEY; noted at New Hucknall Colliery, 18th October 1971; to Pye Hill Colliery, Jacksdale, 25th March 1982; noted in green livery, Pye Hill Colliery, 28th March 1982; seen in green livery with number D2132 and name LESLEY, Pye Hill Colliery, 15th July 1983; despatched on low-loader to C.F. Booth Ltd, Rotherham, for scrap, 27th November 1984; scrapped, 28th November 1984; Pye Hill Colliery closed 9th August 1985.

D2133   Swindon   1960   82A   7/69   P   D2133

new 23rd February 1960; withdrawn, 6th July 1969; sold to British Cellophane Ltd, Bridgwater, Somerset, and moved July 1969; noted in blue livery with no number, Bridgwater, March 1970; went under own power to Bristol (Bath Road) Depot, for an engine overhaul, 1979; returned to Bridgwater; seen in blue livery with number BCL D2133, Bridgwater, 17th September 1982; re-sold to West Somerset Railway, Minehead, and moved by road on 10th July 1996; noted in BR green livery with number D2133, West Somerset Railway, August 2008, June 2015 and May 2022.

D2134 Swindon 1960 82A 7/76 P D2134

03134 new 29th February 1960; withdrawn, 4th July 1976; sold to Bird's Commercial Motors Ltd, Long Marston, and moved October 1976; re-sold for export, minus its engine; noted southbound on A45, on a low-loader, 11th January 1977; shipped from Harwich Docks, January 1977 (see Appendix C); returned to England and moved to South Yorkshire Railway Preservation Society, Meadowhall, Sheffield, 27th April 1995; noted in blue and white livery and hand-written number D2134, Meadowhall, 27th April 1996; re-sold to Royal Deeside Railway, Banchory; moved to Royal Deeside Railway, 12th December 2000; noted in black livery with number D2134, Royal Deeside Railway, 16th September 2022.

D2135 Swindon 1960 30E 1/76 F 03135

03135 new 21st March 1960; withdrawn, 30th January 1976; sold to Shipbreaking (Queenborough) Ltd; noted in a four-locomotive convoy, Hither Green Yard, 7th July 1976; to Shipbreaking (Queenborough) Ltd, Kent; arrived at Queenborough on 16th July 1976; noted with no engine, Queenborough, 30th October 1976; noted dismantled, (frame/wheels/half its cab), Queenborough, 16th February 1977 and 27th March 1977; not seen on a visit to Queenborough, 15th August 1977; believed to have been used as a source of spares and then scrapped at Queenborough, 1977.

D2138 Swindon 1960 82C 5/69 P D2138

new 5th April 1960; withdrawn, 18th May 1969; sold to NCB; moved to Bestwood Colliery, Nottinghamshire, 23rd October 1970; to Pye Hill Colliery, Jacksdale, April 1971; seen at Pye Hill Colliery, 9th May 1971; seen in original BR green livery with 82C and 'D.2138' on cabside, Pye Hill Colliery, 7th October 1977; to BR Toton Depot, under own power, for tyre turning, June 1978; noted at Toton Depot, 10th and 18th June 1978; returned to Pye Hill Colliery, Jacksdale, under own power, 19th June 1978; seen at Pye Hill Colliery, 5th July 1983; noted in green livery, Pye Hill Colliery, 4th June 1985; Pye Hill Colliery closed 9th August 1985; re-sold to Midland Railway, Butterley, Derbyshire, and moved under its own power, 20th August 1985; noted at Butterley, 10th September 1985; noted in green livery with number D2138, Butterley, July 2016; to Andrew Briddon, Darley Dale, by 25th May 2024.

D2139 Swindon 1960 85A 5/68 P D.2139

new 11th April 1960; withdrawn, 11th May 1968; seen in BR green livery with number D2139, with odometer reading of 47,315 miles, Worcester Depot, 26th July 1968; sold to A.R. Adams & Son; despatched from 85A Worcester Depot, 10th December 1968; to A.R. Adams & Son, Newport; used as hire locomotive (see Appendix A); re-sold to NSF; moved to NSF Coed Ely Coking Plant, Tonyrefail (received number 1), and moved by 30th March 1971; noted at Coed Ely, 30th March 1971; to BR Canton Depot, Cardiff, for repairs, 29th September 1977; displayed at open day, Canton Depot, 1st October 1977; noted at Canton Depot (with engine removed), 24th October 1977; noted at Canton Depot, 3rd November 1977 and 19th March 1978; to Coed Ely Coking Plant, Tonyrefail, 1978; to BREL Swindon Works, for repairs, 31st March 1981; noted at BREL Swindon Works, 19th November 1981; to Coed Ely Coking Plant, Tonyrefail, 2nd March 1982; to Monkton Coking Plant, Hebburn, December 1983; seen in green livery with large number 1 on cabside, Monkton Coking Plant, 20th April 1987; withdrawn in 1987; Monkton Coking Plant closed about December 1990; noted at Monkton Coking Plant, 15th February 1992; re-sold to South Yorkshire Railway Preservation Society, Meadowhall, Sheffield, where it arrived on 24th April 1992;

to Heritage Shunters Trust, Rowsley, March 2002; restored in green livery and returned to working order in 2008; briefly carried fictional number D2000; seen in green livery with number D.2139 and 85A shedplate, Rowsley, 24th July 2021.

D2141   Swindon   1960   87E   7/85   P   03141
03141   new 4th May 1960; rebuilt with cut down cab for working Burry Port & Gwendraeth Valley line; withdrawn, 15th July 1985; sold to C.F. Booth Ltd of Rotherham and despatched by rail, 17th October 1985; ran hot and detached at Severn Tunnel Junction; noted (with no coupling rods) at Severn Tunnel Junction, 3rd November 1985; sale to C.F. Booth Ltd cancelled; noted at Severn Tunnel Junction Depot, 30th June 1986; re-sold to White Wagtail Ltd, c/o DeMulder & Sons, Gun Range Farm, Earl Shilton, near Coventry, and moved by road on 30th June 1986; noted in a plant and equipment storage yard at White Wagtail Ltd, 7th April 1987; re-sold to Cotswold Rail, Fire Service College, Moreton in Marsh, and moved September 2000; noted in BR blue livery with number 03141, Moreton in Marsh, 4th January 2001; to Cotswold Rail, RAF Quedgeley, near Gloucester, March 2001; noted at Quedgeley, 2nd July 2001; re-sold to Dean Forest Railway, Lydney Junction, spring 2002; noted at Lydney Junction, 20th August 2003; to Swansea Vale Railway, 19th June 2005; re-sold to Pontypool & Blaenavon Railway, and moved 29th April 2008; noted in tatty blue livery with cab boarded-up, Pontypool & Blaenavon Railway, 29th October 2016; re-sold to Gwendraeth Valley Railway, Mudlescwm, Kidwelly, and moved about March 2020; noted at Mudlescwm, October 2021.

D2144   Swindon   1961   87E   2/86   P   03144
03144   new 19th January 1961; rebuilt with cut down cab for working Burry Port & Gwendraeth Valley line; repainted green at Landore shed and displayed at open day, October 1985; withdrawn, 21st February 1986; sold to MoD; despatched from BR Landore Depot; to MoD Long Marston, Worcestershire, 24th March 1987; received the name WESTERN WAGGONER; to MoD Grantham Barracks, for display, June 1992; to MoD Bicester, 26th June 1992; to Yorkshire Engine Company, Long Marston, 11th December 1995; to 275 Squadron, MoD Bicester, 5th September 1996; to Wensleydale Railway, Leeming Bar, on long-term loan, by 25th January 2004; used in a MoD track re-laying exercise (officially 'Exercise Turnout 20') by 507 Specialist Team Royal Engineers, at Wensleydale Railway for two weeks in September 2020; noted in BR blue livery with number 03144, Wensleydale Railway, 18th May 2023.

D2145   Swindon   1961   87E   7/85   P   03145
03145   new 12th February 1961; rebuilt with cut down cab for working Burry Port & Gwendraeth Valley line; seen at Landore Depot, 7th April 1985; withdrawn, 15th July 1985; sold to C.F. Booth Ltd of Rotherham and despatched by rail, 17th October 1985; after a few days at Severn Tunnel Junction, it moved to Gloucester, where it ran hot on 23rd October 1985; noted at BR Gloucester (Horton Road) Depot, 12th February 1986, 11th May 1986 and 30th June 1986; noted at Gloucester Old Yard, 1st July 1986; the sale to C.F. Booth Ltd was cancelled; despatched from BR Gloucester Old Yard, on a road low-loader, 11am on 2nd July 1986; to White Wagtail Ltd, c/o DeMulder & Sons, Gun Range Farm, Earl Shilton, near Coventry; noted in a plant and equipment storage yard at White Wagtail Ltd, 7th April 1987; re-sold to Cotswold Rail, Fire Service College, Moreton in Marsh, September 2000; noted at Moreton in Marsh, 4th January 2001 and 2nd July 2001; re-sold to D2578 Locomotive Group, Moreton on Lugg, and moved 6th August 2001; noted in blue livery with number 03145, Moreton on Lugg, 8th October 2017; to Dean Forest Railway, for gala, 31st August 2023; noted in use, 9th and 10th September 2023; to Moreton on Lugg, 14th September 2023.

D2146   Swindon   1961   87E   9/68   F   D2146
new 14th February 1961; withdrawn, 21st September 1968; sold to Bird's Commercial
Motors Ltd; despatched from 82C Swindon Depot, late June 1969; arrived at Bird's
Commercial Motors Ltd, Long Marston, Worcestershire, 3rd July 1969; noted in BR green
livery at Long Marston, 14th September 1969; noted at MoD Long Marston, 2nd July 1971;
used in an Army training exercise, MoD Long Marston (a photograph exists of it being lifted
off the track by Army steam crane number 62003), 16th October 1971; returned to Bird's,
Long Marston; noted at Long Marston, early November 1971; noted in BR green livery with
number D2146, Long Marston, 1st September 1972 and 13th July 1973; noted being used
for spares, 13th July 1973; engine removed by March 1978; noted dismantled on 29th July
1978; remains scrapped, about August 1978.

D2148   Swindon   1960   55C   11/72   P   D2148
new 24th May 1960; withdrawn, 19th November 1972; sold to NCBOE; despatched from
BR Healey Mills Depot, August 1973; to Hargreaves (West Riding) Ltd, NCBOE Bowers
Row Disposal Point, Astley, Yorkshire, August 1973; to Lindley Plant Ltd, NCBOE Gatewen
Disposal Point, Denbighshire, September 1973; to NCBOE Bowers Row Disposal Point,
December 1973; noted at Bowers Row, 19th February 1975; seen in BR blue livery, Bowers
Row, 31st July 1977; noted repainted orange, Bowers Row, 5th May 1980; suffered
collision damage, late April 1983; seen stored in the yard (in orange livery with no number)
with severe damage to the rear of its cab, 11th June 1983; a replacement cab (from 03149)
was purchased from BREL Doncaster Works and fitted on site at Bowers Row, January
1984; noted with orange body and blue cab, Bowers Row, 2nd May 1984; noted in orange
livery with no number, Bowers Row, 13th March 1987; re-sold to Steamport Transport
Museum, Southport, and moved by road on 14th March 1987; Steamport closed in 1998;
re-sold to Ribble Steam Railway, Preston, and moved on 17th April 1999; cosmetically
restored and became a static exhibit; re-entered service in 2004; to EWS Crewe Electric
Depot, for tyre turning, 23rd January 2009; to Ribble Steam Railway, Preston, 15th
February 2009; noted in green livery with number D2148 and 27C shed-plate, Ribble Steam
Railway, 23rd April 2022.

D2150   Swindon   1960   55B   11/72   F   NFT
new 10th June 1960; withdrawn, 19th November 1972; sold to British Salt Ltd; despatched
from York Depot, working 9Z10 under own power, at 18:20 on Friday 11th May 1973
(Freight Advice number 1616); to British Salt Ltd, Middlewich, Cheshire, 11th May 1973;
seen in green livery with no number, British Salt Ltd, 22nd July 1983; noted in badly
corroded condition, British Salt Ltd, 5th April 1993 and 17th January 1997; initially acquired
for preservation by Staffordshire Locomotives and moved to J. & H. Parry & Sons
(Shawbury), The Oaks, Shawbury Heath, Shawbury, Shropshire, for storage, about April
2000; moved to Allelys, Studley, Warwickshire, about September 2000; used as a source
of spares; engine removed and exported to Egypt; remains scrapped on site at Studley,
about July 2001.

D2152   Swindon   1960   87E   10/83   P   D2152
03152 new 22nd June 1960; rebuilt with cut down cab for working Burry Port &
Gwendraeth Valley line; withdrawn, 2nd October 1983; moved to Gloucester, 18th May
1984; moved to Swindon Works, 30th May 1984; despatched from BREL Swindon Works,
by road, 6th March 1986; to Swindon & Cricklade Railway, Blunsden; re-sold to Swindon
Railway Workshop Ltd, April 1988; used as works shunter; displayed at Membury Services
West (on the M4) from about April 1990; returned to Swindon Railway Workshop Ltd,

November 1990; re-sold to Swindon & Cricklade Railway, and moved June 1992; noted in green livery with number D2152, Swindon & Cricklade Railway, May 2019.

D2153   Swindon   1960   51L   11/75   F   03153
03153   new 27th June 1960; withdrawn, 8th November 1975; noted at Thornaby Depot, 22nd April 1976; sold to Shipbreaking (Queenborough) Ltd; despatched by rail from Tyne Yard to Stranraer, 18th May 1976; moved by road to Shipbreaking (Queenborough) Ltd, Cairnryan Port yard; exported, May 1976 (see Appendix C).

D2156   Swindon   1960   52A   11/75   F   03156
03156   new 8th August 1960; withdrawn, 22nd November 1975; noted at Gateshead Depot on 19th April 1976; sold to Shipbreaking (Queenborough) Ltd; despatched by rail from Tyne Yard to Stranraer, 18th May 1976; moved by road to Shipbreaking (Queenborough) Ltd, Cairnryan Port yard; exported, May 1976 (see Appendix C).

D2157   Swindon   1960   50C   12/75   F   03157
03157   new 16th August 1960; withdrawn, 29th December 1975; sold to Shipbreaking (Queenborough) Ltd; noted in a four-locomotive convoy, Hither Green Yard, 7th July 1976; to Shipbreaking (Queenborough) Ltd, Kent; arrived at Queenborough on 16th July 1976; noted with no rods, Queenborough, 14th February 1977; exported from Sheerness Docks, March 1977 (see Appendix C).

D2158   Swindon   1960   NC   7/87   P   D2158 / MARGARET
03158   **ANN**
new 22nd August 1960; withdrawn, 5th July 1987; moved to March Depot, 3rd August 1987; despatched from March Depot, 25th January 1992; to Knights of Old Ltd, Old, Northamptonshire, for storage; sold to Peak Rail, Darley Dale, and moved 3rd January 1997; seen with number 03158 and name MARGARET ANN, Darley Dale, 12th December 2003; re-sold to Ecclesbourne Valley Railway, Wirksworth, and moved 7th September 2004; to Lincolnshire Wolds Railway, Ludborough, 16th June 2009; re-sold to Great Central Railway (Nottingham), Ruddington, Nottingham, and moved 12th August 2009; re-sold to Titley Junction Station, near Kington, Herefordshire, and moved 3rd July 2014; noted in green livery with number D2158, Titley Junction Station, May 2017; to Mangapps Railway Museum, Burnham-on-Crouch, 29th October 2019.

D2162   Swindon   1960   BD   5/89   P   03162
03162   new 21st September 1960; last worked at Birkenhead on 9th March 1989; sold to Wirral Borough Council and moved for storage to BR Chester Depot, 30th March 1989; officially withdrawn, 23rd May 1989; despatched from BR Chester Depot to Llangollen Railway, on long-term loan, October 1989; noted in blue livery with number 03162, Llangollen Railway, March 2005 and June 2016; withdrawn for major overhaul, 2022; re-wheeled on 27th January 2024.

D2164   Swindon   1960   30A   1/76   F   ?
03164   new 7th October 1960; withdrawn, 23rd January 1976; sold to Shipbreaking (Queenborough) Ltd; despatched from Stratford Depot, 5th July 1976; noted in a four-locomotive convoy, Hither Green Yard, 7th July 1976; to Shipbreaking (Queenborough) Ltd, Kent; arrived at Queenborough on 16th July 1976; noted in BR blue livery with number 03164, and no rods, Queenborough, 14th February 1977; exported from Sheerness Docks, March 1977 (see Appendix C).

D2170 Swindon 1960 BD 4/89 P D2170
03170 new 21st November 1960; last worked at Birkenhead Docks on 9th March 1989; to BR Chester Depot, for storage, 30th March 1989; withdrawn, period-beginning 24th April 1989; to BR Longsight Depot, for repairs, 27th April 1989; sold to Otis Euro Transrail Ltd, Salford, and moved 9th May 1989; seen in BR blue with number 03170, Otis Euro Transrail Ltd, 17th November 1990; re-sold to Harry Needle Railroad Company, December 1999; to Fragonset, Derby, for certification, April 2000; returned to Otis Euro Transrail Ltd, on hire; to Barrow Hill Engine Shed, Staveley, 28th September 2000; to Battlefield Line, Shackerstone, 13th August 2001; noted in blue livery with number 03170, Battlefield Line, May 2007; re-sold to Epping Ongar Railway, Essex, and moved 17th September 2010; noted in blue livery with number 03170, Epping Ongar Railway, 14th September 2019; noted in green livery with number D2170, Epping Ongar Railway, 4th May 2024.

D2176 Swindon 1961 ZC 5/68 F D2176
new 15th December 1961; withdrawn, 11th May 1968; sold to George Cohen, Sons & Co Ltd, Cransley, Northamptonshire, and moved October 1968; used as yard shunter; noted at Cransley, 28th September 1969; noted in green livery with number D2176, Cransley, 21st June 1970; noted at Cransley, 30th October 1971; scrapped, November 1971.

D2178 Swindon 1962 81F 9/69 P D2178
new 9th January 1962; withdrawn, 13th September 1969; a BR letter dated 3rd November 1969 offered D2178 for sale at £2,500, delivered to nearest station; sold to A.R. Adams & Son; despatched from 81F Oxford Depot, January 1970; to A.R. Adams & Son, Newport, Monmouthshire; used as a hire locomotive (see Appendix A); re-sold to National Smokeless Fuels Ltd, Coed Ely Coking Plant, Tonyrefail, May 1974; given number 2; noted at Coed Ely, 18th June 1974 and 20th April 1975; to BREL Swindon Works, for repairs, by July 1979; noted at BREL Swindon Works, 28th July 1979, 12th August 1979 and 30th October 1979; noted at BR Severn Tunnel Junction Depot (on way back to Coed Ely Coking Plant, Tonyrefail), 1st November 1979; noted in blue livery with number 2, Coed Ely, 22nd April 1984; all track lifted at Coed Ely, July-August 1985; noted disused at Coed Ely, 15th August 1985; sold to Caerphilly Railway Preservation Society, Caerphilly, and moved 12th November 1985; noted in blue livery, Caerphilly, 28th May 1986; seen in blue livery with number 2, Caerphilly, 5th April 1991; site closed in 1996; re-sold to Gwili Railway, Bronwydd Arms, and moved 21st October 1996; to St Philip's Marsh Depot, Bristol, for tyre turning, 7th December 2011; returned to Gwili Railway, Bronwydd Arms, 14th December 2011; noted in green livery with number D2178, Gwili Railway, 7th July 2016 and 22nd September 2021.

D2179 Swindon 1962 NC 7/87 P 03179 / CLIVE
03179 new 18th January 1962; withdrawn (in Network South East livery), 5th July 1987; moved to Isle of Wight, by ferry, 30th June 1988; reinstated to departmental stock, August 1988; rebuilt with cut-down cab, to enable it to pass through a low tunnel in Ryde; repainted in NSE colours, December 1988; withdrawn, 29th October 1993; spent five years in store; sold to West Anglia & Great Northern Company, Electric Maintenance Depot, Hornsey; left Isle of Wight, by ferry, 8th June 1998; arrived at Hornsey, 12th June 1998; ran to King's Cross Station, under own power, where named CLIVE after a depot employee of 43 years standing; to Nene Valley Railway, Wansford, for gala, 29th February 2008; noted in purple livery, Wansford, 2nd March 2008; returned to Hornsey, early March 2008; noted in white, blue and red livery with number 03179, Hornsey, October 2009; re-sold to Rushden Historical Transport Society, Rushden, Northamptonshire, and moved on 19th July 2016; noted in First Capital Connect purple livery with number 03179, Rushden, 22nd July 2016.

D2180   Swindon   1962   NC   3/84   P   03180
03180   new 1st February 1962; withdrawn, 31st March 1984; sold to Mayer Newman Ltd; despatched from BR Norwich Crown Point Depot; to Mayer Newman Ltd, Snailwell, Cambridgeshire, 26th July 1984; noted in blue livery, Snailwell, 30th September 1986; still working in January 1988; later used as a source of spares for 03012 and 03020; re-sold to South Yorkshire Railway Preservation Society, Meadowhall, Sheffield, South Yorkshire, where it arrived on 21st December 1991; placed in store, with no engine; to Battlefield Line, Shackerstone, for storage, 2nd August 2001; noted in blue livery with number 03180, Battlefield Line, April 2008; privately purchased in 2011; to Heritage Shunters Trust, Rowsley, 14th November 2011; stored unserviceable, needing new engine; a major overhaul commenced in 2022.

D2181   Swindon   1962   87E   5/68   F   PRIDE OF GWENT
new 21st February 1962; to Worcester Depot, for storage, 17th July 1967; withdrawn, 11th May 1968; seen in BR green livery with number D2181, with odometer reading of 28,375 miles, Worcester Depot, 26th July 1968; noted at Worcester Depot, 29th October 1968; sold to A.R. Adams & Son; despatched from 85A Worcester Depot, 10th December 1968; to A.R. Adams & Son, Newport, Monmouthshire; used as a hire locomotive (see Appendix A); re-sold to Gwent Coal Distribution Centre, Newport, Monmouthshire, and noted there on 23rd January 1970; noted at Gwent Coal Distribution Centre, by now in red livery, 17th August 1970; noted at Gwent Coal Distribution Centre, painted red with name PRIDE OF GWENT, 21st December 1972; noted at BR Ebbw Junction Depot, for repairs, 1977; returned to Gwent Coal Distribution Centre; replaced by Hudswell Clarke D1186, December 1986; re-sold to Marple & Gillott Ltd, Attercliffe, Sheffield, for scrap, December 1986; seen in red livery with number D2181, being cut up, Marple & Gillott Ltd, 29th January 1987.

D2182   Swindon   1962   87E   5/68   P   D2182
new 5th March 1962; to Worcester Depot, for storage, 17th July 1967; withdrawn, 11th May 1968; seen at Worcester Depot, 26th July 1968; noted at Worcester Depot, 29th October 1968; sold to A.R. Adams & Son; despatched from 85A Worcester Depot, 29th November 1968; to A.R. Adams & Son, Newport, Monmouthshire; used as a hire locomotive (see Appendix A); re-sold to Lindley Plant Ltd, Gatewen Disposal Point, Denbighshire, and moved there in September 1973; noted with number 3/3, Gatewen, 7th August 1980; to NCBOE Bennerley Disposal Point, December 1981; to NCBOE Wentworth Stores, Harley, near Rotherham, 18th March 1982; seen with number 3/3, in olive green livery with no BR number, Wentworth Stores, 12th February 1983 and 3rd April 1983; to NCBOE Bennerley Disposal Point, Ilkeston, 6th May 1983; noted with number 3/3, Bennerley, 2nd June 1983; to NCBOE Coalfield Farm Disposal Point, Hugglescote, Leicestershire, July 1983; noted at Hugglescote, 17th August 1985; re-sold to Warwick District Council, Victoria Park, Leamington Spa, and moved 20th April 1986; noted with number 3/3, Victoria Park, 26th April 1986; noted in green livery, Victoria Park, 4th October 1986; re-sold to Gloucestershire Warwickshire Railway, Toddington, and moved 11th January 1993; seen at GWR Toddington, 20th April 1994; noted in green livery with number D2182, Gloucestershire Warwickshire Railway, May 2009, 16th January 2017, and 27th September 2019.

D2184   Swindon   1962   87E   12/68   P   D2184
new 9th April 1962; to Worcester Depot, for storage, 24th November 1968; withdrawn, 28th December 1968; noted at Worcester Depot, 4th May 1969; sold to Co-operative; despatched from Worcester Depot, 20th August 1969; to Co-operative Wholesale Society Ltd, Coal Concentration Depot, Southend-on-Sea, Essex; noted at CWS Southend, 7th

October 1969; noted in green livery with number D2184, CWS Southend, 30th April 1979 and 25th May 1980; re-sold to Colne Valley Railway, Castle Hedingham, Essex, and moved 17th October 1986; noted in green livery with number D2184, Colne Valley Railway, 13th July 1988; noted in black livery with number D2184, Colne Valley Railway, March 2010, and 30th August 2015.

D2185   Swindon   1962   85A   12/68   F   NFT
new 3rd May 1962; withdrawn, 28th December 1968; sold to Bird's; despatched from 85A Worcester Depot, 21st May 1969; to Bird's Commercial Motors Ltd, Long Marston, Worcestershire; noted in BR green livery with number D2185, Long Marston, 31st May 1969; noted at Long Marston, 12th September 1970; to Abercarn Tinplate Works, April 1971; noted repainted in yellow livery, with BR number painted over, Abercarn, 9th October 1971; to Bird's (Swansea) Ltd, 40 Acre Site, Cardiff, by 16th July 1972; noted at 40 Acre Site, 22nd July 1973; seen at 40 Acre Site, 25th January 1974; moved to Bird's new site at Tremorfa, Cardiff, autumn 1977; scrapped, about June 1978.

D2186   Swindon   1962   81F   9/69   F   D2186
new 23rd May 1962; to store, about March 1969; reinstated, May 1969; hired to a private firm for track lifting contract between Thame and Morris Cowley; contract completed by 12th September 1969; withdrawn, 13th September 1969; a BR letter dated 3rd November 1969 offered D2186 for sale at £2,500, delivered to nearest station; sold to A.R. Adams & Son; despatched from 81F Oxford Depot, 6th February 1970; arrived at A.R. Adams & Son, Newport, Monmouthshire, 8th February 1970; retained its BR livery and number whilst with Adams; used as a hire locomotive (see Appendix A); scrapped, January 1981.

D2187   Swindon   1962   82C   5/68   F   NFT
new 8th March 1962; to Worcester Depot, for storage, 28th November 1967; withdrawn, 11th May 1968; sold to Bird's Commercial Motors Ltd, Long Marston, Worcestershire, and moved September 1968; noted working at Long Marston, 14th September 1969; noted at Long Marston (repainted in yellow livery, with BR number painted over), 25th January 1970; noted on 7th October 1972; noted being used for spares, 13th July 1973; noted dismantled on 29th July 1978; remains scrapped, September 1978.

D2188   Swindon   1961   83B   5/68   F   D2188
new 20th March 1961; withdrawn, 11th May 1968; sold to Bird's; despatched from 82C Swindon Depot, September 1968; to Bird's Commercial Motors Ltd, Long Marston, Worcestershire; noted at Long Marston, 14th September 1969; noted in BR green livery with number D2188, Long Marston, 30th March 1970 and 13th July 1973; scrapped, February 1978.

D2189   Swindon   1961   6A   3/86   P   NPT
03189   new 31st March 1961; withdrawn, 16th March 1986; moved to March Depot, for storage, 27th July 1987; put up for sale in 1991; sold to Steamport; despatched from 31B March Depot; arrived at Steamport, Southport, 18th December 1991; Steamport closed in 1998; re-sold to Ribble Steam Railway, Preston, and moved 17th April 1999; fitted with replacement engine, 2006; noted in green undercoat with no number, Ribble Steam Railway, 23rd April 2022.

D2192   Swindon   1961   82C   1/69   P   D2192 / TITAN
new 5th May 1961; withdrawn, 25th January 1969; sold to Dart Valley Railway, Devon, and moved 25th August 1970; seen at Dart Valley Railway, 21st June 1973; to Torbay & Dartmouth Railway, Paignton, 24th July 1977; noted in blue livery, named ARDENT,

Torbay & Dartmouth Railway, 22nd April 1984; to South Devon Railway Trust, Buckfastleigh, mid-1991; noted in blue livery, South Devon Railway, 25th July 1991; to Dartmouth Steam Railway, Devon, after 25th July 1991; noted in green livery with number D2192, 16th October 1993; noted in black livery with number D2192 and name TITAN, Dartmouth Steam Railway, 14th May 2012 and September 2013.

D2193   Swindon   1961   82C   1/69   F   2
new 15th May 1961; withdrawn, 25th January 1969; noted at Worcester Depot, 4th May 1969; sold to A.R. Adams & Son; despatched from Worcester Depot, September 1969; to A.R. Adams & Son, Newport, Monmouthshire; used as a hire locomotive (see Appendix A); scrapped, January 1981.

D2194   Swindon   1961   85A   9/68   F   D2194
new 23rd May 1961; noted at Hereford, 13th May 1968; seen at Worcester Depot, 26th July 1968; withdrawn, 21st September 1968; noted at Worcester Depot, 2nd February 1969; sold to Bird's; despatched from Worcester Depot, May 1969; to Bird's Commercial Motors Ltd, Long Marston, Worcestershire; noted at Long Marston, 14th September 1969; noted in BR green livery with number D2194, Long Marston, 30th March 1970 and 13th July 1973; noted dismantled on 29th July 1978; scrapped, about August 1978.

D2195   Swindon   1961   82A   9/68   F   D10
new 5th June 1961; withdrawn, 21st September 1968; noted at Worcester Depot, 12th April 1969 and 4th May 1969; sold to Llanelly Steel; despatched from Worcester Depot; to Llanelly Steel Co Ltd, Carmarthenshire (sold per R.E. Trem Ltd, Finningley, Doncaster), June 1969; noted at Llanelly Steel, 5th July 1969; noted with number D2195, Llanelly Steel, 4th October 1969; seen in green livery with number D10, Llanelly Steel, 10th June 1980; scrapped, September 1981.

D2196   Swindon   1961   8H   6/83   P   NPT
03196   new 8th June 1961; withdrawn, 12th June 1983; sold to R.O. Hodgson Ltd, Carnforth, Lancashire, and moved 15th June 1983; noted at R.O. Hodgson Ltd, 23rd October 1983; to Steamtown, Carnforth, on hire, autumn 1991; seen in BR blue livery with number 03196, with name JOYCE on bonnet top, Carnforth, 2nd October 1993; re-sold to Steamtown, July 1996; noted in blue livery with number 03196, Carnforth, July 2008; noted in green livery with no number, Steamtown, 21st January 2023; to Aysgarth Station, Aysgarth, 24th May 2023.

D2197   Swindon   1961   NC   6/87   P   03197
03197   new 15th June 1961; withdrawn, 26th June 1987; to Leicester Depot, for storage, 20th July 1987; put out to tender, August 1988; sold to Harry Needle Railroad Company, 1989; remained at Leicester Depot where used for spares; moved to private location, August 1991; to Lavender Line, Isfield, for storage, 31st July 2001; re-sold to Mangapps Railway Museum, Burnham-on-Crouch, 2010; to Sonic Rail Ltd, Burnham-on-Crouch, Essex, for overhaul, 13th December 2010; to Mangapps Railway Museum, 24th August 2012; noted in blue livery with number 03197, Mangapps Railway Museum, August 2016; to Isle of Wight Railway, 25th September 2016; to Mid-Norfolk Railway, Dereham, 3rd October 2018; to Mangapps Railway Museum, August 2019.

D2199   Swindon   1961   12C   6/72   P   D2199
new 29th June 1961; withdrawn, 18th June 1972; sold to NCB; to BREL Doncaster Works, for overhaul and fitting with air brakes, summer 1973; seen at BREL Doncaster Works, 16th February 1974; moved to Rockingham Colliery, Birdwell, Barnsley, February 1974; noted

at Rockingham Colliery, 7th August 1974; seen repainted in dark blue livery with no BR number and lettered ROCKINGHAM COLLIERY 1, Rockingham Colliery, 27th March 1978 and 8th October 1978; to Barrow Colliery, Worsborough, Barnsley, early January 1979; seen at Barrow Colliery, 16th April 1979; to Houghton Main Colliery, Barnsley, about June 1979; noted at Houghton Main Colliery, 14th September 1979; to Royston Drift Mine, Barnsley, 14th August 1980; to Barrow Colliery, Worsborough, 8th July 1981; to Royston Drift Mine, 23rd March 1982; seen in dark blue livery with no BR number and lettered ROCKINGHAM COLLIERY 1, at Royston, 27th August 1983; moved by low-loader into non rail connected Royston Machinery Stores, September 1986; seen at Royston Machinery Stores, 15th November 1986; put up for sale in early 1987; re-sold to South Yorkshire Railway Preservation Society, Attercliffe, Sheffield, and moved on 12th August 1987; to South Yorkshire Railway Preservation Society, Meadowhall, Sheffield, 12th September 1988; restored to working order, in green livery, late 1988; displayed in green livery with number D2199, BR Tinsley Depot, Sheffield, open day, 29th September 1990; to RMS Locotec Ltd, for use at Eurotunnel, Cheriton Terminal, on hire, 14th March 1997; returned to South Yorkshire Railway Preservation Society, Meadowhall, Sheffield, 12th September 1997; to RMS Locotec Ltd, Dewsbury, on hire, about January 2001; returned to South Yorkshire Railway Preservation Society, 2001; to Hanson Quarry Products, Machen, near Newport, on hire, 5th February 2001; to Heritage Shunters Trust, Rowsley, 6th April 2006; seen in BR green livery with number D2199, Heritage Shunters Trust, Rowsley, 9th March 2007 and 3rd September 2017.

D2371   Swindon   1958   52A   11/87   P   03371
03371   new as Departmental No.92 working at Chesterton Junction, 31st December, 1958; to capital stock and renumbered D2371, 30th July 1967; withdrawn, 26th November 1987; put out to tender, August 1988; moved to Tyneside Central Freight Depot, Gateshead, for storage; left Tyneside Central Freight Depot, by low-loader, 27th October 1988; briefly stored in haulier's yard; arrived at A.J. Wilkinson, Rowden Mill Station, near Bromyard, Worcestershire, 10th November 1988; restored; re-sold to Dartmouth Steam Railway, Devon, and moved there on 2nd February 2015; noted in green livery with number D2371, Dartmouth Steam Railway, July 2016; noted in blue livery with number 03371, Dartmouth Steam Railway, 26th January 2019 and 14th October 2023.

D2373   Swindon   1961   9D   5/68   F   No.1 / DAWN
new 2nd August 1961; withdrawn, 11th May 1968; noted stored at Bolton Depot, 31st July 1968; sold to NCB; despatched from Bolton Depot, September 1968; to NCB Manvers Main Coal Preparation Plant, Wath upon Dearne, Rotherham; this was the first of many ex-BR shunters at Manvers Main; noted named JIM, Manvers Main, 1st December 1968; seen named JIM at Manvers Main workshops, 9th March 1969; had undergone a sex change when noted renamed DAWN, Manvers Main, 1st November 1970; seen in yellow livery, with 'D.2373', white rose emblem No.1, and DAWN on cabside, Manvers Main, 27th November 1976; put into store, June 1981; scrapped on site by Ernest Nortcliffe & Son Ltd of Parkgate, Rotherham, March 1982.

D2381   Swindon   1961   16C   6/72   P   D2381
new 27th October 1961; withdrawn, 18th June 1972; to BREL Derby Works, early March 1973; noted at BREL Derby Works, 4th March 1973; overhauled and repainted; noted at Etches Park Depot, Derby, 9th April 1973; to Flying Scotsman Enterprises, Market Overton, Rutland, by rail, 13th April 1973; to Steamtown Railway Museum, Carnforth, 19th March 1976; seen at Steamtown, 24th April 1976; seen in dark green livery with 03381 numberplate (a number it never carried on BR), Carnforth, 2nd October 1993; Steamtown

closed after the 1997 season and site became West Coast Railway Company; noted in green livery with number D2381 painted on cabside and 03381 on radiator, 1st April 1994; noted in green livery with number D2381 (OOU for many years in steam shed), Carnforth, 15th April 2023.

D2397   Doncaster   1961   NC   7/87   F   03397
03397   new 15th September 1961; withdrawn, 5th July 1987; moved to March Depot, 27th July 1987; sold to The Vic Berry Company; despatched from March Depot; to Leicester Stabling Point, 16th March 1989; tripped to The Vic Berry Company, Leicester, 23rd March 1989; noted in blue livery with number 03397, The Vic Berry Company, 9th April 1989; used for spares; remains scrapped, January 1991.

D2399   Doncaster   1961   NC   7/87   P   03399
03399   new 21st October 1961; withdrawn, 5th July 1987; moved to March Depot, 22nd July 1987; sold to Mangapps Railway Museum; despatched from March Depot, by road, 22nd March 1989; to Mangapps Railway Museum, Burnham-on-Crouch; noted in blue livery with number 03399, Mangapps, June 2012 and August 2016; to Isle of Wight Railway, Havenstreet, for a gala, 26th September 2018; returned to Mangapps Railway Museum; noted in blue livery with number 03399, Mangapps, 29th June 2019.

SECTION 2:

Drewry Car Co Ltd 0-6-0 diesel mechanical locomotives built by Vulcan Foundry Ltd, numbered D2200-D2214, and introduced 1952. Fitted with a Gardner 8L3 engine developing 204bhp at 1200rpm, five speed gearbox, and driving wheels of 3ft 3in diameter. Later classified TOPS Class 04.

D2203   DC   2400   1952   ZC   12/67   P   NPT
         VF   D145
delivered new (number 11103) to March Depot, 27th June 1952; to the Wisbech and Upwell Tramway, 19th July 1952; fitted with side skirting, front and rear cowcatchers and a speed governor; quickly transferred to Yarmouth Docks tramway system; withdrawn from Crewe Works, 16th December 1967; sold to Hemel Hempstead Lightweight Concrete Co Ltd, Cupid Green, Hertfordshire, and moved 12th March 1968; noted working BLS rail tour along branch, 10th May 1969; noted in red livery with no number, Cupid Green, 7th February 1971; noted working LCGB rail tours along branch, 12th May 1973 and 14th June 1975; connection to BR severed in 1979; re-sold to Embsay & Bolton Abbey Railway, and moved 8th February 1982; seen in green livery with number D2203, Embsay & Bolton Abbey Railway, 9th October 1999 and 20th July 2008; noted in black livery with no number, Embsay & Bolton Abbey Railway, 9th September 2022; put on sale for £11,000, 27th February 2023; to Mangapps Railway Museum, Burnham-on-Crouch, 29th March 2023.

D2204   DC   2485   1953   55F   10/69   F   D5
         VF   D211
new as number 11105, 6th March 1953; withdrawn, 26th October 1969; sold to Briton Ferry Steel Co Ltd, Glamorgan, and moved March 1970 (sold via W. & F. Smith Ltd, Ecclesfield, Sheffield); noted in yellow livery, Briton Ferry, 23rd May 1970; initially given number D11 but later renumbered D5; noted dismantled in 1974; remains scrapped, September 1979.

D2205	DC	2486	1953	51L	7/69	P	D2205
	VF	D212					

new as number 11106, 16th March 1953; withdrawn, 7th July 1969; sold to Tees & Hartlepool Port Authority, Middlesbrough Docks, and moved July 1970; initially numbered MD1; seen repainted yellow, with T&HPA logo and number 1 on cabside, with no BR number, Middlesbrough Docks, 15th June 1979; to T&HPA, Grangetown Docks, September 1980; seen in yellow livery, Grangetown Docks, 15th May 1982; noted at Grangetown Docks, 17th March 1983; re-sold to Kent & East Sussex Railway, Tenterden, and moved there on 21st August 1983; carried fictional number 11223; re-sold to West Somerset Railway, Minehead, and moved on 18th November 1989; noted in black livery with number 11223, West Somerset Railway, Minehead, 14th July 1991; to Somerset & Avon Railway, Radstock, 2nd February 1994; to West Somerset Railway, Minehead, July 1996; re-sold to Heritage Shunters Trust, Rowsley, and moved 14th October 2012; returned to working order, in BR green livery, in 2016; noted in green livery with number D2205, Rowsley, 25th May 2019.

D2207	DC	2482	1953	ZC	12/67	P	D2207
	VF	D208					

new as number 11108, 23rd February 1953; withdrawn from Crewe Works, 16th December 1967; moved to Crewe South Depot by 26th December 1967; sold to Hemel Hempstead Lightweight Concrete Co Ltd, Cupid Green, Hertfordshire, and moved 1st February 1968; noted in red livery with no number, Cupid Green, 7th February 1971; re-sold to North Yorkshire Moors Railway and moved there in September 1973; seen in green livery with number D2207, NYMR Goathland, 1st June 1991; seen working breakdown train, same livery, NYMR Grosmont, 7th December 1991; seen, same livery, NYMR Levisham, 11th June 1994; to RMS Locotec Ltd, Dewsbury, for overhaul, 22nd February 2005; to RMS Locotec Ltd, Wakefield, for further repairs, 29th June 2006; to North Yorkshire Moors Railway, 31st January 2007; noted in green livery with number D2207, NYMR, 30th December 2018; usually works as pilot at carriage and wagon works, Pickering.

D2208	DC	2483	1953	5A	7/68	F	D2208
	VF	D209					

new as number 11109, 23rd February 1953; withdrawn, 27th July 1968; sold to NCB; despatched from Crewe South Depot, 10th November 1968; to NCB Manvers Main Coal Preparation Plant, Wath upon Dearne, Rotherham; noted at Manvers Main, 15th December 1968; to Cortonwood Colliery, Wombwell, March 1969; noted at Cortonwood Colliery, 30th March 1969; to Cadeby Colliery, Conisbrough, April 1969; noted repainted with no number, Cadeby Colliery, 4th May 1969; to Silverwood Colliery, Thrybergh, September 1970; noted at Silverwood Colliery, 4th February 1971; noted dismantled to just frame, wheels and cab, Silverwood Colliery, 27th July 1975; these remains cut up by a scrap dealer from Worksop, August 1978; cab and several other parts noted in a scrapyard adjacent to the BR stabling point, Worksop, 6th May 1979.

D2209	DC	2484	1953	8J	7/68	F	No.16 / TRACEY
	VF	D210					

new as number 11110, 2nd March 1953; withdrawn, 27th July 1968; sold to NCB; despatched from Allerton Depot, 10th November 1968; to NCB Manvers Main Coal Preparation Plant, Wath upon Dearne, Rotherham; noted at Manvers Main CPP, 15th December 1968; seen repainted in rich green livery and named ERNEST, Manvers Main CPP, 9th March 1969 and 2nd May 1970; had undergone a sex change when noted renamed TRACEY, Manvers Main CPP, 1st November 1970; to Manvers Coking Plant, 11th September 1972; to Manvers CPP, 20th September 1972; to Manvers Coking Plant, 30th April 1973; to Manvers CPP, 15th May 1973; to Manvers Coking Plant, 10th July 1973; to Manvers CPP, 19th September 1973; to Kiveton Park Colliery, 13th July 1974; seen working, in turquoise livery, with white rose emblem No.16, Kiveton Park Colliery, 12th February 1977; noted disused, 4th October 1980; used for spares from September 1981; reduced to frame, wheels and cab, 17th October 1982; wheels transferred to D2328, March 1983; remains scrapped on site by Brinsworth Metals Ltd, 19th August 1985.

D2211	DC	2509	1954	16C	7/70	F	WILF CLEMENT
	VF	D243					

new as number 11112, 24th September 1954; noted at Derby Depot, 29th September 1968; withdrawn, 19th July 1970; sold to Powell Duffryn Fuels Ltd; despatched from BR Derby Depot, August 1970; delivered to Carmarthen Station; then tripped to Powell Duffryn Fuels Ltd, NCBOE Coed Bach Disposal Point, Kidwelly, August 1970; named WILF CLEMENT; noted at Kidwelly, 16th July 1978; re-sold to Rees Industries Ltd, Saron Works, Pencoed, Llanelli, and moved there on 3rd August, 1978; scrapped, about November 1980.

D2213	DC	2529	1954	8H	8/68	F	D2213
	VF	D257					

new as number 11114, 10th October 1954; withdrawn, 17th August 1968; sold to NCB; moved to Manvers Main Coal Preparation Plant, Wath upon Dearne, Rotherham, September 1969; seen at Manvers Main, 2nd May 1970; used for spares, by July 1975; noted in green livery with number D2213, with no engine, dumped at Manvers Workshops, 7th July 1975; seen at Manvers Workshops, 27th November 1976; remains scrapped, February 1978.

SECTION 3:

Drewry Car Co Ltd 0-6-0 diesel mechanical locomotives built by Vulcan Foundry Ltd and Robert Stephenson & Hawthorns Ltd, numbered D2215-D2273, and introduced 1955. Fitted with a Gardner 8L3 engine developing 204bhp at 1200rpm, five speed gearbox, and driving wheels of 3ft 6in diameter. Later classified TOPS Class 04.

D2216	DC	2539	1955	30A	5/71	F	3
	VF	D265					

new as number 11122, 5th July 1955; withdrawn, 9th May 1971; sold to Shipbreaking (Queenborough) Ltd, Kent, and moved June 1972; noted at Queenborough 11th July 1972 and 9th August 1972; exported from Sheerness Docks, August 1972 (see Appendix C).

D2219	DC	2542	1955	8H	4/68	F	D2219
	VF	D268					

new as number 11125, 17th August 1955; withdrawn, 6th April 1968; sold to Barnsley District Coking Co Ltd; despatched from Crewe South Depot, by rail, October 1968; to Barnsley District Coking Co Ltd, Barrow Coking Plant, Barnsley, October 1968; seen in BR green livery with number D2219, Barrow Coking Plant, 28th June 1969; to NCB Barrow Colliery, Barnsley, on loan, August 1969; returned to Barrow Coking Plant; scrapped on site by Geeson's of Pentrich, May 1977.

D2225   DC   2548   1955   8F   3/69   F   D2225 / DEBRA
**   VF   D274**

new as number 11131, 18th October 1955; withdrawn, 15th March 1969; seen at 8F
Springs Branch Depot, 23rd August 1969; sold to NCB; moved to Manvers Main Coal
Preparation Plant, Wath upon Dearne, Rotherham, January 1970; seen at Manvers Main
CPP, 2nd May 1970; seen with name DEBRA, Manvers Main CPP, 1st November 1970; to
Manvers Coking Plant, 5th June 1972; to Manvers CPP, 13th June 1972; to Wath Main
Colliery, 24th June 1974; to Manvers Main CPP, July 1974; to Wath Main Colliery, for
storage, 8th December 1976; seen stored in shed, Wath Main Colliery, 6th March 1978;
seen still stored in shed, Wath Main Colliery, 10th February 1984; offered for sale for
preservation, February 1984; no takers, so scrapped on site by Wath Skip Hire Ltd of
Rotherham, July 1985.

D2228   DC   2551   1955   8F   7/68   F   D2228 / 4
**   VF   D277**

new as number 11134, 27th November 1955; seen at 8F Springs Branch Depot, Wigan,
9th July 1968; withdrawn, 13th July 1968; sold to Bowaters UK Paper Co Ltd, Sittingbourne,
Kent, and despatched from 8F Springs Branch Depot, 17th October 1968; suffered a hot
box en-route and spent four months at Rugeley; arrived at Bowaters, 17th February 1969;
noted in BR green livery with number D2228, Bowaters, 7th July 1969; allocated number 4
at Bowaters; noted 26th March 1979; scrapped about April 1979.

D2229   DC   2552   1955   52A   12/69   P   D2229
**   VF   D278**

new as number 11135, 28th November 1955; withdrawn, 7th December 1969; moved to
Thornaby Depot, for storage, April 1970; sold to NCB; despatched from Thornaby Depot,
28th August 1970; to NCB Brookhouse Colliery, Beighton; allocated plant number 521/52;
to Orgreave Colliery, by 28th July 1971; to Brookhouse Colliery, early March 1972; seen at
Brookhouse Colliery, 12th March 1972; to Orgreave Colliery about October 1973; to
Brookhouse Colliery, 12th July 1974; seen at Brookhouse Colliery, 27th March 1976; seen
in BR green livery, with number D2229 and 'white rose emblem' No.5, Brookhouse Colliery,
3rd November 1978; given major mechanical service, by two fitters from Baguley-Drewry,
April 1980; repainted in green livery, with 'Ex British Rail D.2229' on cabside and home-
made plate 'Pride of the NCB' on radiator grille, June 1980; seen at Brookhouse Colliery,
5th November 1981; rail traffic ceased at Brookhouse Colliery, early 1983; stored in the
shed; to Manton Colliery, 28th March 1983; seen at Manton Colliery, 23rd November 1983;
seen disused, as pit converted to MGR system, Manton Colliery, 16th October 1989; re-
sold to South Yorkshire Railway Preservation Society, Meadowhall, Sheffield, and moved
there on 25th May 1990; noted in green livery, Meadowhall, 28th May 1990; to Heritage
Shunters Trust, Rowsley, 12th March 2002; cosmetically restored in black livery, 2002;
seen with number D2229, Heritage Shunters Trust, Rowsley, 12th December 2003;
restoration began in 2019; repainted in green livery with number D2229, 2021.

D2231   DC   2555   1956   16C   6/69   F   No.8001
**   VF   D281**

new as number 11150, 19th January 1956; noted at Derby Works, 26th January 1969;
noted at 16C Derby Depot, 29th March 1969; withdrawn, 28th June 1969; sold to R.E. Trem
& Co, Finningley, Doncaster, 1969; noted at Etches Park, Derby, 27th March 1970; to
Steelbreaking & Dismantling Co, Chesterfield, for storage, about April 1970; seen at
Steelbreaking & Dismantling Co, Chesterfield, 21st April 1970; noted at Steelbreaking &

Dismantling Co, Chesterfield, 14th June 1970; exported (per Trem) at unknown date thereafter (see Appendix C).

D2238	DC	2562	1956	8H	7/68	F	D2238 / CAROL
	VF	D288					

new as number 11157, 10th May 1956; withdrawn, 27th July 1968; sold to NCB; moved to Manvers Main Coal Preparation Plant, Wath upon Dearne, Rotherham, 10th November 1968; noted at Manvers Main, 1st December 1968; seen repainted in rich green livery and named TOM, Manvers Main, 9th March 1969; had undergone a sex change when noted renamed CAROL, Manvers Main, 1st November 1970; to Manvers Main Coking Plant, Wath upon Dearne, about 1971; to Coventry Home Fire Plant, Keresley, on loan (as cover while Hunslet 6658 was at Hunslet Engine Company, Leeds, for repairs), 15th June 1974; returned to Manvers Main Coking Plant, Wath upon Dearne, 5th December 1975; seen repainted in yellow livery, Manvers Main Coking Plant, 18th February 1978; to Coventry Home Fire Plant, Keresley, on loan for about six months in 1979; returned to Manvers; Manvers Main Coking Plant closed, 2nd December 1980; seen disused at Manvers Main Coking Plant, 30th January 1981; scrapped on site by Ernest Nortcliffe & Son Ltd of Parkgate, Rotherham, about June 1982.

D2239	DC	2563	1956	75C	9/71	F	NFT
	VF	D289					

new as number 11158, 18th May 1956; withdrawn, 30th September 1971; sold to NCB; despatched from Selhurst Depot; to Dodworth Colliery, Barnsley, August 1972; noted in tatty BR green livery with number D2239, Dodworth Colliery, 31st August 1972; repainted green with no number, April 1973; seen at Dodworth Colliery, 22nd March 1974; seen in green livery with no number, with odometer reading of 22,141 miles, Dodworth Colliery, 27th August 1983; noted at Dodworth Colliery, 17th September 1985; sold to C.F. Booth Ltd, Rotherham, for scrap, and moved 20th March 1986; seen in green livery with no number, C.F. Booth Ltd, 21st March 1986; chassis only remained, 27th March 1986; remains scrapped, late March 1986.

D2240	DC	2564	1956	BC	4/68	F	D2240
	VF	D290					

new as number 11159, 3rd June 1956; withdrawn, 20th April 1968; sold to A. King & Sons Ltd; despatched from Birkenhead Mollington Street Depot; to A. King & Sons Ltd, Norwich, 20th July 1968; noted at Norwich Depot, possibly for repairs, on various dates between 5th October 1968 and 6th October 1970; moved to A. King & Sons Ltd, Norwich, about 6th October 1970; it MAY have been unsuitable due to sharp curvature of tracks; scrapped, December 1970.

D2241	DC	2565	1956	30E	5/71	F	D2241
	VF	D291					

new as number 11160, 15th June 1956; withdrawn, 23rd May 1971; stored in the yard at Colchester Depot; sold to George Cohen, Sons & Co Ltd, Cransley, Northamptonshire, and moved September 1971; seen in green livery with number D2241, Cransley, 23rd June 1973; noted working on various dates to 1975; scrapped on 10th and 11th November 1976.

D2242	DC	2572	1956	55H	10/69	F	?
	RSH	7858					

new as number 11212, 26th October 1956; withdrawn, 26th October 1969; sold to R.E. Trem & Co of Finningley, Doncaster; consigned to C.F. Booth Ltd of Rotherham, for storage, early June 1970; seen at C.F. Booth Ltd on various dates between 13th June 1970 and 21st August 1971; to Shipbreaking (Queenborough) Ltd, Kent, at unknown date thereafter; noted at Queenborough, 23rd March 1972 and 9th May 1972; exported from Sheerness Docks, May 1972 (see Appendix C).

D2243	DC	2575	1956	51L	7/69	F	MD2
	RSHN	7862					

new as number 11213, 26th October 1956; withdrawn, 7th July 1969; sold to Tees & Hartlepool Port Authority, Middlesbrough Docks, and moved July 1970; numbered MD2; withdrawn after an accident in 1972; used for spares, 1972; scrapped in March 1973.

D2244	DC	2576	1956	55F	6/70	F	5
	RSHN	7863					

new as number 11214, 12th November 1956; withdrawn, 29th June 1970; seen at 55F Bradford Depot, 28th June 1970; sold to A.R. Adams & Son, Newport, Monmouthshire, July 1970; despatched from 55F Bradford Depot, August 1970; delayed en-route as ran hot at Belper; delivered to Adams, August 1970; numbered 5; used by Adams as a hire locomotive (see Appendix A); noted being scrapped on 5th January 1981.

D2245	DC	2577	1956	50D	12/68	P	No.2
	RSHN	7864					

new as number 11215, 12th November 1956; withdrawn, 28th December 1968; sold to Derwent Valley Railway; despatched from 50D Goole Depot, May 1969; delivered to Derwent Valley Railway Company, Layerthorpe, York; given number 2; re-sold to Battlefield Line, Shackerstone, Leicestershire, and moved 17th May 1978; noted at Shackerstone, still numbered 2, 16th September 1979; seen with number 11215, Shackerstone, 26th February 1989; noted in green livery with number D2245, Shackerstone, 17th November 2012; to Derwent Valley Light Railway Society, Murton, York, 17th July 2013; to Shackerstone, 22nd July 2013; re-sold to Derwent Valley Light Railway Society, and moved 30th May 2014; noted in green livery with number D2245, Derwent Valley Light Railway Society, 1st April 2017; renumbered No.2, October 2023.

D2246	DC	2578	1956	55G	7/68	P	D2246
	RSHN	7865					

new as number 11216, 6th December 1956; withdrawn, 8th July 1968; noted at BR Knottingley Depot, 30th August 1968; sold to Coal Mechanisation Ltd; despatched from BR Knottingley Depot, January 1969; noted in a freight train at Three Bridges, 17th January 1969; arrived at Coal Mechanisation Ltd, Crawley Coal Concentration Depot, Sussex, 19th January 1969; noted at Crawley CCD, 20th January 1969; noted in green livery with no number, Crawley, 27th April 1980; to Coal Mechanisation Ltd, Tolworth Coal Depot, Surrey, by 12th August 1982; named BLUEBELL; noted in green livery with no number, Tolworth, 30th January 1984; to British Coal, West Drayton Landsale Depot, 19th November 1990; noted at West Drayton Landsale Depot, 11th January 1991; re-sold to South Yorkshire Railway Preservation Society, Meadowhall, Sheffield, and left on 16th December 1994; arrived at Meadowhall, 19th December 1994; to Elsecar Steam Railway, near Barnsley, on hire, 20th April 1995; returned to South Yorkshire Railway Preservation Society, 21st August 1996; re-sold to South Devon Railway, Buckfastleigh, in August 2000; arrived at Buckfastleigh, 9th January 2001; repainted and numbered 11216, May 2001; noted in BR green livery with number D2246, South Devon Railway, 25th June 2010 and 31st July 2019.

D2247 DC   2579   1956   55B   11/69 F   D6
       RSHN   7866

new as number 11217, 14th December 1956; withdrawn, 9th November 1969; to Ford Motor Co Ltd, Dagenham, for use as a standby loco, December 1969; noted at Dagenham, 11th February 1970; returned to BR York, April 1970; sold to Briton Ferry Steel Co Ltd, Glamorgan (via W. & F. Smith Ltd, Ecclesfield, Sheffield) June 1970; noted with number D12, 26th August 1970; renumbered D6 about 1971; works closed on 18th November 1978; used for spares, February 1979; remains scrapped, September 1979.

D2248 DC   2580   1957   55F   6/70   F   2243 / No.18 / SUE
       RSHN   7867

new as number 11218, 8th January 1957; withdrawn, 29th June 1970; seen at BR Bradford Depot, 22nd August 1970; sold to NCB; despatched from Bradford Depot to Manvers Main Coal Preparation Plant, Wath upon Dearne, Rotherham, August 1970; noted at Manvers Main, 22nd February 1971; to Maltby Colliery, Rotherham, about September 1971; seen in BR green livery with number D2248, Maltby Colliery, 15th July 1972; during a repaint into turquoise livery at Maltby Colliery in 1975 the incorrect number 2243 was applied; seen with number 2243, with 'white rose emblem' No.18, and with nameplate SUE mounted on bonnet front, Maltby Colliery, 18th January 1976; noted disused at Maltby Colliery, 12th December 1986; scrapped on site by Carol & Good Ltd of Thurcroft, near Rotherham, April 1987.

D2249 DC   2581   1957   30E   12/70 F   D2249
       RSH   7868

new as number 11219, 17th January 1957; withdrawn, 27th December 1970; sold to Shipbreaking (Queenborough) Ltd; despatched from Colchester Depot; to Shipbreaking (Queenborough) Ltd, Kent, 21st September 1971; noted at Queenborough, 25th September 1971; worked for only a short period; later used for spares; reduced to its chassis and part of the cab, 22nd October 1972; not seen on a visit of 28th July 1973 and believed scrapped.

D2258 DC   2602   1957   16C   9/70   F   D2258 / 4-2
       RSHD   7879

new as number 11228, 8th October 1957; noted at 16C Derby Depot, 29th March 1969; withdrawn, 5th September 1970; sold to NCB (Opencast Executive); moved to Hargreaves (West Riding) Ltd, Bennerley Disposal Point, Ilkeston, January 1971; noted at Bennerley, 10th August 1974; to BR Toton Depot, for repairs, December 1974; returned to Bennerley Disposal Point, Ilkeston, January 1975; noted at Bennerley, 28th March 1979; seen in blue livery with numbers D2258 and 4-2, Bennerley, 5th April 1983; to NCBOE Wentworth Stores, Harley, near Rotherham, 27th February 1984; seen at Wentworth Stores, 14th September 1984; put out to tender, July 1986; re-sold to C.F. Booth Ltd, Rotherham, for scrap, 2nd September 1986; noted at C.F. Booth Ltd, 7th September 1986; allocated Booth's plant number 91017; seen in blue livery with numbers D2258 and 4-2, C.F. Booth Ltd, 8th January 1987; scrapped, mid-January 1987.

D2259 DC   2603   1957   73F   12/68 F   D2259 / 5
       RSHD   7889

new as number 11229, 8th October 1957; withdrawn, 22nd December 1968; sold to Bowaters UK Paper Co Ltd, Sittingbourne, Kent, and moved February 1969; noted in BR green livery with number D2259, Bowaters, 6th September 1969 and 4th October 1969;

allocated number 5 at Bowaters; noted being used for spares, Ridham Dock, 30th October 1976; remains scrapped on site by Smeeth Metal Co Ltd, January 1978.

D2260   DC   2604   1957   55F   10/70 F   THOMAS HARLING
**   RSHD   7890**

new 10th December 1957; seen at 55F Bradford Depot, 22nd August 1970; withdrawn, 24th October 1970; sold to NCBOE; to Tilsley & Lovatt Ltd, Trentham, for repairs, 11th March 1971; to Powell Duffryn Fuels Ltd, NCBOE Mill Pit Disposal Point, Cefn Cribbwr, July 1971; noted in PD blue and white livery with no number but nameplate THOMAS HARLING above side panels, Mill Pit, 30th July 1971 and 9th July 1980; to Cwm Mawr Disposal Point, Tumble, 3rd November 1981; to Coed Bach Disposal Point, Kidwelly, December 1981; scrapped on site by Rees Industries Ltd of Llanelli, June 1983.

D2262   DC   2606   1957   51A   9/68   F   7
**   RSHD   7892**

new 21st November 1957; withdrawn, 23rd September 1968; sold to Ford Motor Co Ltd, Dagenham, and moved March 1969; allocated Ford number 7; noted with number 7, Dagenham, 6th September 1969; suffered collision damage in 1975; subsequently used for spares; noted with no engine, Dagenham, August 1976; remains scrapped on site, July 1978.

D2267   DC   2611   1958   50D   12/69 F   1
**   RSHD   7897**

new 20th January 1958; withdrawn, 27th December 1969; sold to Ford Motor Co Ltd, Dagenham, and moved January 1970; allocated Ford number 6; noted with number 6, Dagenham, 17th September 1971; to BREL Swindon Works, for rebuild, 19th May 1977; noted in blue livery with number 6, Swindon Works, 21st May 1977; noted in blue livery with Ford emblem on cab side and number 1 on bonnet side and front buffer beam, Swindon Works A Shop, 18th July 1977; to Ford Motor Co Ltd, Dagenham, 8th November 1977; noted with number 1, Ford, Dagenham, 19th August 1978; noted in blue livery with Ford logo and number 01 on cab back, Dagenham, 30th September 1990; noted dismantled (no engine), 17th October 1996; re-sold in April 1997; moved to East Anglian Railway Museum, Wakes Colne, Essex, 24th September 1998; to North Norfolk Railway, Sheringham, 16th February 2000; used for spares; remains scrapped, April 2003.

D2270   DC   2614   1958   55B   2/68   F   D9
**   RSHD   7912**

new 4th March 1958; withdrawn, 13th February 1968; sold to Briton Ferry Steel Co Ltd, Llanelly, Glamorgan; sold per R.E. Trem Ltd, Finningley, Doncaster; despatched from Stourton Depot; seen at Ickles Sidings, Rotherham, with odometer reading of 45,482 miles, 1st August 1968; arrived at Llanelly later in August 1968; allocated number D9; noted freshly repainted yellow and lost its BR number, Briton Ferry Steel Co Ltd, 7th July 1969; the works closed on 18th November 1978; used for spares, February 1979; remains scrapped, September 1979.

D2271   DC   2615   1958   55F   10/69 P   D2271
**   RSHD   7913**

new 14th March 1958; seen at 55F Bradford Depot, 30th August 1969; withdrawn, 26th October 1969; sold to C.F. Booth Ltd, Rotherham (sold per R.E. Trem Ltd, Finningley, Doncaster), and moved May 1970; seen at C.F. Booth Ltd on various dates between 29th August 1970 and 3rd April 1972; privately purchased for preservation and moved to

Thomas Hill Ltd, Kilnhurst, for storage, 27th July 1972; seen at Thomas Hill Ltd, 22nd December 1972 and 30th June 1973; to Normanton Barracks, Derby, for storage, 7th September 1973; to Midland Railway, Butterley, 10th May 1975; seen at Butterley, 5th August 1975; seen repainted in maroon livery with number 2271, Butterley, 15th July 1978; re-sold to West Somerset Railway, Minehead, 15th May 1982; noted in blue livery with number D2271, West Somerset Railway, 2nd September 1991; noted in light blue livery with number D2271, West Somerset Railway, 23rd August 2004; to South Devon Railway, Buckfastleigh, 1st November 2018; noted in green livery, Buckfastleigh, 31st July 2019.

| D2272 | DC | 2616 | 1958 | 55F | 10/70 | P | 2272 / ALFIE |
| | RSHD | 7914 | | | | | |

new 28th March 1958; seen at 55F Bradford Depot, 28th June 1970; withdrawn, 24th October 1970; sold to British Fuel Company, Coal Concentration Depot, Blackburn, and moved March 1971; seen repainted in black livery with number 2272, Blackburn, 7th April 1979 and 20th July 1980; seen repainted in turquoise livery, with number 2272 on one cabside and D2272 on other side, and named ALFIE, Blackburn, 22nd May 1985; became spare loco, January 1997; re-sold to South Yorkshire Railway Preservation Society, Meadowhall, Sheffield, and moved 1st May 1997; to Lavender Line, Isfield, for storage, 7th August 2001; to Heritage Shunters Trust, Rowsley, early February 2004; seen with number 2272, Heritage Shunters Trust, Rowsley, 9th March 2007, noted awaiting restoration, 2nd September 2019.

SECTION 4:

Drewry Car Co Ltd 0-6-0 diesel mechanical locomotives built by Robert Stephenson & Hawthorns Ltd, numbered D2274-D2340, and introduced 1959. Fitted with a Gardner 8L3 engine developing 204bhp at 1200rpm, five speed gearbox, and driving wheels of 3ft 7in diameter. Later classified TOPS Class 04. Departmental DS1173, built in 1947, was transferred to capital stock as number D2341 in 1967 and completed Class 04, but some particulars were different from those of the main batch.

| D2274 | DC | 2620 | 1959 | 8J | 5/69 | F | D2274 / No.17 |
| | RSHD | 7918 | | | | | |

new 10th July 1959; withdrawn, 17th May 1969; sold to NCB; despatched from 8J Allerton Depot, 24th June 1969; to NCB Maltby Colliery, Rotherham; seen with number D2274, Maltby Colliery, 5th July 1969; seen freshly repainted maroon, with no number, Maltby Colliery, 20th September 1969; seen with white rose emblem No.17, with number D2274 reapplied, Maltby Colliery, 18th January 1976; ceased working in 1976; seen in maroon livery with number D2274, dumped behind screens, dismantled for spares, no wheels, Maltby Colliery, 25th May 1978 and 8th April 1980; remains scrapped on site, September 1980.

| D2276 | DC | 2622 | 1959 | 30A | 8/69 | F | D2276 |
| | RSHD | 7920 | | | | | |

new 9th August 1959; withdrawn, 10th August 1969; noted in BR green livery with number D2276, 30A Stratford Depot, 13th June 1970; sold to A.R. Adams & Son; despatched from Stratford Depot; to A.R. Adams & Son, Newport, Monmouthshire, July 1970 (see Appendix A).

D2279	DC	2656	1960	30A	5/71	P	D2279
	RSHD	8097					

new 15th February 1960; withdrawn, 23rd May 1971; sold to CEGB; despatched from Stratford Depot under its own power; to CEGB Rye House Power Station, Hoddesdon, Hertfordshire, 1st October 1971; allocated number 2; re-sold to East Anglian Railway Museum, Wakes Colne, Essex, and moved January 1981; noted in black livery with number 11249, East Anglian Railway Museum, July 2009; to Peak Rail, Rowsley, for repairs, 31st January 2014; to Andrew Briddon, Darley Dale, for repairs, 29th May 2015; to haulage yard, near Ashbourne, for storage, 14th June 2018; to East Anglia Railway Museum, Wakes Colne, Essex, 4th July 2018; repainted in green livery with number D2279, by August 2019.

D2280	DC	2657	1960	30E	3/71	P	D2280
	RSHD	8098					

new 25th February 1960; withdrawn, 14th March 1971; sold to Ford Motor Co Ltd; despatched from 30E Colchester Depot, under its own power, 28th June 1971; to Ford Motor Co Ltd, Dagenham; allocated Ford number 1; noted with number 1, Dagenham, 17th September 1971; to BREL Swindon Works, for rebuild, 8th July 1977; to Ford Motor Co Ltd, Dagenham, 25th November 1977; renumbered 2, November 1977; noted with number 2, Ford, Dagenham, 19th August 1978 and 10th March 1992; re-sold to a buyer in Essex and moved to a private location, April 1997; to East Anglian Railway Museum, Wakes Colne, Essex, 24th September 1998; to North Norfolk Railway, Sheringham, 16th February 2000; noted in black livery with no number, North Norfolk Railway, September 2007 and 2nd September 2012; re-sold to Gloucestershire Warwickshire Railway, Toddington, and moved 22nd March 2018; used as yard shunter; repainted in BR green livery with number D2280, May 2024.

D2281	DC	2658	1960	30E	10/68	F	D2281
	RSHD	8099					

new 27th February 1960; withdrawn, 20th October 1968; sold to Briton Ferry Steel Co Ltd, February 1969 (sold per R.E. Trem Ltd, Finningley, Doncaster); despatched on a road low-loader from Colchester Depot, 14th April 1969; to Briton Ferry Steel Co Ltd, Glamorgan, noted in BR green livery, Briton Ferry, 23rd May 1970; noted being used for spares, 26th August 1970; remains scrapped, August 1971.

D2282	DC	2659	1960	30E	12/70	F	D2282
	RSH	8100					

new 28th February 1960; initially allocated to Ipswich Depot where it carried cowcatchers and skirting for use on the Ipswich Docks tramway system; withdrawn from Colchester Depot, 27th December 1970; sold to Shipbreaking (Queenborough) Ltd; despatched from Colchester Depot; to Shipbreaking (Queenborough) Ltd, Kent, September 1971; noted at Queenborough, 25th September 1971; used for spares; noted on 22nd October 1972; not seen on a visit of 28th July 1973 and believed that the remains were scrapped.

D2284	DC	2661	1960	30E	4/71	P	D2284
	RSHD	8102					

new 25th March 1960; withdrawn, 11th April 1971; noted in BR green livery with number D2284, Colchester Depot, 2nd May 1971; sold to NCB; noted in a freight train in Colchester Yard, 29th June 1971; arrived at NCB North Gawber Colliery, Mapplewell, Barnsley, 16th July 1971; seen at North Gawber Colliery, 22nd March 1974 and 24th June 1975; to Grimethorpe Colliery, 30th January 1976; noted at Grimethorpe Colliery, 1st August 1976 and 25th January 1978; to Woolley Colliery, Darton, early March 1978; seen at Woolley Colliery, 2nd July 1978; seen on blocks with wheels out, at rear of shed, Woolley Colliery, 16th April 1979, 12th January 1980 and 16th May 1981; seen at Woolley Colliery, 6th August 1983; became surplus when 'merry-go-round' working introduced in 1985; donated by NCB to South Yorkshire Railway Preservation Society, Chapeltown, and moved 2nd August 1985; seen at Chapeltown, 25th August 1985; to South Yorkshire Railway Preservation Society, Attercliffe, Sheffield, December 1986; to South Yorkshire Railway Preservation Society, Meadowhall, Sheffield, 12th September 1988; noted in green livery, Meadowhall, 27th August 1989; to Heritage Shunters Trust, Rowsley, March 2002; returned to use, May 2003; seen with number D2284, Heritage Shunters Trust, Rowsley, 12th December 2003; noted in BR green livery with number D2284, Rowsley, 7th September 2019.

| D2289 | DC | 2669 | 1960 | 70D | 9/71 | P | LONATO SPA |
| | RSHD | 8122 | | | | | |

new 28th April 1960; noted in blue livery with number D2289, BR Eastleigh Depot, 25th September 1970; withdrawn, 30th September 1971; sold to Shipbreaking (Queenborough) Ltd, Kent, and moved April 1972; exported from Sheerness Docks, May 1972 (see Appendix C); privately purchased and returned to England; arrived at Western Docks, Dover, 12th June 2018; to Heritage Shunters Trust, Rowsley, 13th June 2018; noted in red livery with no number, Rowsley, 2nd September 2018.

| D2293 | DC | 2673 | 1960 | 73F | 4/71 | F | D2293 |
| | RSHD | 8126 | | | | | |

new 4th June 1960; withdrawn, 30th April 1971; moved to Eastleigh, for storage, September 1971; sold to Shipbreaking (Queenborough) Ltd; despatched from BR Eastleigh Depot, 16th March 1972; arrived at Shipbreaking (Queenborough) Ltd, Kent, 18th March 1972; noted at Queenborough, 23rd March 1972 and 9th August 1972; used for spares; noted in dismantled state, 22nd October 1972; not seen on a visit of 28th July 1973 and believed the remains scrapped.

| D2294 | DC | 2674 | 1960 | 70D | 2/71 | F | 01 |
| | RSHD | 8127 | | | | | |

new 4th June 1960; withdrawn, 28th February 1971; noted at Eastleigh Depot, 4th July 1971; sold to Shipbreaking (Queenborough) Ltd; despatched from BR Eastleigh Depot, 16th March 1972; arrived at Shipbreaking (Queenborough) Ltd, Kent, 18th March 1972; used as yard shunter; noted at Queenborough, 9th May 1972; noted repainted in light blue livery with no number, Queenborough, 9th August 1972, 28th July 1973, 16th February 1977, and 2nd September 1978; seen in blue livery with number 01, Queenborough, 13th September 1979; noted out of use, Queenborough, 4th May 1984 and 24th July 1985; noted stripped to a shell, 7th September 1985; scrapped on site, early October 1985 (was gone by 17th).

| D2295 | DC | 2675 | 1960 | 70D | 4/71 | F | ? |
| | RSHD | 8128 | | | | | |

new 24th June 1960; withdrawn, 30th April 1971; noted at Eastleigh Depot, 4th July 1971; sold to Shipbreaking (Queenborough) Ltd; despatched from BR Eastleigh Depot; noted in Woking Yard, 16th March 1972; arrived at Shipbreaking (Queenborough) Ltd, Kent, 18th March 1972; noted at Queenborough, 23rd March 1972 and 9th May 1972; exported from Sheerness Docks, May 1972 (see Appendix C).

D2298   DC   2679   1960   52A   12/68   P   D2298
**   RSHD   8157**

new 11th October 1960; to Gateshead Depot, 7th July 1968; withdrawn, 28th December 1968; sold to Derwent Valley Railway; despatched from Gateshead Depot; to Derwent Valley Railway Company, Layerthorpe, York, 16th April 1969; seen in lined green livery, and named LORD WENLOCK (after the company's first chairman), Derwent Valley Railway, 20th June 1979; worked the last passenger train on the DVLR, a BRSRS special, with 'DVR Farewell' headboard, 27th September 1981; advertised for sale, late September 1981; noted at Derwent Valley Railway, 16th August 1982; re-sold to Quainton Railway Society, near Aylesbury, Buckinghamshire, and arrived 22nd October 1982; noted with no name or number, Quainton Railway Society, 10th October 1983; noted in blue livery with no number, Quainton Railway Society, May 2014; noted in green livery with number D2298, Quainton Railway Society, April 2017; to East Lancashire Railway, Bury, for gala, 25th May 2017; returned to Quainton Railway Society, 7th June 2017; noted in green livery with number D2298, Quainton Railway Society, July 2019.

D2299   DC   2680   1960   52A   1/70   F   D2299 / DIANA
**   RSHD   8158**

new 11th October 1960; withdrawn, 11th January 1970; moved to Thornaby Depot, April 1970; sold to NCB; despatched from Thornaby Depot, 9th July 1970; to NCB Bestwood Colliery, Nottinghamshire; to Hucknall Colliery, 7th August 1970; noted in BR green livery, Hucknall Colliery, 20th March 1976; to Calverton Colliery, 14th November 1977; noted at Calverton Colliery, 7th April 1978; to Hucknall Colliery, 23rd August 1978; named JONAH; to Calverton Colliery, 1980; noted in BR green livery with number D2299, Calverton Colliery, 20th July 1980; to Hucknall Colliery, 1980; noted at Hucknall Colliery, 28th August 1980; Hucknall Colliery closed on 5th February 1982; noted with name DIANA, Hucknall Colliery, 15th August 1982; re-sold to R.E. Trem Ltd of Finningley, Doncaster, who removed the engine; remains to C.F. Booth Ltd, Rotherham, for scrap, February 1984; scrapped, week-ending 17th February 1984.

D2300   DC   2681   1960   8J   5/69   F   D2300
**   RSHD   8159**

new 24th October 1960; withdrawn, 17th May 1969; sold to NCB; despatched from Allerton Depot to NCB Shireoaks Colliery, 25th June 1969; noted with number D2300, Shireoaks Colliery, 18th June 1970; given number 30; seen at Shireoaks Colliery, 1st January 1971; to Steetley Colliery, on loan, 12th September 1974; returned to Shireoaks Colliery, 18th November 1974; seen with BTC 3537 of 1976 registration plate, Shireoaks Colliery, 28th August 1977; to Manton Colliery, 18th October 1978; noted at Manton Colliery, 17th April 1979; seen in blue livery with no BR number, Manton Colliery, 6th April 1980; out of use by late 1983; scrapped on site by Hoyland Dismantling Co Ltd, August 1986.

D2302   DC   2683   1960   16C   6/69   P   D2302
**   RSHD   8161**

new 31st October 1960; withdrawn, 23rd June 1969; sold to British Sugar Corporation Ltd, and delivered to Woodston Factory, Peterborough, August 1969; to BSC Allscott Factory, Shropshire, early October 1969; noted in BR green livery with number D2302, Alscott, 4th October 1969 and 8th October 1969; re-sold to G.G. Papworth Ltd, Queen Adelaide Rail Distribution Centre, Ely, Cambridgeshire, and moved 12th July 1983; seen in faded green livery with number D2302, G.G. Papworth Ltd, 13th October 1986; site taken over by Potter Group, 1991; seen repainted in yellow livery with number D2302, Ely, 9th March 1992; re-sold to Harry Needle Railroad Company; moved to McCall's yard, Meadowhall, 25th September 1993; moved into South Yorkshire Railway Preservation Society, Meadowhall, Sheffield, 27th September 1993; noted in yellow livery with number D2302, Meadowhall, June 1999; to Rutland Railway Museum, Cottesmore, on loan, 16th March 2001; to Barrow Hill Engine Shed, Staveley, 18th May 2004; noted in green livery with number D2302, Barrow Hill Engine Shed, March 2009; re-sold to D2578 Locomotive Group, Moreton on Lugg, and moved there on 16th November 2011; noted in green livery with number D2302, Moreton on Lugg, October 2012 and September 2023.

D2304 DC   **2685**   **1960**   **51A**   **2/68**   **F**   **D8**
RSHD   **8163**

new 4th November 1960; withdrawn, 13th February 1968; sold to C.F. Booth Ltd, Rotherham, and seen there 23rd July 1968; seen in BR green livery with number D2304, with odometer reading of 30,946 miles, Ickles Sidings, Rotherham, 1st August 1968; re-sold to Llanelly Steel Co Ltd, Carmarthenshire (sold per R.E. Trem Ltd, Finningley, Doncaster) and moved in August 1968; noted at Llanelli, 5th July 1969; noted with number D8, Llanelli, 4th October 1969; noted being used for spares, November 1972; remains scrapped, May 1977.

D2305 DC   **2686**   **1960**   **51A**   **2/68**   **F**   **D9**
RSHD   **8164**

new 10th November 1960; withdrawn, 13th February 1968; sold to C.F. Booth Ltd, Rotherham, and seen there 23rd July 1968; seen in BR green livery with number D2305, with odometer reading of 25,817 miles, Ickles Sidings, Rotherham, 1st August 1968; re-sold to Llanelly Steel Co Ltd, Carmarthenshire (sold per R.E. Trem Ltd, Finningley, Doncaster) and moved in August 1968; allocated number D9; noted at Llanelli, 5th July 1969; noted with number D2305, Llanelli, 4th October 1969; noted in tatty BR green with number D9, Llanelli, 4th May 1978; seen disused, Llanelli, 10th June 1980; scrapped, September 1981.

D2306 DC   **2687**   **1960**   **51L**   **2/68**   **F**   **D6**
RSHD   **8165**

new 17th November 1960; withdrawn, 13th February 1968; sold to Llanelly Steel Co Ltd; noted on a low-loader at Cardiff, 11th July 1968; arrived at Llanelly Steel Co Ltd, Carmarthenshire (sold per R.E. Trem Ltd, Finningley, Doncaster), 11th July 1968; noted at Llanelly Steel, 5th July 1969; noted with number D6, Llanelly Steel, 4th October 1969; noted disused with no rods and number D6, Llanelly Steel, 4th May 1978; seen dumped with no centre wheels, in green livery with no BR number, with number D6, Llanelly Steel, 10th June 1980; scrapped, September 1981.

D2307 DC   **2688**   **1960**   **51L**   **2/68**   **F**   **D7**
RSHD   **8166**

new 28th November 1960; withdrawn, 13th February 1968; sold to Llanelly Steel Co Ltd; despatched from Thornaby Depot by road; arrived at Llanelly Steel Co Ltd, Carmarthenshire (sold per R.E. Trem Ltd, Finningley, Doncaster), 11th July 1968; noted at Llanelly Steel, 5th July 1969; noted with number D7, Llanelly Steel, 4th October 1969; out of use by July 1978; scrapped, October 1979.

D2308   **DC**   **2689**   **1960**   **51A**   **2/68**   **F**   **D8**
             RSHD   **8167**

new 6th December 1960; withdrawn, 13th February 1968; sold to C.F. Booth Ltd, Rotherham; seen at C.F. Booth Ltd, 23rd July 1968; re-sold to Briton Ferry Steel Co Ltd, Glamorgan (sold per R.E. Trem Ltd, Finningley, Doncaster); seen in BR green livery with number D2308, with odometer reading of 27,297 miles, Ickles Sidings, Rotherham, 1st August 1968; seen passing through Rotherham on a low-loader, 24th August 1968; arrived at Briton Ferry in late August 1968; noted still carrying its BR number, Briton Ferry Steel Co Ltd, 7th July 1969 and 14th September 1971; to Duport Steel Works Ltd, Llanelli, 25th October 1979; used for spares; scrapped, May 1980.

D2310   **DC**   **2691**   **1960**   **52A**   **2/69**   **P**   **04110**
             RSHD   **8169**

new 22nd December 1960; withdrawn, 1st February 1969; sold to Coal Mechanisation Ltd; noted at Gateshead Depot, 13th April 1969; despatched from Gateshead Depot, 22nd April 1969; journey protracted due to hot boxes en-route at Wellingborough, and later at Bedford; arrived at Coal Mechanisation Ltd, Tolworth Coal Depot, Surrey, 19th May 1969; noted in green livery with number D2310, Tolworth, 5th June 1969; noted in green livery with number D2310, and COLMEC on cab side, Tolworth, 3rd August 1973; later repainted maroon; noted at Tolworth, 9th February 1980; re-sold to Harry Needle Railroad Company; to South Yorkshire Railway Preservation Society, Meadowhall, Sheffield, 14th September 1994; to Battlefield Line, Shackerstone, on loan, 3rd October 2001; noted in black livery with number D2310, Battlefield Line, May 2007; noted in BR blue livery with number 04110, Battlefield Line, 26th July 2008; re-sold to Battlefield Line, about October 2011; noted with number 04110, Battlefield Line, October 2019; noted in green livery, Battlefield Line, 12th June 2021.

D2317   **DC**   **2698**   **1961**   **52A**   **9/69**   **F**   **No.10**
             RSHD   **8176**

new 22nd February 1961; withdrawn, 30th September 1969; sold to NCB; to Manvers Main Coal Preparation Plant, Wath upon Dearne, Rotherham, 30th December 1969; seen at Manvers Main, 2nd May 1970; to Cortonwood Colliery, Wombwell, 5th May 1970; noted at Cortonwood Colliery, 14th August 1970; given number 10 in 1973; seen at Cortonwood Colliery, 26th June 1976; noted with 'white rose emblem' No.10, on loan, New Stubbin Colliery, Rawmarsh, July 1976; returned to Cortonwood Colliery, about August 1976; seen in original BR green livery, with 'white rose emblem' No.10, and with RE 945 of 1953 registration plate, Cortonwood Colliery, 19th February 1978, 2nd March 1982 and 17th August 1985; colliery closed, 25th October 1985; scrapped on site by Wath Skip Hire Ltd of Rotherham, July 1986.

D2322   **DC**   **2703**   **1961**   **52A**   **8/68**   **F**   **D2322 / No.24**
             RSHD   **8181**

new 28th March 1961; withdrawn, 8th August 1968; sold to NCB; to Orgreave Colliery, February 1969; noted at Orgreave Colliery, 3rd April 1969; used mainly at Orgreave Colliery although it sometimes worked along a private NCB branch line to the nearby Treeton Colliery where it could occasionally be seen; seen at Orgreave Colliery, 6th July 1969; seen in original BR green livery, with BR number, with 'white rose emblem' No.24, and BTC registration plate 3277 of 1967, Orgreave Colliery, 19th January 1975; seen at Orgreave Colliery, 26th April 1980; to Kiveton Park Colliery, by rail under its own power, 29th April 1980; seen in original BR green livery, with BR number, and with 'white rose emblem'

No.24, Kiveton Park Colliery, 2nd September 1982; noted working, 25th April 1983; noted with no bonnet panels and no engine, Kiveton Park Colliery, 30th September 1983; noted as frame, wheels and cab only, 16th March 1985; remains scrapped on site by 28th November 1985.

D2324	DC	2705	1961	55B	7/68	P	2324 / JUDITH
	RSHD	8183					

new 26th April 1961; withdrawn, 8th July 1968; stored at York; sold to G.W. Talbot Ltd, Coal Concentration Depot, Aylesbury, Buckinghamshire, and moved January 1969; noted with number 2324, Aylesbury, 12th October 1969; noted in green livery with number 2324, Aylesbury, August 1970; noted with no number, Aylesbury, 13th July 1980; re-sold to Redland Roadstone Ltd, Barrow upon Soar, Leicestershire, after 18th November 1989; re-sold to Harry Needle Railroad Company; moved to South Yorkshire Railway Preservation Society, Meadowhall, Sheffield, 29th March 1995; to Lavender Line, Isfield, for storage, about June 2001; to Barrow Hill Engine Shed, Staveley, 27th March 2006; noted in light green livery with number 2324, BHES, July 2007; to Heritage Shunters Trust, Rowsley, 1st October 2008; seen in green livery with no number, Heritage Shunters Trust, Rowsley, 12th December 2008; to Nemesis Rail, Burton upon Trent, early January 2015; noted in light green livery with number 2324, Nemesis, 2nd July 2023.

D2325	DC	2706	1961	50D	7/68	P	D2325
	RSHD	8184					

new 29th April 1961; withdrawn, 8th July 1968; sold to NCB; despatched from 50D Goole Depot, December 1968; to NCB Norwich Coal Concentration Depot, December 1968; repainted in light green livery, with number D2325 on one side only, early April 1969; noted at Norwich CCD, 5th August 1979; the CCD closed in October 1986 and all track was lifted by 26th October 1986; to Tannick Commercial Repairs, Norwich, for storage, October 1986; re-sold to John Jolly, Bridgewick Farm, Dengie, Southminster, Essex, 19th March 1987; to Mangapps Railway Museum, Burnham-on-Crouch, 19th March 1989; noted in green livery with number D2325, Mangapps, June 2012 and May 2017.

D2326	DC	2707	1961	52A	8/68	F	D2326
	RSHD	8185					

new 2nd June 1961; withdrawn, 29th August 1968; sold to NCB; despatched from 52A Gateshead Depot, February 1969; noted at Thornaby Depot, 13th April 1969; delivered to NCB Manvers Main Coal Preparation Plant, Wath upon Dearne, Rotherham; acquired for spares only; noted at Manvers Main, 4th May 1969; seen at Manvers Main, 2nd May 1970; being used for spares in 1971; seen (heavily dismantled remains) alongside Manvers Main Workshops, 25th November 1973; remains scrapped on site, Autumn 1975.

D2327	DC	2708	1961	52A	8/68	F	No.12 / 521/12
	RSHD	8186					

new 29th May 1961; withdrawn, 14th August 1968; sold to NCB; noted at Thornaby Depot, 13th April 1969; to Manton Main Colliery, about April 1969; to Dinnington Colliery, 9th August 1971; noted with erroneous number 521/12, Dinnington Colliery, 1st June 1972; numbers 521/72 and 521/12 were the correct numbers for D2327 and D2328 respectively; to Elsecar Central Workshops, 3rd May 1973; returned to Dinnington Colliery, 1973; seen at Dinnington Colliery, 5th April 1974; to Elsecar Central Workshops, 15th November 1974; to Dinnington Colliery, 20th January 1975; seen in turquoise livery with no BR number, with 'white rose emblem' No.12 and erroneous number 521/12, Dinnington Colliery, 5th March 1976; seen with RE 952 of 1976 registration plate, Dinnington Colliery, 20th April 1979 and

10th November 1980; seen (long disused with jackshaft removed), Dinnington Colliery, 23rd November 1983; to Coopers (Metals) Ltd, Brightside, Sheffield, for scrap, 5th January 1984; scrapped, early February 1984.

D2328   DC   2709   1961   52A   9/68   F   No.31
**   RSHD   8187**

new 6th June 1961; withdrawn, 16th September 1968; stored in Thornaby Depot roundhouse; noted at Thornaby Depot, 13th April 1969; sold to NCB; despatched from Thornaby Depot, 6th June 1969; to NCB Dinnington Colliery; noted at Dinnington Colliery, 13th June 1969; seen with number D2328, Dinnington Colliery, 13th June 1970 and 29th April 1972; its correct plant number 521/12 was not carried; to Steetley Colliery, April 1973; given number 31; noted at Steetley Colliery, 28th May 1973; to BREL Doncaster Works, for tyre turning, 2nd February 1977; returned to Steetley Colliery, 10th February 1977; to Shireoaks Colliery, by April 1982; to Kiveton Park Colliery, 13th May 1982; seen in green livery with no BR number, and 'white rose emblem' No.31, Kiveton Park Colliery, 2nd September 1982; received wheels from D2209, March 1983; by low-loader to Cortonwood Colliery, Wombwell, 18th July 1985; seen repainted in turquoise livery with number No.31, Cortonwood Colliery, 17th August 1985; noted at Cortonwood Colliery, 31st March 1986; scrapped on site by Wath Skip Hire Ltd of Rotherham, July 1986.

D2329   DC   2710   1961   52A   7/68   F   D2329
**   RSHD   8188**

new 21st June 1961; withdrawn, 29th July 1968; sold to R.E. Trem Ltd of Finningley, Doncaster, via BR tender number 17/230/521T/66; re-sold to Derwent Valley Railway Company, Layerthorpe, York, January 1969; used for spares; remains scrapped, April 1970.

D2332   DC   2713   1961   52A   6/69   F   D.2332 / LLOYD
**   RSHD   8191**

new 10th July 1961; withdrawn, 8th June 1969; sold to NCB; to Manvers Main Coal Preparation Plant, Wath upon Dearne, Rotherham, January 1970; seen in BR green livery with number D2332, Manvers Main, 2nd May 1970; seen in yellow livery, Manvers Main, 14th August 1970; noted with name LLOYD, Manvers Main, 1st November 1970; to Manvers Main Coking Plant, on loan, 1st February 1974; returned to MMCPP, 24th April 1974; to Cadeby Colliery, Conisbrough, 28th August 1975; seen in yellow livery with 'D.2332' and LLOYD on cabside, Cadeby Colliery, 28th September 1975; to Thurcroft Colliery, 14th June 1976; noted at Thurcroft Colliery, 23rd April 1977; seen at Thurcroft Colliery, 2nd October 1979; to Shireoaks Colliery, 29th June 1981; seen at Shireoaks Colliery, 11th July 1981; to Thurcroft Colliery, 3rd September 1982; seen at Thurcroft Colliery, 27th January 1983; to Dinnington Colliery, 19th July 1985; noted in yellow livery with name LLOYD, Dinnington Colliery, 18th September 1985; put out to open tender, May 1986; there were no takers so scrapped on site, July 1986.

D2333   DC   2714   1961   52A   9/69   F   3 / P1062C
**   RSHD   8192**

new 20th July 1961; withdrawn, 4th September 1969; sold to Ford Motor Co Ltd; despatched from 52A Gateshead Depot, December 1969; journey protracted due to running hot at Northallerton, where seen in sidings on 30th December 1969; stored at York; delivered to Ford Motor Co Ltd, Dagenham, 2nd March 1970; allocated Ford number 8; suffered collision damage, May 1971; noted stored outside repair shop, 4th June 1971; to BR Stratford Depot, to repair collision damage, September 1971; returned to Ford,

Dagenham; to BREL Swindon Works, for rebuild, 3rd May 1977; to Ford Motor Co Ltd, Dagenham, by road, 23rd January 1978; renumbered 3 and P1062C, January 1978; noted with number 3, Ford, Dagenham, 19th August 1978; noted jacked-up with no wheels, 26th October 1986; scrapped, early 1990.

D2334	DC	2715	1961	51A	7/68	P	D2334
	RSHD	8193					

new 25th July 1961; withdrawn, 8th July 1968; seen at 51A Darlington Depot, 16th August 1968; moved to Thornaby Depot for storage; noted at Thornaby Depot, 13th April 1969; sold to NCB; despatched from Thornaby, 2nd June 1969; to NCB Manvers Main Coal Preparation Plant, Wath upon Dearne, Rotherham, 2nd June 1969; allocated plant number 521/31; seen in BR green livery, with number D2334, Manvers Main, 7th June 1969; to Thurcroft Colliery, 8th October 1969; noted at Thurcroft Colliery, 2nd August 1970; given number 33; seen at Thurcroft Colliery, 5th April 1974; seen up on blocks (no wheels) behind shed, in original BR green livery, with number D2334 plus 'white rose emblem' No.33, Thurcroft Colliery, 8th March 1978; seen with number 33, Thurcroft Colliery, 23rd November 1983; to Dinnington Colliery, 19th June 1985; noted at Dinnington Colliery, 29th June 1985; to Maltby Colliery, Rotherham, 24th February 1986; seen with number 33, Maltby Colliery, 24th October 1986; re-sold to South Yorkshire Railway Preservation Society, Meadowhall, Sheffield, where arrived 12th November 1988; noted in green livery, Meadowhall, 27th August 1989; noted in green livery, with number 33, Meadowhall, 4th May 1992; to Knights of Old Ltd, Old, Northamptonshire, 28th September 1993; re-sold to Churnet Valley Railway, Cheddleton, Staffordshire, and moved 10th July 1994; noted in green livery with number D2334, Churnet Valley Railway, 11th June 2016; re-sold to a private individual and moved to Mid-Norfolk Railway, Dereham, 7th January 2017.

D2335	DC	2716	1961	51A	7/68	F	No.2
	RSHD	8194					

new 27th July 1961; withdrawn, 8th July 1968; seen at 51A Darlington Depot, 16th August 1968; moved to Thornaby Depot, for storage, December 1968; noted at Thornaby Depot, 13th April 1969; sold to NCB; despatched from Thornaby Depot, 2nd June 1969; to NCB Manvers Main Coal Preparation Plant, Wath upon Dearne, Rotherham, 2nd June 1969; seen in BR green livery, with number D2335, Manvers Main, 7th June 1969; to Maltby Colliery, Rotherham, late June 1969; seen with number D2335, Maltby Colliery, 5th July 1969 and 20th September 1969; seen in maroon livery with number No.2, Maltby Colliery, 15th July 1972; seen out of use, being used for spares, with no engine, in faded maroon livery and number No.2, Maltby Colliery, 13th July 1979 and 8th April 1980; remains scrapped on site, September 1980.

D2336	DC	2717	1961	51A	7/68	F	D2336
	RSHD	8195					

new 24th August 1961; withdrawn, 15th July 1968; seen at 51A Darlington Depot, 16th August 1968; moved to Thornaby Depot, for storage, December 1968; noted at Thornaby Depot, 13th April 1969; sold to NCB; despatched from Thornaby Depot, 2nd June 1969; to NCB Manvers Main Coal Preparation Plant, Wath upon Dearne, Rotherham, 2nd June 1969; seen in BR green livery, with number D2335, Manvers Main, 7th June 1969 and 2nd May 1970; later used for spares; noted with no engine, dumped outside Manvers Workshops, 7th July 1975; remains scrapped, about February 1978.

D2337	DC	2718	1961	51A	7/68	P	D2337
	RSHD	8196					

new 30th August 1961; withdrawn, 15th July 1968; seen at 51A Darlington Depot, 16th August 1968; moved to Thornaby Depot, for storage, December 1968; noted at Thornaby Depot, 13th April 1969; sold to NCB; despatched from Thornaby Depot, 2nd June 1969; to NCB Manvers Main Coal Preparation Plant, Wath upon Dearne, Rotherham, 2nd June 1969; seen in BR green livery, with number D2337, Manvers Main, 7th June 1969; repainted in NCB yellow, about July 1971; regularly worked the internal NCB branch to Barnburgh Main Colliery; seen in yellow livery, with no BR number, named DOROTHY, and with 'white rose emblem' number 3, Barnburgh Main Colliery, 24th March 1976; seen at Manvers Main, 17th September 1986 and 28th November 1987; re-sold to members of South Yorkshire Railway Preservation Society, Attercliffe, Sheffield, and moved on 22nd February 1988; noted in yellow livery, Attercliffe, 27th March 1988; to South Yorkshire Railway Preservation Society, Meadowhall, Sheffield, 15th September 1988; noted in yellow livery, Meadowhall, 27th August 1989; to Heritage Shunters Trust, Rowsley, 15th March 2002; restored, and repainted in BR green in October 2003; seen with number D2337, Heritage Shunters Trust, Rowsley, 12th December 2008; noted in BR green livery with number D2337, Rowsley, 18th August 2018.

D2340	DC	2593	1956	55F	10/68	F	D1
	RSHD	7870					

originally a demonstration locomotive; purchased by British Railways and entered service at Darlington on 9th March 1962; withdrawn, 13th October 1968; sold to Briton Ferry Steel Co Ltd, Glamorgan, and moved April 1969 (sold per R.E. Trem Ltd, Finningley, Doncaster); noted in BR green livery with number D2340, Briton Ferry Steel Co Ltd, 7th July 1969 and 26th August 1970; later given number D10; subsequently renumbered D1; noted in yellow livery with number D1, Briton Ferry Steel Co Ltd, 1st October 1976; used for spares in February 1979; remains scrapped, September 1979.

SECTION 5:

Andrew Barclay, Sons & Co Ltd built 0-4-0 diesel mechanical locomotives, numbered D2410-D2444, and introduced 1958 for yard shunting around the BR Scottish Region. Fitted with a Gardner 8L3 engine developing 204bhp at 1200rpm; five speed gearbox, and driving wheels of 3ft 7in diameter. Later classified TOPS Class 06.

D2420	AB	435	1959	RSD	1984	P	06003

06003 new 14th February 1959; it was the second-last locomotive of the class in service when it was withdrawn as number 06003 on 1st February 1981; transferred to BR Departmental stock, 5th March 1981; moved to Reading Signal Depot, 9th April 1981; renumbered 97804 on 19th April 1981; became out of use when Signal Depot closed in 1984; moved to Reading Depot; withdrawn, June 1985; to Old Oak Common Depot, for open day, September 1985; sold to C.F. Booth Ltd; despatched from 81A Old Oak Common Depot, by road, 25th September 1986; to C.F. Booth Ltd, Rotherham; allocated Booth's plant number 91018; seen in BR blue livery with number 97804 and two-way arrow logo, C.F. Booth Ltd, 17th October 1986; re-sold to members of South Yorkshire Railway Preservation Society, Attercliffe, Sheffield, and moved on 9th March 1987; seen at South Yorkshire Railway Preservation Society, Attercliffe, 10th March 1987; to South Yorkshire Railway Preservation Society, Meadowhall, Sheffield, 14th September 1988; noted in blue livery, BR Tinsley Depot open day, 29th September 1990; returned to South Yorkshire Railway Preservation Society; to Crewe Works open day, 2nd May 1997; to Battlefield Line, Shackerstone, on loan, 9th May 1997; returned to South Yorkshire Railway Preservation Society, 23rd October 1997; to Barrow Hill Engine Shed, Staveley, 11th June 1999; to Rutland Railway Museum, Cottesmore, on loan, 9th September 1999; to Barrow Hill Engine Shed, Staveley, 13th September 2002; to UK Coal, Widdrington Disposal Point, Northumberland, on hire, 18th July 2003; to Barrow Hill Engine Shed, Staveley, 28th June 2006; to Heritage Shunters Trust, Rowsley, 13th May 2008; seen in blue livery with number 06003, Heritage Shunters Trust, Rowsley, 12th December 2008; to Museum of Science & Industry, Manchester, on loan, 14th November 2011; to Heritage Shunters Trust, Rowsley, 31st January 2013; withdrawn from service, September 2013; overhauled, repainted green and renumbered D2420; returned to use, September 2019; seen in green livery with number D2420, Rowsley, 24th July 2021; repainted in BR blue with number 06003, about August 2023.

D2432   AB   459   1960   65A   12/68   F   NFT
new 22nd February 1960; withdrawn, 15th December 1968; noted at Eastfield Depot, 20th April 1969; sold to R.E. Trem Ltd, Finningley, Doncaster; moved direct to Shipbreaking (Queenborough) Ltd, Kent, being noted on low-loader at Gosforth, Newcastle upon Tyne, 19th May 1969; noted on low-loader at Stamford, 20th May 1969; noted on low-loader at M2 Medway Service Area, 20th May 1969; arrived at Queenborough, 20th May 1969; used as yard shunter; repainted in green livery (different shade to BR green) while at Queenborough; noted in green livery with no number, Queenborough, 24th January 1971, 20th September 1974 and 14th February 1977; suffered a bent crankshaft in a derailment; exported from Sheerness Docks, March 1977 (see Appendix C).

SECTION 6:

Hudswell, Clarke & Co Ltd built 0-6-0 diesel mechanical locomotives, numbered D2510-D2519, and introduced 1961. Fitted with a Gardner 8L3 engine developing 204bhp at 1200rpm; four speed gearbox, and driving wheels of 3ft 6in diameter. No TOPS classification.

D2511   HC   D1202   1961   12C   12/67   P   D2511
new 28th August 1961; withdrawn, 30th December 1967; sold to NCB; despatched from 12C Barrow Depot and noted en-route at Carnforth, 4th May 1968; to NCB Brodsworth Colliery, Doncaster, May 1968; noted with numbers D2511 and BRM5477, Brodsworth Colliery, 9th November 1968 and 27th April 1969; noted at Brodsworth Colliery, 11th April 1976; damaged in a collision with a Class 40 diesel, Brodsworth Colliery, 1977; re-sold to Keighley & Worth Valley Railway, Haworth, and moved 8th October 1977; seen with numbers D2511 and BRM5477, Haworth 28th May 1978; noted with collision damage to its cab, Haworth, 13th July 1979; repaired, repainted, and entered service in 1980; noted in green livery with number D2511, Haworth, 21st March 1992, 10th October 2003 and 12th June 2022; now sole remaining member of its class.

D2513   HC   D1204   1961   12C   8/67   F   D2513
new to Barrow shed on 5th October 1961; withdrawn, 19th August 1967; sold to NCB; despatched from Barrow Depot, December 1968; to NCB Cadeby Colliery, Conisbrough; noted at Cadeby Colliery, 14th April 1969; seen in BR green livery with number D2513,

Cadeby Colliery, 6th November 1970 and 28th September 1975; scrapped on site, October 1975.

D2518   HC   D1209   1961   5A   2/67   F   D2518 / 3219-016
new 22nd November 1961; withdrawn, 25th February 1967; noted at BR Crewe South Depot, 19th March 1967 and 1st August 1967; sold to NCB; despatched from Crewe South Depot, August 1967; to NCB Hatfield Colliery, Doncaster, August 1967; noted at Hatfield Colliery, 22nd October 1967 and 9th June 1968; later used for spares; written off, 3rd May 1974; remains scrapped on site by NCB, 1975.

D2519   HC   D1210   1961   5A   7/67   F   D2519 / 3219-017
new 30th November 1961; to Crewe South Depot, for storage, and noted there 16th April 1967; withdrawn, 29th July 1967; sold to NCB in July 1967; noted at Crewe South Depot, 2nd August 1967; despatched from Crewe South Depot, February 1968; to NCB Hatfield Colliery, Stainforth, Doncaster; noted at Hatfield Colliery, 9th June 1968; seen in original BR green livery with number D2519, Hatfield Colliery, 2nd May 1971, 14th August 1973 and 23rd May 1981; re-sold to Keighley & Worth Valley Railway, Haworth, and moved 2nd April 1982; used for spares for D2511; remains to Marple & Gillott Ltd, Attercliffe, Sheffield, for scrap, 27th March 1985; scrapped, April 1985.

SECTION 7:

Hunslet Engine Co Ltd built 0-6-0 diesel mechanical locomotives, numbered D2550-D2618, and introduced 1955 for use around Eastern and Scottish Regions of BR. Fitted with a Gardner 8L3 engine developing 204bhp at 1200rpm, four speed gearbox, and driving wheels of 3ft 4in diameter (D2550-D2573) and 3ft 9in (D2574-D2618). Later classified TOPS Class 05.

D2554   HE   4870   1956   70H   9/83   P   D2554
05001   new 13th May 1956; noted in store, Fratton Depot, 14th August 1966; transferred to the Isle of Wight, for use during the electrification of the Island Line, 1966; thereafter retained for use on engineer's trains; given number 05001 (the only 05 to gain a TOPS number), August 1974; noted in BR blue livery with number 05001, 27th September 1975; transferred to Departmental stock as number 97803, 1st January 1981; seen in blue livery with number 97803, Sandown, 6th June 1981; noted out of use at Ryde, 15th August 1983; withdrawn, September 1983; sold to Isle of Wight Steam Railway, 11th June 1984; moved to Havenstreet, 23rd August 1984; noted in green livery with number D2554, Isle of Wight Steam Railway, July 1987; seen in green livery with number D2554, and with 70H shedplate, Isle of Wight Steam Railway, 17th September 1992; noted in green livery with number D2554, Isle of Wight Steam Railway, October 2016; was given nickname 'Nuclear Fred' by railway staff and a nameplate was later fitted on the front, by 26th May 2018.

D2561   HE   4999   1957   8F   8/67   F   D3
new as 11164, 19th February 1957; withdrawn, 26th August 1967; sold to Llanelly Steel Co Ltd, Carmarthenshire, and moved March 1968 (sold per R.E. Trem Ltd, Finningley, Doncaster); noted at Llanelly Steel, 5th July 1969; noted in green livery with number D3, Llanelly Steel, 4th October 1969; noted being used for spares, Llanelly Steel, 18th July 1970; remains scrapped, October 1972.

D2568　HE　　　5006　1957　8F　　8/67　F　D2568

new as 11171, 20th September 1957; withdrawn, 26th August 1967; sold to Briton Ferry Steel Co Ltd, Glamorgan, and moved 9th May 1968 (sold per R.E. Trem Ltd, Finningley, Doncaster); engine removed and bodywork scrapped, about May 1969; the dismantled frame (cut into two pieces) was noted near the locomotive shed, Briton Ferry Steel Co Ltd, 7th July 1969 and 18th July 1972; remains subsequently scrapped.

D2569　HE　　　5007　1957　8C　　8/67　F　D6

new as 11172, 20th September 1957; withdrawn, 26th August 1967; sold to Briton Ferry Steel Co Ltd; despatched from 8C Speke Junction Depot, May 1968; to Briton Ferry Steel Co Ltd, Glamorgan (sold per R.E. Trem Ltd, Finningley, Doncaster); given number D6; engine removed and parts of bodywork scrapped, about May 1969; frame and cab only noted near the locomotive shed, Briton Ferry Steel Co Ltd, 7th July 1969; dismantled remains noted on 23rd May 1970; remains subsequently scrapped.

D2570　HE　　　5008　1957　8F　　7/67　F　NFT

new as 11173, 30th October 1957; withdrawn, 22nd July 1967; sold to Briton Ferry Steel Co Ltd; despatched from 8C Speke Junction Depot, March 1968; to Briton Ferry Steel Co Ltd, Glamorgan (sold per R.E. Trem Ltd, Finningley, Doncaster); used for spares; noted in dismantled state, in yellow livery with no number, Briton Ferry Steel Co Ltd, 7th July 1969; remains scrapped, June 1971.

D2578　HE　　　5460　1958　62A　7/67　P　D2578

new 18th November 1958; withdrawn, 7th July 1967; sold to Hunslet Engine Co Ltd, Leeds, December 1967; rebuilt as Hunslet 6999 of 1968; re-sold to H.P. Bulmer Ltd, Cider Manufacturers, Hereford, July 1968; received the name CIDER QUEEN at the opening of the new Bulmer rail facility, November 1968; was registered to run on BR metals, BTC number 3393; worked a brake van tour from Hereford Station to Moreton-on-Lugg RAOC depot, 8th July 1971; seen in green livery with cast number 2 plate and CIDER QUEEN nameplate, H.P. Bulmer Ltd, 27th May 1974; noted at Bulmer's, 12th February 1984 and 16th September 1990; re-sold to D2578 Locomotive Group (formed early 2001), May 2001; moved to D2578 Locomotive Group, Moreton on Lugg, 6th August 2001; during a repaint it was temporarily in engineer's grey livery with fictitious number 05101, spring 2004; noted with number 05101, 11th April 2004; later repainted green with number D2578; noted in green livery with number D2578, Moreton on Lugg, 21st October 2012 and September 2023.

D2587　HE　　　5636　1959　62C　12/67　P　D2587

new 10th November 1959; withdrawn, 30th December 1967; sold to The Hunslet Engine Co Ltd, Leeds, and moved September 1968; noted at The Hunslet Engine Company, 10th November 1968; overhauled (Hunslet 7180 of 1969) and fitted with a 384hp engine; re-sold to the CEGB and moved to Chadderton Power Station, September 1969; seen repainted in orange livery with number 2, Chadderton, 3rd March 1979; became redundant when Chadderton closed; to CEGB Kearsley Power Station, for storage, 3rd November 1981; noted stored at Kearsley, 13th May 1982; eventually declared surplus; re-sold to East Lancashire Railway, Bury, and moved in March 1983; seen in orange livery with number 2, East Lancashire Railway, Bury, 20th August 1983; noted in green livery, East Lancashire Railway, Bury, 2nd September 1990; re-sold to South Yorkshire Railway Preservation Society, Meadowhall, Sheffield, 30th August 1997; placed in store; to Lavender Line, Isfield, for storage, about June 2001; to Heritage Shunters Trust, Rowsley, early February 2004; repainted in BR green in 2004; returned to use, September 2005; seen with number

D2587, Rowsley, 9th March 2007; to Barrow Hill Engine Shed, Staveley, for storage, 14th November 2011; to Heritage Shunters Trust, Rowsley, 15th February 2013; seen in green livery with number D2587 and 56A shedplate, Rowsley, 18th June 2022.

D2595   HE   5644   1960   62A   6/68   P   D2595

new 30th January 1960; withdrawn, 8th June 1968; sold to The Hunslet Engine Co Ltd, Leeds, and moved September 1968; noted in erecting shop, The Hunslet Engine Company, 10th November 1968; overhauled (Hunslet 7179 of 1969) and fitted with a 384hp engine; re-sold to CEGB and moved to Chadderton Power Station, September 1969; seen repainted in orange livery with number 1, Chadderton, 3rd March 1979; to CEGB Kearsley Power Station, 24th September 1981; noted stored (rail traffic ceased) at Kearsley, 13th May 1982; re-sold to East Lancashire Railway, Bury, and moved there in March 1983; seen in orange livery with number 1, East Lancashire Railway, Bury, 20th August 1983; noted in green livery with number D2595, East Lancashire Railway, Bury, 17th June 1989; re-sold to Steamport Transport Museum, Southport, Merseyside, and moved by road, 19th October 1989; Steamport closed in 1998 and the rolling stock was moved to the new Ribble Steam Railway, Preston; D2595 moved on 31st March 1999; repainted in fictitious BR rail blue, 2008; noted in BR blue livery with number D2595, Ribble Steam Railway, 31st August 2012 and 23rd April 2022.

D2598   HE   5647   1960   50D   12/67   F   SAM

new 30th June 1960; withdrawn, 30th December 1967; seen at 50D Goole Depot, 28th January 1968; sold to NCB; despatched from 50D Goole Depot, 17th May 1968; delivered to NCB Rossington Colliery, Doncaster; noted at Rossington Colliery, 9th November 1968; noted with name SAM, Rossington Colliery, 7th December 1968; to Askern Colliery, Doncaster, July 1971; noted at Askern Colliery, 29th June 1974; noted disused after suffering fire damage, Askern Colliery, 11th November 1974; to Lambton Engine Works, for assessment, February 1975; scrapped, May 1975.

D2599   HE   5648   1960   50D   12/67   F   3219-013

new 8th July 1960; withdrawn, 30th December 1967; noted in the shed at Hatfield Colliery, Doncaster (possibly on trial or on approval), 22nd October 1967; seen at 50D Goole Depot, 28th January 1968; sold to NCB; despatched from 50D Goole Depot, May 1968; delivered to NCB Hickleton Colliery, Doncaster, May 1968; to Frickley Colliery, South Elmsall, September 1968; noted in green livery with number D2599, Frickley Colliery, 20th September 1968 and 28th June 1970; noted with additional number F.SE.357, Frickley Colliery, 15th November 1975; to Askern Colliery, Doncaster, 16th June 1976; seen in green livery with number D2599, Askern Colliery, 23rd April 1977 and 18th June 1978; noted on blocks with wheels out, Askern Colliery, 18th May 1980; was included on a NCB list of plant for disposal dated February 1981; scrapped on site by R.D. Geeson Ltd of Ripley, May 1981.

D2600   HE   5649   1960   50D   12/67   F   D7

new 13th July 1960; withdrawn, 30th December 1967; sold to Briton Ferry Steel Co Ltd; despatched from 50D Goole Depot, 29th April 1968; to Briton Ferry Steel Co Ltd, Glamorgan (sold per R.E. Trem Ltd, Finningley, Doncaster); given number D7; noted freshly repainted and lost its BR number, Briton Ferry Steel Co Ltd, 7th July 1969; withdrawn at end of 1970; scrapped, June 1971.

D2601   HE   5650   1960   50D   12/67   F   D5
new 20th July 1960; withdrawn, 30th December 1967; seen at 50D Goole Depot, 28th January 1968; sold to Llanelly Steel Co Ltd; despatched from Goole Depot, 29th April 1968; delivered to Llanelly Steel Co Ltd, Carmarthenshire (sold per R.E. Trem Ltd, Finningley, Doncaster), 3rd May 1968; noted at Llanelly Steel, 5th July 1969; noted with number D5, Llanelly Steel, 4th October 1969; out of use by August 1976; scrapped, September 1979.

D2607   HE   5656   1960   6G   12/67   F   D2607
new 18th October 1960; withdrawn, 30th December 1967; sold to NCB; moved to Dinnington Colliery, July 1968; noted at Dinnington Colliery, 21st July 1968 and 9th September 1968; to Steetley Colliery about October 1968; noted at Steetley Colliery, 9th November 1968; seen at Steetley Colliery, 1st January 1971; to Shireoaks Colliery, on loan, May 1971; returned to Steetley Colliery, August 1971; noted at Steetley Colliery, 28th May 1973; to NCB Fence Workshops, Woodhouse Mill, Sheffield, for overhaul, 28th May 1974; returned to Steetley Colliery, 30th October 1974; to Shireoaks Colliery, on loan, 17th May 1975; returned to Steetley Colliery, July 1975; to Treeton Colliery, on loan, 3rd December 1975; returned to Steetley Colliery, December 1975; noted at Steetley Colliery, 21st February 1976; seen at BR Doncaster Depot, for tyre turning, 20th August 1977; returned to Steetley Colliery about September 1977; noted at Steetley Colliery, 6th November 1977; seen in original BR green livery with number D2607, with BTC 2881 of 1976 registration plate, Steetley Colliery, 28th August 1979; noted at Steetley Colliery, 19th April 1980; to Shireoaks Colliery, on loan, 26th August 1980; returned to Steetley Colliery, 16th January 1981; to Treeton Colliery, on loan, April 1981; returned to Steetley Colliery, July 1981; noted dumped, no side rods, Steetley Colliery, 25th April 1983; to Coopers (Metals) Ltd, Brightside, Sheffield, for scrap, 12th June 1984; scrapped, by 4th July 1984.

D2611   HE   5660   1961   50D   12/67   F   D2611 / YM1835
new 16th January 1961; withdrawn, 30th December 1967; seen at 50D Goole Depot, 28th January 1968; sold to NCB; despatched from 50D Goole Depot, May 1968; delivered to NCB Yorkshire Main Colliery, Edlington, Doncaster; an NCB employee stated it had been purchased from BR for £1,000; noted in BR green livery at Yorkshire Main Colliery, 9th June 1968; seen at Yorkshire Main Colliery, 22nd August 1970; noted at Yorkshire Main Colliery, 27th July 1975; written off, 3rd August 1976; scrapped on site, about December 1976.

D2613   HE   5662   1961   50D   12/67   F   D2613 / BRM5481
new 21st February 1961; noted (presumably on-hire), Hatfield Colliery, 3rd June 1967; noted on-hire, Thorne Colliery Washery, 29th August 1967; withdrawn, 30th December 1967; sold to NCB; despatched from 50D Goole Depot, May 1968; delivered to NCB Hatfield Colliery, May 1968; to NCB Brodsworth Colliery, Woodlands, Doncaster, 1968; noted with number BRM5481, Brodsworth Colliery, 9th November 1968; seen in BR green livery with numbers D2613 and BRM5481, Brodsworth Colliery, 2nd May 1971; to Bentley Colliery, 1974; noted at Bentley Colliery, 14th March 1975 and 15th August 1976; written off, 6th April 1977; scrapped on site by Walter Heselwood Ltd of Sheffield, June 1977.

D2616   HE   5665   1961   50D   12/67   F   D2616
new 17th March 1961; noted at Hatfield Colliery, Doncaster (possibly on trial or on approval), 22nd October 1967; withdrawn, 30th December 1967; seen at 50D Goole Depot, 28th January 1968; sold to NCB; despatched from 50D Goole Depot, May 1968; delivered to NCB Hatfield Colliery, Stainforth, Doncaster; noted at Hatfield Colliery, 9th June 1968;

seen in green livery with number D2616, Hatfield Colliery, 2nd May 1971 and 26th June 1971; dismantled in April 1973; scrapped, June 1973.

D2617　　HE　　　　5666　　1961　　62C　　　12/67　F　D2617
new 28th March 1961; withdrawn, 30th December 1967; noted at Dunfermline Depot, July 1968; sold to Hunslet Engine Co Ltd, Leeds, and moved September 1968; used for spares in rebuilds of D2587 and D2595 (which see); seen at Hunslet Engine Co, 21st November 1969 and 28th June 1970; remains scrapped on site by Robinson & Birdsell in late April/early May 1971.

SECTION 8:

North British Locomotive Co Ltd built 0-4-0 diesel hydraulic locomotives, numbered D2708-D2780, and introduced 1957. Fitted with a North British/M.A.N. W6V 17.5/22A engine developing 225bhp at 1100rpm, and driving wheels of 3ft 6in diameter. No TOPS classification.

D2720　　NB　　　　27815　　1958　　64H　　　8/67　F　NFT
new, 24th June 1958; withdrawn, 3rd August 1967; sold to James N. Connell Ltd; despatched from Leith Central; to James N. Connell Ltd, Coatbridge, Lanarkshire, about May 1968; noted at Connell, 2nd June 1968; noted with no number, 12th June 1971; scrapped, July 1971.

D2726　　NB　　　　27821　　1958　　ZN　　　2/67　F　NFT
new 30th July 1958; transferred to Wolverton Works, where arrived on 3rd March 1966; withdrawn, 11th February 1967; sold to R.E. Trem Ltd, Finningley, Doncaster, and despatched from Wolverton Works direct to Shipbreaking (Queenborough) Ltd, Kent, September 1967; used as yard shunter; noted at the end of the pier, with no engine, Queenborough, 12th September 1971; the engine had been removed and sent to R.E. Trem Ltd, Finningley, Doncaster; remains scrapped on site, October 1971.

D2736　　NB　　　　27831　　1958　　65A　　　3/67　F　D2736
new 20th November 1958; withdrawn, 4th March 1967; seen at 65A Eastfield Depot, 2nd April 1967; sold to Bird's; despatched from Eastfield Depot, Glasgow, July 1967; to Bird's Commercial Motors Ltd, Long Marston, Worcestershire; to Bird's (Swansea) Ltd, Pontymister Works, Risca, August 1967; noted at Risca, 17th December 1967; to Bird's (Swansea) Ltd, 40 Acre Site, Cardiff, 25th February 1968; noted with engine removed, 40 Acre Site, 28th April 1968; noted at 40 Acre Site, 24th October 1968 and 6th October 1969; scrapped about October 1969.

D2738　　NB　　　　27833　　1958　　65A　　　6/67　F　NFT
new 28th November 1958; seen at 65A Eastfield Depot, 2nd April 1967; withdrawn, 24th June 1967; sold to Andrew Barclay, Sons & Co Ltd, Kilmarnock, and moved October 1967; one of four North British shunters purchased by Andrew Barclay for refurbishment and resale; rebuilt 1968; noted at Kilmarnock, 21st April 1969; sold to NCB; to Killoch Colliery, Ochiltree, about July 1969; noted fitted with 'Andrew Barclay rebuild 1968' cast plate, Killoch Colliery, 4th April 1970; seen in faded BR green livery, with no number, Killoch Colliery, 29th May 1978; re-sold to Alex Smith Metals of Ayr, by March 1979; scrapped on site, January 1980.

D2739   NB   27834   1958   65A   3/67   F   D2739
new 11th December 1958; put in store at Kipps Depot, 26th November 1966; withdrawn, 4th March 1967; sold to Bird's; despatched from Kipps Depot, July 1967; to Bird's Commercial Motors Ltd, Long Marston, Worcestershire; scrapped, September 1969.

D2757   NB   28010   1960   65A   7/67   F   NFT
new 24th May 1960; withdrawn, 3rd July 1967; sold to Bird's; despatched from Eastfield Depot; noted at Abergavenny Goods Yard, 4th November 1967; arrived at Bird's (Swansea) Ltd, Pontymister Works, Risca, 9th November 1967; noted at Pontymister, 10th March 1968; to Bird's (Swansea) Ltd, 40 Acre Site, Cardiff, February 1968; noted at 40 Acre Site on various dates between 17th February 1969 and 18th August 1971; scrapped, later in 1971; the cab was noted, still laid in the yard on 20th October 1972.

D2763   NB   28016   1960   65A   6/67   F   NFT
new 4th July 1960; seen at 65A Eastfield Depot, 2nd April 1967; withdrawn, 24th June 1967; sold to Andrew Barclay, Sons & Co Ltd, Kilmarnock, and moved October 1967; one of four North British shunters purchased by Andrew Barclay for refurbishment and resale; rebuilt 1968; noted at Andrew Barclay, Kilmarnock, 21st April 1969; re-sold to BSC Landore Foundry, Swansea, and moved there by September 1969; it carried no number; scrapped, April 1977.

D2767   NB   28020   1960   65A   6/67   P   D2767
new 17th August 1960; seen at 65A Eastfield Depot, 2nd April 1967; withdrawn, 24th June 1967; sold to Andrew Barclay, Sons & Co Ltd, Kilmarnock, and moved October 1967; one of four North British shunters purchased by Andrew Barclay for refurbishment and resale; rebuilt 1968; noted at Andrew Barclay, Kilmarnock, 21st April 1969; re-sold to Burmah Oil Co Ltd, Stanlow, Cheshire, and arrived on 24th April 1969; noted with no number, Stanlow, 14th August 1970 and 6th April 1982; re-sold to East Lancashire Railway, Bury, and moved there by road on 12th June 1983; seen in green livery with number D2767 plus small numberplate 76.067, East Lancashire Railway, Bury, 20th August 1983; to Manchester Metrolink, on hire, 25th October 1991; returned to East Lancashire Railway, Bury, late 1991; noted in green livery with number D2767, Bury, 12th June 1995; re-sold to Scottish Railway Preservation Society, Bo'ness, and moved 25th July 2001; noted in green livery with number D2767, Bo'ness, April 2007 and 26th July 2015.

D2774   NB   28027   1960   65A   6/67   P   D2774
new 11th October 1960; seen at 65A Eastfield Depot, 2nd April 1967; withdrawn, 24th June 1967; sold to Andrew Barclay, Sons & Co Ltd, Kilmarnock, and moved October 1967; one of four North British shunters purchased by Andrew Barclay for refurbishment and resale; rebuilt 1968; to NCB Killoch Colliery, Ochiltree, on hire, about July 1969; noted fitted with 'Andrew Barclay rebuild 1968' cast plate, Killoch Colliery, 4th April 1970; returned to Andrew Barclay, Sons & Co Ltd, 1971; re-sold to NCB; to Celynen North Colliery, Newbridge, early April 1971; noted at Celynen North, 31st May 1972; to Hafodyrynys Colliery, Pontypool, about 1974; noted at Hafodyrynys Colliery, 25th April 1975; to BR Canton Depot, Cardiff, for tyre turning, March 1976; noted at BR Canton Depot, 19th March 1976; noted back at Hafodyrynys Colliery, 26th March 1976; to Celynen North Colliery, Newbridge, April 1976; to Celynen South Colliery, Abercarn, 4th September 1976; noted in green livery with no number, Celynen South Colliery, June 1977; to Celynen North Colliery, Newbridge, March 1978; noted at Celynen North Colliery, Newbridge, 21st March 1978; to Celynen South Colliery, early 1980; noted at Celynen South Colliery, 1st June 1980; seen in green livery with no number, Celynen South Colliery, 24th March 1982; to BR Canton

Depot, Cardiff, for tyre turning, 17th May 1982; to NCB Mountain Ash Works, 27th May 1982; noted at Mountain Ash Colliery, 23rd December 1982; to Celynen South Colliery, Abercarn, by 12th March 1983; noted disused at Celynen South, 12th August 1985; last rail traffic at Celynen South, 23rd October 1985; noted disused, Celynen South, 5th April 1986 and 16th February 1987; re-sold to East Lancashire Railway, Bury, by 26th May 1987; noted in green livery with number D2774, Bury, 12th June 1995; re-sold to Strathspey Railway, Aviemore, and moved 2nd May 2001; noted in green livery with number D2774, Boat of Garten, 7th June 2024.

D2777   NB   28030   1960   65A   3/67   F   D2777
new 21st November 1960; stored at Kipps Depot from 26th November 1966; withdrawn, 4th March 1967; sold to Bird's; despatched from Kipps Depot; to Bird's Commercial Motors Ltd, Long Marston, Worcestershire, July 1967; to Bird's (Swansea) Ltd, Pontymister Works, Risca, about 9th November 1967; noted at Pontymister, 10th March 1968; used for spares; remains scrapped, May 1968.

SECTION 9:

Yorkshire Engine Co Ltd built 0-4-0 diesel hydraulic locomotives, numbered D2850-D2869, and introduced 1960. Fitted with a Rolls-Royce C6NFL engine developing 179bhp at 1800rpm, and driving wheels of 3ft 6in diameter. Later classified TOPS Class 02.

D2853   YE   2812   1960   8J   6/75   P   02003 / PETER
02003   ex-works, YE Meadowhall Works, 25th October 1960; withdrawn, 1st June 1975; noted in green livery with number 02003, Allerton Depot, 30th July 1975; sold to L.C.P. Fuels Ltd; despatched from Allerton Depot; to L.C.P. Fuels Ltd, Shut End Works, West Midlands, 19th November 1975; named PETER; noted in green livery with name PETER, Shut End, 26th March 1982; was named after Peter Pitt of Lunt, Comley & Pitt; seen in blue livery with name PETER, Shut End, 25th September 1984; seen in shed, Shut End, 27th September 1987; displayed (in green livery with number 02003) at BR Bescot Depot, Walsall, open day, 9th October 1988; returned to Shut End, October 1988; noted at Shut End, 15th November 1995; re-sold to HNRC and moved to South Yorkshire Railway Preservation Society, Meadowhall, Sheffield, 18th April 1997; to Rutland Railway Museum, Cottesmore, for storage, 22nd June 2001; to HNRC, Barrow Hill Engine Shed, Staveley, 1st September 2003; noted in blue livery with number 02003, Barrow Hill Engine Shed, 5th October 2003; to Appleby Frodingham Railway Society, Scunthorpe, on loan, 24th March 2006; noted in green livery with number D2853, Appleby Frodingham, June 2011; to Barrow Hill Engine Shed, 28th January 2016; seen with number D2853, Barrow Hill Engine Shed, 5th May 2016; seen in green livery with number 02003, Barrow Hill Engine Shed, 12th March 2022; advertised for sale by HNRC, October 2022.

D2854   YE   2813   1960   8J   2/70   P   D2854
ex-works, 4th November 1960; noted at 16C Derby Depot, 29th March 1969; withdrawn as 'surplus – wear and tear', 28th February 1970; sold per R.E. Trem Ltd, Finningley, Doncaster, for C.F. Booth Ltd; despatched from Allerton Depot; to C.F. Booth Ltd, Rotherham, 29th August 1970; seen at C.F. Booth Ltd, 29th August 1970; repaired and used as a yard shunter; seen in green livery with number D2854, C.F. Booth Ltd, 22nd November 1972; fitted with a new engine, early March 1983; seen working, C.F. Booth Ltd, 16th March 1983; disused needing repairs, 1985; seen in faded BR green livery, long disused, C.F. Booth Ltd, 17th April 1988; re-sold to South Yorkshire Railway Preservation

Society, Attercliffe, Sheffield, and moved in May 1988; to South Yorkshire Railway Preservation Society, Meadowhall, Sheffield (next door to where the loco was built), 14th September 1988; noted in green livery, Meadowhall, 27th August 1989; to Middleton Railway, Leeds, on loan, 10th September 1994; returned to South Yorkshire Railway Preservation Society, 15th October 1994; to Supertram, Nunnery Depot, Sheffield, on hire, 7th November 1994; returned to South Yorkshire Railway Preservation Society, 27th May 1995; to Heritage Shunters Trust, Rowsley, March 2002; seen in green livery with number D2854, Rowsley, 12th December 2003 and 24th July 2021.

D2856   YE   2815   1960   8J   6/75   F   02004
02004   ex-works, 2nd December 1960; withdrawn, 1st June 1975; noted in green livery with number 02004, Allerton Depot, 30th July 1975; sold to Redland Roadstone Ltd; despatched from Allerton Depot; to Redland Roadstone Ltd, Buddon Wood Quarry, Mountsorrel, Leicestershire, 25th September 1975; seen at Mountsorrel, 4th January 1976; used for spares; noted (cab, frame and wheels only), dumped in the works lorry park, Buddon Wood Quarry, Mountsorrel, on various dates from 30th September 1976 to May 1986; scrapped on site by The Vic Berry Company of Leicester, later in 1986.

D2857   YE   2816   1960   8J   4/71   F   NFT
ex-works, 14th December 1960; withdrawn, 18th April 1971; sold to Bird's; despatched from Allerton Depot, October 1971; to Bird's Commercial Motors Ltd, Long Marston, Worcestershire; noted at Long Marston, 28th October 1971 and 1st May 1973; noted with name SIR ALFRED, Long Marston, 20th June 1980; seen in green livery with no name or number, Long Marston, 26th September 1984; seen in light blue livery with no number, Long Marston, 24th July 1990 and 31st March 1992; scrapped on site, June 1992.

D2858   YE   2817   1960   9A   2/70   P   D2858
ex-works, 21st December 1960; seen stored at Newton Heath Depot, 1st January 1970; withdrawn as 'surplus to requirements', 28th February 1970; sold to Hutchinson; despatched from Newton Heath Depot, about 4th August 1970; to Hutchinson Estate & Dock Co (Widnes) Ltd, Widnes; noted at Hutchinson Estate, 14th August 1970; to Fisons Fertilisers, Widnes, (who took over loco and sidings), September 1978; noted in green livery, Widnes, 12th September 1979; re-sold to Lowton Metals Ltd, Haydock, and moved 5th March 1981; noted at Lowton Metals, 18th July 1983 and 5th September 1986; Lowton Metals closed at the end of December 1986; re-sold to Butterley Engineering Ltd, Ripley, Derbyshire, and moved December 1986; seen in maroon livery with no number, Butterley, 7th June 1987, 26th February 1989 and 26th July 1999; re-sold to Midland Railway, Butterley, Derbyshire, and moved 14th June 2002; seen in green livery with number D2858, Butterley, 30th April 2011.

D2860   YE   2843   1961   8J   12/70 P   D2860
ex-works, 13th September 1961; new to Fleetwood shed, 16th September 1961; withdrawn, 13th December 1970; stored at Allerton; selected to become part of the National Collection; despatched from Allerton Depot, March 1973; noted at Norwood Junction, 21st March 1973; noted at Three Bridges, 26th March 1973; arrived at Curator of Historical Relics, BR Preston Park, Brighton, late March 1973; stored; to National Railway Museum, York, 26th November 1977; to Thomas Hill (Rotherham) Ltd, Kilnhurst, for overhaul and repaint, 14th September 1978; to National Railway Museum, York, 3rd January 1979; used as museum pilot; seen with 'Thomas Hill restored 1978' plate, National Railway Museum, York, 8th March 1980; noted in green livery, National Railway Museum, York, 2nd September 1981; to Gloucestershire Warwickshire Railway, Toddington, for gala, 9th July

2009; to National Railway Museum, York, mid-July 2009; noted in green livery with number D2860, National Railway Museum, York, 29th January 2014.

D2862   YE   2845   1961   10D   12/69   F   ND3 / 63000359
ex-works, 16th October 1961; seen at BR Lostock Hall Depot, 23rd August 1969; withdrawn, 14th December 1969; sold to Tilsley & Lovatt Ltd; despatched from Lostock Hall Depot; to Tilsley & Lovatt Ltd, Trentham, Staffordshire, March 1970; overhauled and repainted; sold to NCB; to Norton Colliery, Staffordshire, January 1971; allocated number ND3; to Chatterley Whitfield Colliery, Tunstall, week-ending 25th September 1971; to Norton Colliery, week-ending 15th October 1971; noted in blue livery with number 63000359, Norton Colliery, June 1976; scrapped, about April 1979.

D2865   YE   2848   1961   50D   3/70   F   NFT
ex-works, 7th November 1961; seen at 50D Goole Depot, 30th August 1969; withdrawn as 'surplus to requirements', 21st March 1970; seen in sidings adjacent to Goole Station, 31st May 1970; sold to APCM, Kilvington, Nottinghamshire, and moved in September 1970; noted at Kilvington, 16th April 1973; seen in bright yellow livery with no BR number, Kilvington, 2nd May 1979 and 17th April 1981; rail traffic ceased, mid-1983; noted disused, Kilvington, 9th November 1984; to Blue Circle Industries PLC, Beeston Cement Terminal, Nottingham, for storage, after April 1984; to The Vic Berry Company, Leicester, for scrap, December 1984; noted at Berry, 29th December 1984; scrapped, June 1985.

D2866   YE   2849   1961   9A   2/70   P   NPT
ex-works, 10th November 1961; withdrawn as 'surplus to requirements', 28th February 1970; noted at BR Newton Heath Depot, 25th July 1970; sold to W.H. Arnott, Young & Co Ltd; despatched from Newton Heath Depot, about August 1970; to W.H. Arnott, Young & Co Ltd, Dalmuir, Dunbartonshire; allocated number AY1021; to BREL Glasgow Works, for repairs, April 1977; returned to W.H. Arnott, Young & Co Ltd, Dalmuir; noted in yellow livery with no number, Dalmuir, 2nd April 1981; became disused, about 1985; re-sold to Caledonian Railway, Brechin, and arrived on 17th October 1987; restored to working order, 1987; noted in yellow livery with number AY1021, Brechin, 3rd April 1988; re-sold to South Yorkshire Railway Preservation Society, Meadowhall, Sheffield, and arrived on 22nd January 1996; to Heritage Shunters Trust, Rowsley, March 2002; seen in blue livery at Heritage Shunters Trust, Rowsley, 12th December 2003.

D2867   YE   2850   1961   6A   9/70   P   D2867
ex-works, 17th November 1961; to Tunnel Portland Cement Co Ltd, Penyffordd, on hire, August 1970; withdrawn, 12th September 1970; sold to Redland Roadstone Ltd; despatched from Chester Depot; to Redland Roadstone Ltd, Budden Wood Quarry, Mountsorrel, Leicestershire, and arrived 23rd October 1970; the quarry was rail-linked to Barrow upon Soar rail loading terminal and, although the locomotive was mostly found at Mountsorrel, it could be seen at either location; at some stage (date not known) the rail link was replaced by a conveyor, and thereafter the locomotive was permanently based at Barrow upon Soar; noted in light green livery with no number and with name DIANE on one side only, Barrow upon Soar, 8th June 1980; seen in green livery with no number and named DIANE, Barrow upon Soar, 3rd April 1986; re-sold to Harry Needle Railroad Company; to South Yorkshire Railway Preservation Society, Meadowhall, Sheffield, 31st March 1995; to Battlefield Line, Shackerstone, on long-term loan, 27th July 2001; put up for sale by HNRC, October 2022; sold to Northamptonshire Ironstone Railway Trust; unloaded at NIRT, with help of JF 4220001, 1st June 2023.

D2868   YE   2851   1961   10D   11/69 P D2868

ex-works, 24th November 1961; withdrawn, 14th November 1969; seen at BR Lostock Hall Depot, 15th August 1970; sold to Lunt, Comley & Pitt Ltd, Shut End Works, Staffordshire, and moved mid-October 1970; named SAM after the fuel yard foreman; seen disused, in blue livery with name SAM, Shut End, 25th September 1984; seen in shed, Shut End, 27th September 1987; noted at Shut End, 15th November 1995; re-sold to South Yorkshire Railway Preservation Society, Meadowhall, Sheffield, and moved 25th April 1997; to Lavender Line, Isfield, for storage, June 2001; noted in faded blue livery, Lavender Line, 15th July 2001; to Barrow Hill Engine Shed, Staveley, 28th January 2004; to Heritage Shunters Trust, Rowsley, 13th May 2008; seen in green livery with number D2868, Heritage Shunters Trust, Rowsley, 12th December 2008; to Museum of Science & Industry, Manchester, on loan, 14th November 2011; noted in green livery with number D2868, Manchester, January 2012; to Heritage Shunters Trust, Rowsley, 1st February 2013; to Barrow Hill Engine Shed, Staveley, 8th October 2018; re-sold to Barrow Hill Engine Shed Society, 2019; green livery with number D2868, Barrow Hill Engine Shed, 12th March 2022.

SECTION 10:

Hunslet Engine Co Ltd built 0-4-0 diesel mechanical locomotives, numbered D2950-D2952, and introduced 1954. Fitted with a Gardner 6L3 engine developing 153bhp at 1200rpm, four speed gearbox, and driving wheels of 3ft 4in diameter. No TOPS classification.

D2950   HE   4625   1954   50D   12/67 F D4

new as number 11500 to Ipswich Depot, 3rd December 1954; cowcatchers and motion guards fitted for use in Ipswich docks; renumbered D2950, April 1958; withdrawn, 30th December 1967; sold to Llanelly Steel Co Ltd; despatched from Goole Depot, 29th April 1968; to Llanelly Steel Co Ltd, Carmarthenshire (sold per R.E. Trem Ltd, Finningley, Doncaster); arrived at Llanelly Steel, 2nd May 1968; numbered D4 in May 1968; noted in green livery with number D4, Llanelly Steel, 4th October 1969 and 4th May 1978; re-sold for preservation and moved to Thyssen Ltd, Old Castle Depot, Llanelli, May 1980; stored pending future restoration; noted with number D4, Thyssen Ltd, 8th July 1980; the project foundered and the locomotive was scrapped on site by Gwillym Jones & Son, spring 1983; the engine was retained and later used in a trawler.

SECTION 11:

Andrew Barclay, Sons & Co Ltd built 0-4-0 diesel mechanical locomotives, numbered D2953-D2956, and introduced 1956. Fitted with a Gardner 6L3 engine developing 153bhp at 1200rpm, four speed gearbox, and driving wheels of 3ft 2in diameter. BR Departmental locomotive number 81 became the second D2956 after the first one was withdrawn. Later classified TOPS Class 01 (a class numbering just five examples).

D2953   AB   395   1955   30A   6/66 P D2953

new as number 11503 to Stratford Depot, 24th January 1956; renumbered D2953, October 1960; noted at March Depot, 11th April 1965; withdrawn, 19th June 1966; sold to Thames Matex Ltd; despatched from Stratford Depot; to Thames Matex Ltd, West Thurrock, Essex, June 1966 (the first BR capital stock diesel shunter sold for industrial service); to BP Refinery (Kent) Ltd, Grain, Kent, on loan, 1967; returned to Thames Matex; noted in blue livery with no number, Thames Matex, 28th May 1971 and 16th September 1971; to Shell

Mex & BP Ltd, Purfleet, on loan, at various times; noted recently repainted in green livery with number D2953, Thames Matex, 14th February 1977; out of use with gearbox problems, about 1980; re-sold to John Wade; to South Yorkshire Railway Preservation Society, Chapeltown, 15th December 1985; to South Yorkshire Railway Preservation Society, Attercliffe, Sheffield, December 1986; seen repainted in blue livery with number D2953, Attercliffe, 18th July 1987; to South Yorkshire Railway Preservation Society, Meadowhall, Sheffield, 15th September 1988; noted in green livery, Meadowhall, 27th August 1989; to Heritage Shunters Trust, Rowsley, March 2002; repaired and put to use in March 2003; seen in green livery with number D2953, Rowsley, 24th July 2021.

D2956　AB　　　398　1956　36A　　5/66　P　D2956
new as number 11506, 30th March 1956; withdrawn, 29th May 1966; sold to A. King & Sons Ltd, Norwich, and moved July 1966; the second capital stock diesel shunter sold into industrial service; noted at A. King & Sons Ltd, Norwich, 5th August 1979 and 25th September 1980; to A. King & Sons Ltd, Snailwell, Cambridgeshire, September 1981; noted disused, in faded green livery with number A48L, Snailwell, 31st December 1983; withdrawn from service due to defective transmission, 1984; donated to East Lancashire Railway, Bury, and moved by road, 30th July 1985; initially painted black and given bogus number 01003; noted in black livery with number 01003, East Lancashire Railway, 31st August 1986; number 01003 was later removed; used for many years as Castlecroft Yard pilot; noted in black livery with number 11506, East Lancashire Railway, 30th June 1996 and 7th July 1998; noted in green livery with number D2956, East Lancashire Railway, 26th March 2002, 6th July 2005 and 1st July 2008; noted in black livery with number D2956, East Lancashire Railway, April 2011 and July 2014; noted in black livery with no number, 8th March 2015 and 28th August 2015; stored serviceable in Baron Street shed, July 2024.

D2956　AB　　　424　1958　36A　　11/67　F　D5
new to BTC Departmental Board, July 1958; after despatch from Andrew Barclay's it ran hot and was repaired at Dumfries Depot, July 1958; new to Peterborough Engineers Department, August 1958; worked as Departmental locomotive number 81; was at Chesterton Junction by May 1960; moved to Doncaster Depot, 2nd July 1967; placed in running stock, 8th July 1967; was surprisingly given the same number as the previously withdrawn D2956 (see above); withdrawn, 12th November 1967; stored at Doncaster Depot; sold to Briton Ferry Steel Co Ltd, Glamorgan, and moved 19th March 1968 (sold per R.E. Trem Ltd, Finningley, Doncaster); given number D5; scrapped, August 1969.

SECTION 12:

Ruston & Hornsby Ltd built 0-4-0 diesel mechanical locomotives (Ruston's class 165DS), numbered D2957-D2958, and introduced 1956. Fitted with a Ruston 6VPHL engine developing 165bhp at 1250rpm, four speed gearbox, and driving wheels of 3ft 4in diameter. No TOPS classification.

D2958　RH　　　390777 1956　30A　　1/68　F　NFT
new to Immingham Depot as number 11508, 28th May 1956; transferred to Stratford Depot, late 1956; mainly used as depot pilot; withdrawn, 14th January 1968; noted at Stratford Depot, 21st January 1968; sold to C.F. Booth Ltd; despatched from Stratford Depot, 30th April 1968; to C.F. Booth Ltd, Rotherham, May 1968 (sold per R.E. Trem Ltd, Finningley, Doncaster); seen in BR green livery with number D2958, with odometer reading of 3,816 miles, C.F. Booth Ltd, 27th June 1968; used as a yard shunter; seen working, in BR green

livery but with BR number painted over, C.F. Booth Ltd, 22nd November 1972 and 5th July 1975; noted in green livery, with tarpaulin over engine compartment, 18th January 1981; withdrawn in 1981; seen at C.F. Booth Ltd, 20th September 1984; scrapped, October 1984.

SECTION 13:

Ruston & Hornsby Ltd built 0-6-0 diesel electric locomotives (Ruston's class LSSE), numbered D2985-D2998, and introduced 1962 for working at Southampton Docks where they replaced the 'USA' steam locomotives. Fitted with a six-cylinder Paxman 6RPHL engine developing 275bhp at 1360rpm, and driving wheels of 3ft 6in diameter. Later classified TOPS Class 07.

D2985   RH   480686 1962   70D   7/77   P   07001
07001   delivered by rail to Eastleigh, 9th June 1962; to Southampton Docks, June 1962; renumbered 07001 in January 1974; withdrawn, 2nd July 1977; sold to Tilsley & Lovatt Ltd; despatched from BR Eastleigh Depot, 4th April 1978; noted on a low loader on the A34, Litchfield, Hampshire, 4th April 1978; delivered to Tilsley & Lovatt Ltd, Trentham, Staffordshire, 5th April 1978; overhauled and repainted; re-sold to Peakstone Ltd, Holderness Limeworks, Peak Dale, Derbyshire, and moved 30th May 1978; seen in yellow livery with no number, Peak Dale, 27th September 1979, 24th April 1984 and 23rd April 1986; out of use from 1986; re-sold to Harry Needle Railroad Company, 26th April 1989; to South Yorkshire Railway Preservation Society, Meadowhall, Sheffield, June 1989; noted in yellow livery, Meadowhall, 27th August 1989; to Mayer Parry Ltd, Snailwell, on hire, 28th April 1993; returned to South Yorkshire Railway Preservation Society, 22nd October 1997; to Barrow Hill Engine Shed, Staveley, 28th June 1999; repainted in HNRC yellow livery, 2000; was Railtrack registered, December 2000; seen in yellow livery with number 07001, on low-loader on M62, 5th March 2001; to Creative Logistics, Salford, on hire, 5th March 2001; to Barrow Hill Engine Shed, Staveley, 20th July 2009; placed in storage; seen with number 07001, Barrow Hill Engine Shed, 4th May 2011; offered for sale to preservation groups in 2012; re-sold to Heritage Shunters Trust, Rowsley, and moved 20th December 2012; noted in yellow livery, January 2013; repainted in BR blue livery, summer 2013; used on Peak Rail works trains, 2015; seen in BR blue livery with number 07001, Rowsley, 24th July 2021.

D2986   RH   480687 1962   70D   7/77   F   NFT
07002   new to Southampton Docks, 16th June 1962; withdrawn, 2nd July 1977; noted at Eastleigh Depot, 11th February 1978; sold to NCBOE; to Powell Duffryn Fuels Ltd, NCBOE Coed Bach Disposal Point, Kidwelly, Dyfed, April 1978; noted at Coed Bach, 4th May 1978; noted with no number, Coed Bach, 8th July 1980; scrapped on site by T. Davies of Llanelli, September 1982.

D2987   RH   480688 1962   70D   10/76   F   NFT
07003   new to Southampton Docks, 16th June 1962; withdrawn, 4th October 1976; sold to R.E. Trem Ltd, Finningley, Doncaster, and moved there by road low-loader in March 1977; re-sold to British Industrial Sand Ltd, Oakamoor, Staffordshire; noted on a low-loader on the A38 bypass, Alfreton, 23rd June 1978; delivered to British Industrial Sand Ltd, Oakamoor; seen in white livery with no number, with brown camel logo on cabside, disused, Oakamoor, 2nd July 1983; scrapped on site by British Industrial Sand workmen, May 1985.

D2989   RH   480690 1962   70D   7/77   P   07005

07005   new to Southampton Docks, 23rd June 1962; withdrawn, 2nd July 1977; sold to Resco (Railways) Ltd, Woolwich, and moved June 1978; overhauled (works number L106 of 1978); re-sold to ICI Wilton Works, Middlesbrough, and arrived there on 18th July 1979; named LANGBAURGH in November 1979; seen in grey livery with number 07005 and ICI logo on cabside, and with Resco L106 plate, ICI Wilton, 14th June 1988; last worked on 5th July 1987; traction motor removed as spare for 07011; re-sold to Harry Needle Railroad Company, 2000; to Barrow Hill Engine Shed, Staveley, for storage, 21st December 2000; noted in light blue livery with no number, 7th October 2001; to Battlefield Line, Shackerstone, 3rd September 2003; re-sold to Great Central Railway, Loughborough, and moved 21st May 2008; noted in green livery with number D2989, Great Central Railway, Loughborough, September 2013; to Boden Rail Engineering, Washwood Heath, early 2014; to Great Central Railway, Loughborough, 18th January 2018.

D2990   RH   480691 1962   70D   7/77   F   NFT

07006   new to Southampton Docks, 13th July 1962; withdrawn, 2nd July 1977; sold to NCBOE; to Powell Duffryn Fuels Ltd, NCBOE Coed Bach Disposal Point, Kidwelly, Dyfed, April 1978; noted at Coed Bach, 4th May 1978; noted with no number, Coed Bach, 8th July 1980; noted at Coed Bach, 23rd April 1984; suffered fire damage in 1984; scrapped on site by T. Davies of Llanelli, October 1984.

D2991   RH   480692 1962   70D   5/73   P   2991/BRUCE

new to Southampton Docks, 21st July 1962; withdrawn, 6th May 1973; to Eastleigh Works, 1973; taken into departmental service and used as a stationary generator; to Eastleigh Depot open day, 29th May 1983; resumed as stationary generator until 1984, then placed into store; noted in blue livery with number 2991, Eastleigh Works, May 1986; adopted by works staff; repainted green by works staff, October 1986; noted in green livery with number D2991, Eastleigh Works, May 1988; donated to Eastleigh Railway Preservation Society, Eastleigh Works, about September 1988; noted in green livery with number D2991, Eastleigh Works, 28th September 2005; to Knights Rail Services Ltd, Eastleigh Works, as works shunter, about March 2007; given number 07007 (which it never carried in BR service) during a major overhaul, January 2008; resumed working, 19th February 2008; main line registered, April 2008; to Swanage Railway, Dorset, for gala, 6th May 2008; returned to Eastleigh Works, 16th May 2008; noted in blue livery with number 07007, Eastleigh Works, July 2013 and October 2016; repainted, renumbered 2991, and named BRUCE, Eastleigh, March 2023.

D2993   RH   480694 1962   70D   10/76   F   ?

07009   new to Southampton Docks, 21st August 1962; withdrawn, 4th October 1976; sold to Shipbreaking (Queenborough) Ltd, Kent, and moved by low-loader early March 1977; exported from Sheerness Docks, March 1977 (see Appendix C).

D2994   RH   480695 1962   70D   10/76   P   07010

07010   new to Southampton Docks, 5th September 1962; withdrawn, 4th October 1976; noted in BR blue livery with number 07010, Eastleigh Depot, 25th February 1978; sold to Resco (Railways) Ltd; despatched from Eastleigh Depot, June 1978; to Resco (Railways) Ltd, Woolwich; overhauled; re-sold to Winchester & Alton Railway, New Alresford, Hampshire, August 1978; noted at Alresford, 2nd June 1979; re-sold to West Somerset Railway, Minehead, and moved 19th May 1980; noted in green livery with number D2994, West Somerset Railway, September 1986 and 14th July 1991; re-sold to Avon Valley

Railway, Bitton, Gloucestershire, 1st March 1994; noted freshly repainted in BR blue livery with number 07010, Bitton, 12th April 2014; to St Philip's Marsh Depot, Bristol, for tyre turning, 4th June 2014; to Avon Valley Railway, 13th June 2014; seen disused, in blue livery with number 07010, Avon Valley Railway, 16th July 2021; to Peak Rail sidings, Rowsley, 25th January 2024; tripped to Briddon, Darley Dale, for repairs, 29th January 2024.

D2995   RH   480696 1962   70D   7/77   P D2995
07011   new to Southampton Docks, 22nd September 1962; withdrawn, 2nd July 1977; noted in BR blue livery with number 07011, Eastleigh Depot, 18th February 1978; sold to Resco (Railways) Ltd; despatched from 70D Eastleigh Depot; to Resco (Railways) Ltd, Woolwich, June 1978; overhauled (works number L105 of 1978); re-sold to ICI and moved to Billingham Works, Stockton-on-Tees, March 1979; returned to Resco (Railways) Ltd, for repairs, 12th November 1979; to ICI Wilton Works, Middlesbrough, 4th September 1980; named CLEVELAND; seen in grey livery with number 07011 and name CLEVELAND, and with Resco works number L105 plate, ICI Wilton, 14th June 1988; re-sold to Hastings Diesels, St Leonards Depot, Hastings, East Sussex, and moved by road on 17th May 1996; noted at St Leonards Depot, 14th September 1997; to Kent & East Sussex Railway, Tenterden, Kent, 1998; to St Leonards Railway Engineering Ltd, East Sussex, about July 2000; fitted with high-level brake pipes on one end for shunting DEMU vehicles; noted in blue livery with number 07011, St Leonards, 30th June 2011; noted in multi-coloured livery with no number, St Leonards, 8th July 2015 and 15th May 2018; noted in green livery with number D2995, St Leonards, 9th November 2023.

D2996   RH   480697 1962   70D   7/77   P D2996
07012   new to Southampton Docks, 6th October 1962; withdrawn, 2nd July 1977; sold to NCBOE; to Powell Duffryn Fuels Ltd, NCBOE Cwm Mawr Disposal Point, Tumble, Dyfed, April 1978; noted at Cwm Mawr, 4th May 1978; to NCBOE Coed Bach Disposal Point, Kidwelly, about December 1981; noted at Coed Bach, 23rd April 1982; seen in blue and white livery with number 07012, Coed Bach, 2nd April 1991; re-sold to Harry Needle Railroad Company, 1992; noted on a low-loader, northbound on M1 Motorway, 11th December 1992; to South Yorkshire Railway Preservation Society, Meadowhall, Sheffield, 11th December 1992; to Barrow Hill Engine Shed, Staveley, 28th June 1999; to Lavender Line, Isfield, for storage, June 2001; noted in blue and white livery with number 07012, Lavender Line, 15th July 2001; to Barrow Hill Engine Shed, Staveley, 18th July 2006; to Appleby Frodingham Railway Society, Scunthorpe, January 2009; re-sold to a private syndicate; to Barrow Hill Engine Shed, Staveley, 28th January 2016; seen in BR blue livery with number 07012, Barrow Hill Engine Shed, 12th March 2022; seen repainted in BR green with number D2996, Barrow Hill Engine Shed, August 2022.

D2997   RH   480698 1962   70D   7/77   P 07013
07013   new to Southampton Docks, 20th October 1962; withdrawn, 2nd July 1977; noted in BR blue livery with number 07013, Eastleigh Depot, 18th February 1978; sold to Resco (Railways) Ltd, Woolwich, and moved May 1978; overhauled (works number L101 of 1978); re-sold to Dow Chemical Co Ltd, King's Lynn, and moved on 5th October 1978; seen in green livery with no number, King's Lynn, 15th April 1993; re-sold to Harry Needle Railroad Company; to South Yorkshire Railway Preservation Society, Meadowhall, Sheffield, by road, 16th August 1994; to Barrow Hill Engine Shed, Staveley, 28th June 1999; cosmetically restored to BR blue livery; to Heritage Shunters Trust, Rowsley, 24th October 2003; seen with no number, Rowsley, 12th December 2003; noted in BR rail blue livery with number 07013, Rowsley, 28th June 2008; to HNRC, Barrow Hill Engine Shed,

Staveley, 14th November 2011; re-sold to a private syndicate; moved to East Lancashire Railway, Bury, 14th May 2013; stored unserviceable, awaiting restoration, East Lancashire Railway, July 2024.

SECTION 14:

British Railways built 0-6-0 diesel electric locomotives, numbered D3000-D3116, and introduced 1953. Design based on the LMS shunters introduced in 1945. Fitted with an English Electric 6KT engine developing 350bhp at 630rpm, and driving wheels of 4ft 6in diameter. Later classified TOPS Class 08.

D3000   Derby   1952   82A   11/72   P   13000
new (order 6232) as number 13000, 1st November 1952; withdrawn, 19th November 1972; sold to NCB; to Bath Road Depot, Bristol, for pre-delivery service/repairs, early March 1973; to NCB Hafodyrynys Colliery, Pontypool, 19th March 1973; noted at Hafodyrynys Colliery, 20th May 1973; noted in green livery with number D3000, Hafodyrynys Colliery, April 1975; to BR Canton Depot, Cardiff, for repairs and tyre turning, 11th August 1975; noted at Canton Depot, 21st August 1975 and 27th October 1975; to Hafodyrynys Colliery, Pontypool, about November 1975; to Bargoed Colliery, 12th July 1978; to BR Canton Depot, Cardiff, October 1979; noted at Canton Depot, 7th October 1979; to Bargoed Colliery, October 1979; to BR Canton Depot, Cardiff, late May 1980; noted at Canton Depot, 1st June 1980 and 6th July 1980; to Bargoed Colliery, July 1980; to Mountain Ash Colliery, for repairs, 10th July 1981; to Mardy Colliery, 13th November 1981; to Mountain Ash Colliery, December 1981; to Mardy Colliery, May 1982; out of use after rail traffic ceased at Mardy Colliery in August 1986; re-sold to Brighton Railway Museum; despatched on 18th March 1987; arrived at Brighton, 21st March 1987; noted at BR Brighton Depot, 11th September 1987; re-sold to South Yorkshire Railway Preservation Society, Meadowhall, Sheffield, and moved 9th March 1993; to Barrow Hill Engine Shed, Staveley, by road, 7th March 2001; to Appleby Frodingham RPS, Scunthorpe, on loan, 14th April 2008; to Heritage Shunters Trust, Rowsley, 27th January 2011; to Harry Needle Railroad Company, Worksop Depot, 6th July 2020; noted in black livery with number 13000, Worksop Depot, 24th June 2024.

D3002   Derby   1952   82A   7/72   P   13002
new as number 13002, 24th October 1952; withdrawn, 17th July 1972; sold to Foster Yeoman Quarries Ltd, Merehead Stone Terminal, Somerset, and moved in November 1972; noted at BR Westbury Depot, for repairs, February 1976; to BR Bath Road Depot, Bristol, for repairs, 13th April 1976; noted at Bath Road Depot, 19th May 1976; returned to Merehead Stone Terminal, June 1976; given blue livery, number 11 and painted name DULCOTE; to Gloucester Depot, for repairs, late January 1980; noted at Gloucester Depot, 2nd February 1980 and 1st March 1980; returned to Merehead; noted in blue livery, Merehead Stone Terminal, 25th September 1981; re-sold to Plym Valley Railway; left Merehead Stone Terminal in late June 1982 and spent about two weeks stored in Westbury Yard; moved to Plym Valley Railway, Marsh Mills, Plymouth, 9th July 1982; seen in blue livery, with names YEOMAN and DULCOTE and number 11, Marsh Mills, 13th September 1982; noted in blue livery with number 13002, Marsh Mills, June 2012; noted in black livery with number 13002, Marsh Mills, 19th June 2016 and 4th September 2022.

D3003　Derby　　　　　　　1952　82A　　7/72　F　22 / MEREHEAD

new as number 13003, 31st October 1952; withdrawn, 17th July 1972; sold to Foster Yeoman Quarries Ltd, Merehead Stone Terminal, Somerset, and moved in May 1973; given number 22 and name MEREHEAD; to BR Bath Road Depot, Bristol, for repairs, March 1974; returned to Merehead; to BREL Derby Works, for repairs, 30th June 1974; noted at Derby Works, 4th August 1974 and 22nd September 1974; returned to Merehead Stone Terminal, late 1974; noted at Gloucester Depot, 3rd May 1980 and 5th May 1980; seen in blue livery with 22 and MEREHEAD on side, with plate 'Overhauled and modified, Derby 1974', Merehead Stone Terminal, 25th September 1981; donated to Children's Playground, Wanstrow, near Cranmore, Somerset, and moved February 1982; seen preserved, in blue livery with names YEOMAN and MEREHEAD and number 22, Wanstrow, 17th September 1982; noted at Wanstrow, 28th July 1991; scrapped, December 1991.

D3011　Derby　　　　　　　1952　70D　　10/72　F　LICKEY

new (order 6232) as number 13011, 19th December 1952; withdrawn, 1st October 1972; sold to British Leyland Ltd, Longbridge, Birmingham, and moved on 8th January 1973; the sale price was £7,000; given name LICKEY; seen with number 3011, Longbridge, 20th June 1973; to BREL Derby Works, for repair and fitting with air brakes, 15th January 1976; to British Leyland Ltd, Longbridge, April 1976; to Tyseley Diesel Depot, Birmingham, for tyre turning, November 1981; returned to Longbridge; seen in green livery, with LICKEY nameplate and BTC 3078 of 1964 registration plate, Longbridge, 4th December 1984; re-sold to Marple & Gillott Ltd, Attercliffe, Sheffield, for scrap, and arrived on 6th December 1985; seen in green livery, with plate 'British Leyland Motor Corporation 36327' in cab, Marple & Gillott Ltd, 10th December 1985; seen being cut up, 19th December 1985.

D3014　Derby　　　　　　　1952　70D　　10/72　P　D3014

new (order 6232) as number 13014, 24th December 1952; withdrawn, 1st October 1972; sold to NCB; despatched from BR Eastleigh Depot, 17th August 1973; to NCB Merthyr Vale Colliery, Aberfan; noted at Merthyr Vale Colliery, 21st August 1973; to BR Canton Depot, Cardiff, for repairs, December 1974; noted at Canton Depot, for tyre turning, 19th December 1974; to Merthyr Vale Colliery, Aberfan, about January 1975; noted in light green livery with no number, Merthyr Vale Colliery, 1st March 1975; noted at Merthyr Vale Colliery, 21st March 1978; to BR Canton Depot, Cardiff, for repairs, 19th February 1980; to Merthyr Vale Colliery, Aberfan, 21st May 1980; to BR Canton Depot, Cardiff, for repairs, September 1980; to Merthyr Vale Colliery, Aberfan, October 1980; to BR Canton Depot, Cardiff, for change of wheelsets, 3rd October 1981; to Merthyr Vale Colliery, Aberfan, 27th December 1981; to BR Canton Depot, Cardiff, for repairs and repaint, 19th July 1985; noted at Canton Depot, 20th September 1985; to Merthyr Vale Colliery, Aberfan, 28th March 1986; noted at Merthyr Vale Colliery, 6th October 1987; to BR Canton Depot, Cardiff, for repairs, 5th October 1987; to Merthyr Vale Colliery, Aberfan, 30th October 1987; noted with number 3014, Merthyr Vale Colliery, 23rd November 1988; re-sold to Dartmouth Steam Railway; noted on a road low-loader near Cullompton, Devon, 3rd March 1989; arrived at Dartmouth Steam Railway, Devon, 4th March 1989; noted in bright red livery with name VOLUNTEER and number D3014, Dartmouth Steam Railway, 25th July 1991; noted in blue livery with number D3014 and name SAMSON, Dartmouth Steam Railway, September 2007; noted in 'Dart Rail' dark grey livery with number D3014 and name SAMSON, Dartmouth Steam Railway, 22nd June 2017; noted in green livery with number D3014, 30th July 2019.

D3018   Derby   1953   81D   12/91   P   D3018 / HAVERSHAM
08011   new (order 6232) as number 13018, 28th February 1953; withdrawn, 13th December 1991; sold to Chinnor and Princes Risborough Railway, Oxfordshire, and arrived 25th April 1992; renumbered back to D3018; renumbered 13018 by May 1999; noted in black livery with number 13018, Chinnor and Princes Risborough Railway, September 2007; noted in green livery with number D3018 and name HAVERSHAM, Chinnor and Princes Risborough Railway, 30th April 2014 and 4th June 2017.

D3019   Derby   1953   8J   6/73   P   D3019
new (order 6232) as number 13019, 21st March 1953; withdrawn, 10th June 1973; sold to NCBOE; despatched from Allerton Depot, Liverpool, November 1973; journey protracted due to a hot box and stored at Bescot Yard, Walsall, for three weeks; to Powell Duffryn Fuels Ltd, NCBOE Gwaun-cae-Gurwen Disposal Point, West Glamorgan, 13th December 1973; named GWYNETH; to BR Canton Depot, Cardiff, for repairs, 4th June 1978; noted at Canton Depot, 18th November 1978; to Gwaun-cae-Gurwen Disposal Point, 11th December 1978; noted at Gwaun-cae-Gurwen, 3rd April 1979 and 3rd June 1980; seen in blue and white livery with number D3019, Gwaun-cae-Gurwen, 29th October 1982; re-sold to Harry Needle Railroad Company, June 1990; to South Yorkshire Railway Preservation Society, Meadowhall, Sheffield, 5th July 1990; noted in blue and white livery with number D3019, Meadowhall, 4th May 1992 and February 1994; to Battlefield Line, Shackerstone, 30th July 2001; to Tyseley Diesel Depot, Birmingham, for tyre turning, November 2003; re-sold to Cambrian Railway Trust, Llynclys, Oswestry, and moved 20th December 2003; noted in grey livery, Llynclys, 23rd September 2017.

D3022   Derby   1953   41A   9/80   P   D3022
08015   new (order 6232) as number 13022, 9th May 1953; withdrawn, 21st September 1980; to Swindon Works, for storage, 20th October 1980; noted in BREL Swindon Works, 22nd April 1983; sold to Class 08 Society based at Severn Valley Railway; despatched from BREL Swindon Works, 25th May 1983; noted at BR Gloucester Depot, 26th May 1983; arrived at Severn Valley Railway, Bridgnorth, 6th June 1983; was the first Class 08 on the SVR; overhauled; entered traffic as resident Bewdley pilot, spring 1984; seen in green livery with number D3022, Bridgnorth, 11th November 1984; to Bridgnorth, for engine rebuild, 2000; returned to service, summer 2001; used as Kidderminster pilot; to EWS Toton Depot, for repairs, 13th April 2006; returned to Severn Valley Railway, April 2006; to St Philip's Marsh Depot, Bristol, for tyre turning, 7th July 2019; returned to Severn Valley Railway, 11th July 2019; in BR green livery with number D3022, Severn Valley Railway, May 2019.

D3023   Derby   1953   9D   5/80   P   08016
08016   new (order 6232) as number 13023, 30th May 1953; withdrawn, 18th May 1980; sold to Hargreaves Industrial Services Ltd; despatched from Newton Heath Depot, October 1980; to Hargreaves Industrial Services Ltd, NCBOE British Oak Disposal Point, Crigglestone, West Yorkshire; noted at Crigglestone, 14th October 1980; seen in BR blue livery with number 08016, Crigglestone, 8th April 1981; out of use by September 1989; re-sold to South Yorkshire Railway Preservation Society, Meadowhall, Sheffield, and arrived on 24th January 1992; restored in BR green livery; to Heritage Shunters Trust, Rowsley, 4th April 2002; seen with number D3023 and name GEOFF L. WRIGHT, Rowsley, 12th December 2003; repainted into BR blue livery with number 08016, 2004; to Bluebell Railway, Horsted Keynes, on hire for use on their works trains, 27th March 2006; repainted in green livery whilst at Bluebell Railway; to Heritage Shunters Trust, Rowsley, April 2008; repainted back into blue livery, 2008; seen in blue livery with number 08016, Rowsley, 24th July 2021.

D3029 Derby 1953 15A 12/86 P 13029
08021 new (order 6739) as number 13029, 3rd October 1953; withdrawn, 11th December 1986; sold to Birmingham Railway Museum; despatched from Derby RTC, 7th May 1987; to Birmingham Railway Museum, Tyseley; to Tyseley Diesel Depot, for repairs, 3rd December 1988; returned to Birmingham Railway Museum, December 1988; seen with number 13029, Birmingham Railway Museum, 6th June 1993; noted in black livery with number 13029, Tyseley, June 2007 and August 2019.

D3030 Derby 1953 41A 3/85 P LION
08022 new (order 6739) as number 13030, 3rd October 1953; withdrawn, 17th March 1985; moved to Swindon Works, 18th March 1985; sold to Guinness Ltd; despatched from BREL Swindon Works, 5th July 1985; noted at Old Oak Common, 10th July 1985; arrived at Guinness, Park Royal, London, 20th July 1985; repainted in lined black, red and gold livery, with golden harp logo on side, and named LION; to BR Old Oak Common Depot, London, for repairs, early April 1990; noted in black livery, Old Oak Common Depot, 8th April 1990; returned to Guinness Ltd; noted at BR Old Oak Common Depot, London, open day, 18th August 1991; returned to Guinness Ltd; to RFS (Engineering) Ltd, Doncaster, for repairs, 13th December 1993; returned to Guinness Ltd, 11th January 1994; rail traffic ceased, 1995; donated on extended loan to Cholsey & Wallingford Railway, Oxfordshire, and moved by road on 31st August 1997; noted in Guinness black livery with no number and name LION, Cholsey & Wallingford Railway, May 2012 and 4th June 2017.

D3038 Derby 1953 9A 12/72 F 2100/525
new (order 6739) as number 13038, 28th November 1953; withdrawn, 30th December 1972; moved to Newton Heath Depot, for storage; sold to NCB; despatched from BR Newton Heath Depot, 25th October 1973; delivered to NCB Ashington Central Workshops; given plant number 2100/525; to Bates Colliery, Blyth, March 1974; noted at Bates Colliery, 19th September 1977; noted with number 2100/525, disused, Bates Colliery, 5th November 1979; scrapped, June 1980.

D3044 Derby 1954 16A 8/74 P 08032 / MENDIP
08032 new (order 6839) as number 13034, 17th April 1954; withdrawn, 25th August 1974; sold to Foster Yeoman Quarries Ltd; to BREL Derby Works, for overhaul, August 1974; despatched from Derby Works; noted at Bromsgrove, 21st February 1975; noted at Gloucester Depot, 22nd February 1975; arrived at Foster Yeoman Quarries Ltd, Merehead Stone Terminal, Somerset, about 23rd February 1975; given number 33 and name MENDIP, by 1976; noted at Merehead, 11th July 1976; noted in blue livery with number 33 and name MENDIP, Merehead Stone Terminal, 15th June 1979; to BR Gloucester Depot, for repairs; noted at Gloucester Depot, 4th January 1980, 31st January 1980 (with no middle wheels), 10th February 1980, and 1st March 1980; returned to Merehead Stone Terminal, March 1980; noted at Merehead Stone Terminal, 25th September 1981 and 17th June 1982; Foster Yeoman operated as Mendip Rail from 1993; seen in blue livery with number 33 and name MENDIP with plate 'Overhauled and modified. Derby 1975', Merehead Stone Terminal, 7th September 1994; to East Somerset Railway, Cranmore, for repairs, 10th August 2001; noted in Yeoman blue livery with number 08032, Cranmore, 10th November 2001; returned to Merehead Stone Terminal, by 18th November 2001; to East Somerset Railway, Cranmore, by 5th May 2002; returned to Merehead Stone Terminal, late 2002; to East Somerset Railway, Cranmore, April 2003; noted at Cranmore, 1st January 2004; returned to Merehead Stone Terminal, about February 2004; to East Somerset Railway, Cranmore, December 2004; returned to Merehead Stone Terminal; to Whatley Quarry, about March 2006; noted at Whatley Quarry, 28th March 2006; returned

to Merehead Stone Terminal, 9th July 2006; noted at Merehead, 25th July 2006; noted in blue livery with number 08032 and name MENDIP, Merehead open day, 21st June 2008; re-sold to Mid Hants Railway, Ropley, and moved 15th August 2008; to Knights Rail Services, Eastleigh Works, for overhaul, 20th October 2010; received spares from 08826; to Mid Hants Railway, Ropley, 8th February 2012.

D3047   Derby   1954   70D   7/73   F   105
new (order 6839) as number 13047, 28th April 1954; withdrawn, 15th July 1973; to BREL Derby Works, 3rd January 1974; overhauled; modifications included the fitting of buckeye couplers and air-brakes; noted at Derby Works, September 1974; to Syston, for testing, 6th February 1975; hauled to Canada Dock, Liverpool, 17th February 1975; exported to Liberia (see Appendix C).

D3059   Derby   1954   16F   5/80   P   3059 /
08046   BRECHIN CITY
new (order 6839) as number 13059, 28th August 1954; withdrawn, 18th May 1980; moved to Derby, for storage, October 1980; sold to Associated British Maltsters Ltd; despatched from Derby, 21st January 1981; to Associated British Maltsters Ltd (who had taken over the Inverhouse Malt Company's site at Airdrie); to BR Motherwell Depot, for repairs, 7th February 1981; returned to Associated British Maltsters Ltd; seen in brown livery with a pattern of orange side stripes, and yellow rods, with no BR number, Associated British Maltsters Ltd, 11th June 1981; re-sold to Caledonian Railway, Brechin, and moved December 1985; repainted as D3059 and given name BRECHIN CITY; noted in blue livery with number 13059, 16th June 2008; noted in blue livery with number 3059, Brechin, 30th June 2016 and 25th February 2019.

D3067   Darlington   1953   52A   2/80   P   08054 / M414
08054   new as number 13067, 28th October 1953; withdrawn, 16th February 1980; sold to Tilcon Ltd, Swinden Lime Works, Grassington, and moved June 1980; supplied via Thomas Hill (Rotherham) Ltd of Kilnhurst; noted at Grassington, 14th October 1980; seen in blue livery with no number, Grassington, 8th June 1981 and 1st April 1983; re-sold to Embsay & Bolton Abbey Railway, and moved 27th February 2008; noted in blue livery with number 08054, Embsay & Bolton Abbey Railway, July 2008.

D3074   Darlington   1953   40A   6/84   P   UNICORN
08060   new as number 13074, 19th December 1953; withdrawn, 24th June 1984; to Swindon Works, for storage, arriving on 22nd April 1985; sold to Guinness Ltd; despatched from Swindon Works, 5th July 1985; noted at Old Oak Common, 10th July 1985; arrived at Guinness Ltd, Park Royal Works, 20th July 1985; repainted in lined black, red and gold livery, with golden harp logo on side, and named UNICORN; to BR Old Oak Common Depot, London, for repairs, early April 1988; returned to Guinness Ltd; to BR Old Oak Common Depot, London, for repairs, May 1989; returned to Guinness Ltd; to BR Old Oak Common Depot, London, for repairs, May 1990; noted in OOCD, 8th May 1990; returned to Guinness Ltd; noted in black livery, BR Old Oak Common Depot, London, open day, 19th March 1994; returned to Guinness Ltd; rail traffic ceased, 1995; to Cholsey & Wallingford Railway, Oxfordshire, on extended loan, by road on 31st August 1997; noted in Guinness black livery with no number and name UNICORN, Cholsey & Wallingford Railway, May 2012 and 4th June 2017.

D3079   Darlington   1954   55B   12/84 P 13079
08064   new as number 13079, 18th January 1954; withdrawn, 23rd December 1984; sold to NRM; to National Railway Museum, York, 28th October 1985; repainted as 13079 in February 1994; seen in black livery with number 13079, National Railway Museum, York, 29th March 1994 and 16th July 1996; to National Railway Museum, Shildon, 18th August 2004; to National Railway Museum, York, 18th October 2005; to National Railway Museum, Shildon, 18th November 2013; noted in black livery with number 13079, National Railway Museum, Shildon, June 2014 and May 2016; to National Railway Museum, York, 8th June 2016; to National Railway Museum, Shildon, 28th August 2020; seen in black livery with number 13079, Shildon, 19th August 2021.

D3087   Derby   1954   8F   6/73   F NFT
new (order 8240) as number 13087, 30th October 1954; withdrawn, 10th June 1973; sold to CEGB; despatched from Springs Branch Depot, Wigan, 24th October 1973; to CEGB Walsall Power Station; noted in orange livery with number D3087, Walsall, May 1976; to BR Tyseley Diesel Depot, Birmingham, for repairs, November 1981; returned to CEGB Walsall Power Station; scrapped on site by Thos. W. Ward Ltd, May 1983.

D3088   Derby   1954   2F   12/73 F D3088 / 2100/526
new (order 8240) as number 13088, 6th November 1954; withdrawn, 29th December 1973; sold to NCB; despatched from BR Bescot Depot, 11th April 1974; arrived at NCB Ashington Colliery, 19th April 1974; to Ashington Workshops, 20th June 1974; noted at Ashington Workshops, 8th July 1974; to Bates Colliery, Blyth, 6th December 1974; noted at Bates Colliery, 19th September 1977; to Lambton Engine Works, Philadelphia, 20th April 1979; seen in workshops, under overhaul, in original BR green livery with number D3088, Philadelphia, 26th June 1979; to Bates Colliery, Blyth, 6th September 1979; noted repainted in dark blue livery with number 2100/526, Bates Colliery, 5th November 1979; to Lambton Engine Works, Philadelphia, 29th June 1981; noted at Lambton Engine Works, 4th August 1982; to Bates Colliery, Blyth, 15th February 1983; withdrawn with electrical problems, Bates Colliery, by October 1983; rail traffic ceased at Bates Colliery, 26th April 1985; noted at Bates Colliery, 9th October 1985; scrapped on site by C.H. Newton & Co Ltd of Durham, by 4th November 1985; Bates Colliery closed on 25th February 1986.

D3092   Derby   1954   73C   10/72 F 101
new (order 8240) as number 13092, 27th November 1954; withdrawn, 1st October 1972; to BREL Derby Works, November 1972; noted undergoing overhaul, 1st July 1973 and 18th August 1973; modifications included the fitting of buckeye couplers and air-brakes; displayed at media event, BREL Derby Works, April 1974; despatched from BREL Derby Works to Middlesbrough, working as train 9X47, with hauling locomotive and four 350hp shunters (per Freight Advice number 930), at 08:05 on Friday 26th April 1974; exported to Liberia (aboard the MV AVAFORS, a bulk carrier owned by the ore-mining company that owned the Liberian operation) from Middlesbrough Docks, early May 1974 (Appendix C).

D3094   Derby   1954   73F   10/72 F 102
new (order 8240) as number 13094, 11th December 1954; withdrawn, 1st October 1972; to BREL Derby Works, 3rd March 1973; noted undergoing overhaul, 1st July 1973 and 18th August 1973; modifications included the fitting of buckeye couplers and air-brakes; despatched from BREL Derby Works to Middlesbrough, working as train 9X47, with hauling locomotive and four 350hp shunters (per Freight Advice number 930), at 08:05 on Friday 26th April 1974; exported to Liberia (aboard the MV AVAFORS) from Middlesbrough Docks, early May 1974 (see Appendix C).

D3098   Derby   1955   73F   10/72   F   103
new (order 8340) as number 13098, 22nd January 1955; withdrawn, 1st October 1972; to
BREL Derby Works, 3rd March 1973; noted undergoing overhaul, 1st July 1973 and 18th
August 1973; modifications included the fitting of buckeye couplers and air-brakes;
despatched from BREL Derby Works to Middlesbrough, working as train 9X47, with hauling
locomotive and four 350hp shunters (per Freight Advice number 930), at 08:05 on Friday
26th April 1974; exported to Liberia (aboard the MV AVAFORS) from Middlesbrough
Docks, early May 1974 (see Appendix C).

D3099   Derby   1955   73F   10/72   F   NFT
new (order 8340) as number 13099, 22nd January 1955; withdrawn, 8th October 1972;
sold to Shipbreaking (Queenborough) Ltd, Kent, March 1973 and moved about June 1973;
noted at Queenborough, 1st July 1973; later used for spares; noted in green livery with
number D3099, Queenborough, 21st August 1976 and 14th February 1977; bodywork
scrapped by about May 1978; seen as just a frame, English Electric engine and wheels
(with 3099 stamped on rods) dumped in the triangle of lines near the shed, Queenborough,
13th September 1979; these remains scrapped in 1980.

D3100   Derby   1955   75C   10/72   F   104
new (order 8340) as number 13100, 29th January 1955; withdrawn, 1st October 1972; to
BREL Derby Works, 3rd March 1973; noted undergoing overhaul, 1st July 1973 and 18th
August 1973; modifications included the fitting of buckeye couplers and air-brakes;
despatched from BREL Derby Works to Middlesbrough, working as train 9X47, with hauling
locomotive and four 350hp shunters (per Freight Advice number 930), at 08:05 on Friday
26th April 1974; exported to Liberia (aboard the MV AVAFORS) from Middlesbrough
Docks, early May 1974 (see Appendix C).

D3101   Derby   1955   73F   5/72   P   13101
new (order 8340) as number 13101, 5th February 1955; withdrawn, 31st May 1972; sold
to ARC (East Midlands) Ltd, Loughborough, and moved 10th February 1973; used as yard
pilot in their sidings adjacent to Loughborough Station; noted at ARC Loughborough, 16th
June 1979; modified to operate without coupling rods, ARC Loughborough, 31st March
1980; last used in 1981; noted at ARC Loughborough, 9th December 1981; seen in mustard
yellow livery with ARC on bonnet side and number 3101 on cabside, no coupling rods, ARC
Loughborough, 26th September 1984; re-sold to Great Central Railway, Loughborough,
and moved (via lengthy detour due to weight restrictions) on 14th December 1984;
repainted in green livery with number D3101, Great Central Railway, Loughborough,
January 1985; noted in black livery with number D3101, Great Central Railway,
Loughborough, December 2007 and 25th May 2013; noted in green livery with number
13101, Great Central Railway, Loughborough, 5th October 2014 and 1st June 2017.

D3102   Derby   1955   31A   11/77   P   007 / JAMES
08077   new (order 8340) as number 13102, 11th February 1955; withdrawn, 27th
November 1977; stored at Cambridge Depot; sold to Wiggins Teape & Co Ltd, Fort William,
and moved 10th December 1978; noted in Carr Sidings, Doncaster, 10th December 1978;
noted at Wiggins Teape, 12th April 1979; noted broken down, 5th May 1979; to BR Eastfield
Depot, Glasgow, for repairs, May 1979; returned to Wiggins Teape; to BREL Glasgow
Works, for repairs, June 1981; noted at BREL Glasgow Works, 27th June 1981; returned
to Wiggins Teape, 1981; to BR Eastfield Depot, Glasgow, for repairs, 16th March 1984;
returned to Wiggins Teape, 24th March 1984; seen in blue livery with no number, Wiggins
Teape, 2nd October 1984; re-sold to RFS (Engineering) Ltd, Kilnhurst, in part-exchange

for a locomotive obtained by Wiggins Teape; arrived at Kilnhurst, 23rd November 1990; serviced for use in RFS hire fleet; seen in blue livery at Kilnhurst, 3rd February 1991; to ABB, York, on hire, 16th December 1991; returned to RFS (Engineering) Ltd, 17th November 1992; to Roche Products, Dalry, on hire, 17th November 1992; to RFS (Engineering) Ltd, Doncaster, about March 1993; to Teesbulk Handling, Middlesbrough, on hire, 2nd June 1994; returned to RFS (Engineering) Ltd, 13th September 1994; re-sold to Freightliner, December 1996; to Eastleigh East Yard, for storage, 10th December 1996; officially reinstated, December 1996; initially based at Southampton Docks.

D3110   Derby   1955   52A   3/86   F   08085
08085   new (order 8340) as number 13110, 22nd March 1955; withdrawn, 23rd March 1986; moved to Tyne Yard, for storage; sold to RFS (Engineering) Ltd, Doncaster, and moved July 1988; to RFS (Engineering) Ltd, Kilnhurst, July 1989; returned to RFS (Engineering) Ltd, Doncaster, by 4th March 1990; used for spares; seen in BR blue livery with number 08085, Doncaster, 23rd March 1991; remains by road low-loader to C.F. Booth Ltd, Rotherham, for scrap, 24th March 1993; scrapped, 16th April 1993.

SECTIONS 15/16:

Two variants of the popular shunter: Firstly, British Railways built 0-6-0 diesel electric locomotives, numbered D3127-D3136, D3167-D3438, D3454-D3472, D3503-D3611, D3652-D3664, D3672-D3718, D3722-D4048, D4095-D4098, and D4115-D4192, introduced 1953. (Locomotives D3117-D3126 and D3152-D3166 were unclassified.) Fitted with an English Electric 6KT engine developing 350bhp at 680rpm, and driving wheels of 4ft 6in diameter. Later classified TOPS Class 08. Secondly, British Railways built 0-6-0 diesel electric locomotives, numbered D3137-D3151, D3439-D3453, D3473-D3502, D3612-D3651 and D4049-D4094, introduced 1955. Fitted with a Lister-Blackstone ER6T engine developing 350bhp at 750rpm, and driving wheels of 4ft 6in diameter. Later classified TOPS Class 10.

D3140   Darlington   1955   51L   6/68   F   D3140
new as number 13140, April 1955; withdrawn, June 1968; believed purchased by R.E. Trem Ltd, Finningley, Doncaster; NCB Philadelphia finished steam working in February 1969 and D3140 was noted working there, on hire (possibly to cover for a short-term motive power shortage) on 14th March 1969; IRS records show the hire was 10th to 21st March 1969; noted at Thornaby Depot, 13th April 1969; the history of this locomotive thereafter, and its ultimate fate, is not known, and it is included here in the hope that a reader may contact the author with further information.

D3167   Derby   1955   36A   3/88   P   D3167
08102   new (order 9244) as number 13167, 24th August 1955; withdrawn, 14th March 1988; sold to Lincoln City Council, and moved 23rd March 1988; noted at Lincoln, awaiting restoration, 19th July 1988; to BREL Doncaster Works, for overhaul, 12th October 1988; noted in blue livery with number 08102, Doncaster Works, 2nd November 1988; noted in green livery with number D3167, Doncaster Works, March 1989; returned to Lincoln, 5th April 1989; displayed on a short length of track between Central Station and High Street signal box, Lincoln; seen in green livery, Lincoln, 9th March 1991 and 25th February 1993; re-sold to Lincolnshire Wolds Railway, Ludborough, and moved by road low-loader, 8th May 1994; noted in green livery with number D3167, Lincolnshire Wolds Railway, 24th April 2011 and 18th September 2021.

D3174   Derby   1955   31A   7/84   P   D3174 /
08108   DOVER CASTLE

new (order 9244) as number 13174, 23rd September 1955; withdrawn, 8th July 1984; sold to Dower Wood & Co Ltd; despatched from Cambridge Depot, August 1984; to Dower Wood & Co Ltd, Grain Terminal, Newmarket, Suffolk, 8th August 1984; noted in BR blue livery with number 08108, Dower Wood, 3rd April 1985; re-sold to East Kent Railway, Shepherdswell, and moved 3rd August 1991; re-sold to Kent & East Sussex Railway, Tenterden, and moved 13th October 1992; noted in black livery with number D3174 and name DOVER CASTLE, Kent & East Sussex Railway, 7th July 1994 and April 2006; noted under overhaul, 14th May 2016.

D3179   Derby   1955   86A   3/84   F   HO17
08113   new (order 9244) as number 13179, 7th October 1955; withdrawn, 18th March 1984; sold to NCBOE; despatched from Canton Depot, Cardiff, 2nd August 1984; to Powell Duffryn Fuels Ltd, NCBOE, Gwaun-cae-Gurwen, 6th August 1984; repainted blue and white, about 1986; noted in blue and white livery, Gwaun-cae-Gurwen, 15th January 1987; seen with number 08113, Gwaun-cae-Gurwen, 2nd April 1991; re-sold to RMS Locotec Ltd, Dewsbury, and arrived on 21st September 1995; seen in blue and white livery with no number, with PDCP on cabside, Dewsbury, 18th March 1996; to RMS Locotec Ltd, Wakefield, April 2006; used as a source of spares for the hire fleet; remains to Morley Waste Traders Ltd, Leeds, February 2007; scrapped, May 2007.

D3180   Derby   1955   36A   11/83   P   08114 / GOTHAM
08114   new (order 9244) as number 13180, 18th October 1955; withdrawn, 6th November 1983; moved for storage to Swindon Works; sold to Gloucestershire Warwickshire Railway, August 1984; despatched from BREL Swindon Works, 2nd October 1984; arrived at Gloucestershire Warwickshire Railway, 3rd October 1984; repainted as D3180; to Swindon & Cricklade Railway, 14th November 1987; re-sold to Great Central Railway, Loughborough, and moved 22nd January 1991; noted in black livery with number 13180, Great Central Railway, Loughborough, 7th April 1991; noted in green livery with number 13180, Great Central Railway, Loughborough, 4th October 1991; to Great Central Railway (Nottingham), Ruddington, Nottingham, 25th February 1997; noted in green livery with number 13180, Great Central Railway (Nottingham), Ruddington, 31st May 2009 and 17th November 2018; noted in blue livery with number 08114 and name GOTHAM, Ruddington, 10th April 2022; to Epping Ongar Railway, 20th April 2022.

D3183   Derby   1955   82C   12/72   F   D3183

new (order 9244) as number 13183, 24th October 1955; withdrawn, 30th December 1972; sold to NCB; to Bath Road Depot, Bristol, for pre-delivery service/repairs, early March 1973; to NCB Merthyr Vale Colliery, Aberfan, 17th March 1973; noted at Merthyr Vale Colliery, 19th March 1973; to BR Canton Depot, Cardiff, for repairs, May 1975; noted at Canton Depot, 29th May 1975; to Merthyr Vale Colliery, Aberfan, June 1975; noted at Merthyr Vale Colliery, 21st March 1978; to BR Canton Depot, Cardiff, for repairs, 11th September 1980; to Merthyr Vale Colliery, Aberfan, 19th December 1980; seen in original BR green livery with number D3183, Merthyr Vale Colliery, 25th March 1982; to BR Canton Depot, Cardiff, 19th July 1985; noted stored/disused (in green livery) at Canton Depot, 20th September 1985 and 12th August 1986; to Merthyr Vale Colliery, Aberfan, for spares, September 1986; noted at Merthyr Vale Colliery, 30th September 1986; noted being cannibalised, October 1986; remains sold at British Coal auction, held at Tondu, 23rd September 1987; noted still at Merthyr Vale Colliery, 6th October 1987; scrapped by W. Phillips Ltd of Llanelli, December 1987.

D3190   Derby   1955   5A   3/84   P   NPT

08123   new (order 9244) as number 13190, 15th November 1955; withdrawn, 20th March 1984; to Swindon Works, 14th March 1985; sold to Cholsey & Wallingford Railway; despatched from BREL Swindon Works, by road, 7th June 1985; to Cholsey & Wallingford Railway, Oxfordshire; given name GEORGE MASON in a ceremony on 26th October 1985; noted in GWR Brunswick green livery with number 08123, and GWR-style monogram adapted to read CWR, 29th February 1992; noted in green livery with no name and no number, Cholsey & Wallingford Railway, 29th August 2016 and 4th June 2017.

D3201   Derby   1955   40A   9/80   P   13201

08133   new (order 9245) as number 13201, 16th December 1955; withdrawn, 21st September 1980; to Swindon Works, 24th November 1980; sold to Sheerness Steel Co Ltd; despatched from BREL Swindon Works, 6th October 1981; to Sheerness Steel Co Ltd, Sheerness, Kent; noted in use, 4th December 1981; given number 1; to RFS (Engineering) Ltd, Kilnhurst, for repairs, 29th March 1991; returned to Sheerness Steel Co Ltd, 1st May 1991; seen in blue and red livery with numbers 1 and 08133, Sheerness, 7th March 1992; replaced by hired-in loco, 1992; re-sold to a member of the South Yorkshire Railway Preservation Society, Meadowhall, Sheffield, and arrived 30th August 1995; to HNRC, Barrow Hill Engine Shed, Staveley, 19th December 2000; noted at HNRC, 7th October 2001; re-sold to Severn Valley Railway, Bridgnorth, and moved 17th April 2002; underwent lengthy restoration; renumbered D3201 in 2009; entered service in late 2012; renumbered 13201 in 2019; usually works as Kidderminster pilot.

D3225   Darlington   1955   73F   4/77   F   009

08157   new as number 13225, 2nd July 1955; withdrawn, 9th April 1977; sold to Independent Sea Terminals; to Eastleigh Works, for overhaul, 25th June 1977; despatched from BR Eastleigh Works, 28th July 1977; to Independent Sea Terminals, Ridham Dock, Kent; noted with no number, Ridham Dock, 28th April 1979; noted in blue livery with numbers 08157 and 1020, Ridham Dock, 4th May 1984; re-sold to RFS (Engineering) Ltd, Doncaster, and arrived on 7th April 1993; given number 009; seen at RFS (Engineering) Ltd, Doncaster, 28th June 1993; used for spares; to European Metal Recycling, Attercliffe, Sheffield, for scrap, 28th May 1996; scrapped, June 1996.

D3232   Darlington   1956   52A   3/86   P   08164 / PRUDENCE

08164   new as number 13232, 12th January 1956; withdrawn, 23rd March 1986; moved to Tyne Yard, for storage, November 1986; sold to RFS; despatched from BR Tyne Yard, July 1988; to RFS (Engineering) Ltd, Doncaster; noted in grey livery, RFS (Engineering) Ltd, Doncaster, 17th July 1988; given number 002 and name PRUDENCE; to RFS (Engineering) Ltd, Kilnhurst, for weighing, 15th January 1991; returned to RFS (Engineering) Ltd, Doncaster, same day; seen in grey livery with number 002 and nameplate PRUDENCE, RFS (Engineering) Ltd, Doncaster, 23rd March 1991; seen in grey livery with dark grey roof and number 002, with large board on front '08 shunter for ABB, rebuilt at Crewe site', ABB Transportation, York, 23rd April 1996; seen at ABB, York, on hire, 13th May 1996; seen at RFS (Engineering) Ltd, Doncaster, 3rd October 1996; re-sold to East Lancashire Railway, Bury, and arrived 28th January 1998; initially renumbered D3232; noted in BR blue livery with number 08164 and name PRUDENCE, East Lancashire Railway, Bury, November 2019.

D3236   Darlington   1956   55B   3/88   P   D3236

08168   new as number 13236, 7th February 1956; withdrawn, 14th March 1988; sold to ABB Transportation, York, and moved April 1989; repainted as D3236; locomotive included in sale when BREL works privatised; seen numbered D3236, York, 15th February 1995; noted at York, 27th January 1996; sold at public auction, 25th June 1996; re-sold to Battlefield Line, Shackerstone, and moved 12th July 1996; re-sold to Fragonset, Derby, for use as pilot locomotive, 1st July 1999; to Alstom Transport, Old Dalby Test Centre, 10th December 2001; to Battlefield Line, Shackerstone, 22nd July 2004; noted in black livery with number 08168, Battlefield Line, September 2004; to Bluebell Railway, Horsted Keynes, 30th April 2008; to Nemesis Rail, Burton upon Trent, 27th June 2014; seen in black livery with number 13236, Nemesis Rail, 19th August 2014; to Epping Ongar Railway, Essex, for gala, 21st April 2016; to Nemesis Rail, Burton upon Trent, 30th April 2016; repainted in BR green livery with number D3236, May 2020; noted in use as yard shunter, 19th August 2022.

D3238   Darlington   1956   52A   3/86   F   08170
08170   new as number 13238, 15th February 1956; withdrawn, 23rd March 1986; to Tyne Yard, for storage, November 1986; sold to RFS; despatched from BR Tyne Yard, July 1988; to RFS (Engineering) Ltd, Doncaster; noted at RFS (Engineering) Ltd, Doncaster, 17th July 1988; to RFS (Engineering) Ltd, Kilnhurst, 15th July 1989; noted at Kilnhurst, 14th November 1989; used for spares; seen with no engine at Kilnhurst, 28th October 1990 and 23rd March 1991; remains to C.F. Booth Ltd, Rotherham, 11th December 1991; scrapped, December 1991.

D3245   Derby   1956   55B   10/88 F   08177
08177   new (order 9248) as number 13245, 14th May 1956; withdrawn, 30th October 1988; to Goole Depot, for storage; sold to ABB Transportation; despatched from Goole Depot to Doncaster, February 1989; moved from Doncaster to Bescot, 21st February 1989; to ABB Transportation, Crewe, March 1989; locomotive included in sale when BREL works privatised in April 1989; noted in BR blue livery with number 08177, Crewe Works, 7th May 1989; scrapped on site by M.R.J. Phillips (Metals) Ltd, August 1996.

D3255   Derby   1956   85B   12/72 P   D3255
new (order 9248) as number 13255, 8th June 1956; withdrawn, 30th December 1972; sold to NCB; to Bath Road Depot, Bristol, for pre-delivery service/repairs, early March 1973; to NCB Blaenavon Colliery (propelled into exchange sidings by D6973), 19th March 1973; to Bargoed Colliery, 22nd March 1973; noted at Bargoed Colliery, 1st July 1973; to BR Canton Depot, Cardiff, for tyre turning, September 1974; noted at Canton Depot, 26th September 1974; to Bargoed Colliery; to BR Canton Depot, Cardiff, for repairs, 28th December 1975; noted at BR Canton Depot, Cardiff, 28th December 1975 and 6th January 1976; to Bargoed Colliery, about April 1976; to BR Canton Depot, Cardiff, August 1976; noted at BR Canton Depot, 25th August 1976 and 4th September 1976; to Bargoed Colliery, 1976; to BR Canton Depot, Cardiff, for repairs, 18th April 1977; noted at Canton Depot, 4th June 1977; to Bargoed Colliery, August 1977; to BR Canton Depot, Cardiff, March 1978; noted at BR Canton Depot, 19th March 1978; to Bargoed Colliery, about May 1978; to BR Canton Depot, Cardiff, 23rd May 1979; noted at Canton Depot, 9th June 1979; to Bargoed Colliery, about August 1979; to BR Canton Depot, Cardiff, for tyre turning and two-year exam, 6th April 1981; to Mardy Colliery, 27th May 1981; to Mountain Ash Colliery, 28th December 1981; to BR Canton Depot, Cardiff, for repairs, May 1982; noted at BR Canton Depot, Cardiff, 16th May 1982; to Mardy Colliery, 27th May 1982; rail traffic ceased, Mardy Colliery, August 1986; noted at Mardy Colliery, 2nd December 1986; re-sold to Brighton Railway Museum, and moved about May 1987; noted at BR Brighton Depot, 11th September 1987 and 1st October 1999; re-sold to Colne Valley Railway, Castle Hedingham, 9th September 2008; noted in blue livery with no number, Castle Hedingham,

27th September 2008; re-sold to Tim Ackerley, c/o T.W.S. Welders, North Side Works, Leavening, Malton, and moved on 27th August 2009; noted on a short length of track, sheeted over, Leavening, 30th September 2012; noted in blue livery with number D3255, Leavening, April 2016.

D3261   Derby   1956   86A   12/72   P   D3261
new (order 9248) as number 13261, 5th July 1956; withdrawn, 30th December 1972; noted stored at 87E Landore Depot, March 1973; sold to NCB; despatched from 87E Landore Depot, July 1973; to NCB Tower Colliery, Hirwaun, Glamorgan, July 1973; noted at Tower Colliery, 30th November 1973; seen with number 3261, Tower Colliery, 13th May 1975; to BR Canton Depot, Cardiff, for repairs, September 1975; noted at BR Canton Depot, Cardiff, 22nd September 1975; to Tower Colliery, Hirwaun, October 1975; to BR Canton Depot, Cardiff, for repairs, 27th October 1977; despatched from Canton Depot working 9E76, 18:56 on Tuesday 1st November 1977; delivered to Tower Colliery, Hirwaun, 2nd November 1977; seen at Tower Colliery, 24th April 1978; to BR Canton Depot, Cardiff, 28th October 1978; noted at Canton depot, 27th February 1979; noted at Gloucester Old Yard, 4th May 1979; to BREL Swindon Works, for repairs, early May 1979; noted at BREL Swindon Works, 5th June 1979 and 30th October 1979; despatched to Tower Colliery, Hirwaun, 29th November 1979; noted en-route at Gloucester Old Yard, with no side rods, 8th December 1979; seen in green livery with number 3261 and Derby 1956 plate, Tower Colliery, 8th June 1980; was a sale lot at British Coal auction, Tondu, 23rd September 1987; re-sold to Brighton Railway Museum; noted vandalised, Tower Colliery, 23rd October 1988; to Brighton Railway Museum, 11th December 1988; long-stored in the sidings on opposite side of the line to the Preston Park Works; noted in sidings, 1st January 1996; re-sold to Swindon & Cricklade Railway and moved about March 1996; in green livery with number D3261, Swindon & Cricklade Railway, 30th March 2014 and 26th November 2020.

D3265   Derby   1956   86A   9/83   P   13265
08195   new (order 9248) as number 13265, 23rd August 1956; withdrawn, 25th September 1983; to Swindon Works, 12th October 1983; sold to Llangollen Railway; despatched from BREL Swindon Works, 20th March 1986; seen on a low-loader in Birmingham, at 08-30 on 20th March 1986; to Llangollen Railway; noted in green livery with number D3265 and MARK nameplate, Llangollen, 27th March 1989 and 1st January 1992; seen in green livery with number D3265, Llangollen Railway, 23rd April 1999; noted in black livery with number 13265, Llangollen Railway, 28th April 2012 and 20th July 2015.

D3272   Derby   1956   86A   5/89   P   08202
08202   new (order 9248) as number 13272, 6th October 1956; withdrawn, 23rd May 1989; sold to G.G. Papworth Ltd; despatched by road from Canton Depot, Cardiff, 22nd February 1990; off-loaded at Whitemoor Yard; tripped by rail to G.G. Papworth Ltd, Ely, Cambridgeshire, 10th March 1990; seen in yellow livery with number 08202, Ely, 9th March 1992; to The Potter Group Ltd, Knowsley, Merseyside, 9th January 2001; to Gloucestershire Warwickshire Railway, Toddington, for gala and then storage, 7th July 2010; noted in yellow livery with number 08202, Gloucestershire Warwickshire Railway, Toddington, July 2010; to GBRf, Celsa, Cardiff, on hire, late January 2011; to The Potter Group Ltd, Ely, 4th May 2011; re-sold to Chasewater Railway, Staffordshire, and moved 15th January 2014; noted in yellow livery with number 08202, Chasewater Railway, 28th February 2014; re-sold to Avon Valley Railway, Bitton, and moved 30th June 2015; noted in blue livery with number 08202, Avon Valley Railway, 12th March 2016; to Knorr Bremse Rail Services Ltd, Wolverton Works, on hire, 25th January 2018; to Avon Valley Railway, Bitton, 16th May 2018; seen in green livery with number 08202, Avon Valley Railway, 16th

July 2021; to Railway Support Services, Wishaw, 26th March 2024; to Llangollen Railway, on hire, 3rd April 2024.

D3286   Derby   1956   16C   11/80   F   08216

08216   new (order 9248) as number 13286, 30th November 1956; withdrawn, 2nd November 1980; to Swindon Works, for storage, 14th January 1981; sold to Sheerness Steel Co Ltd; despatched from BREL Swindon Works, 22nd April 1983; stored in Acton Yard; to Hoo Junction, 12th May 1983; arrived at Sheerness Steel Co Ltd, Sheerness, Kent, 12th May 1983; given number 2; noted in blue livery with number 08216, Sheerness, 31st August 1986; to RFS (Engineering) Ltd, Kilnhurst, for overhaul, 21st December 1989; seen at Kilnhurst, 3rd April 1990; returned to Sheerness Steel Co Ltd, 5th May 1990; noted at Sheerness, 16th August 1991; seen in blue livery with number 08216, Sheerness, 7th March 1992; withdrawn and parts donated to 08133, 1994; noted semi-dismantled, June 1995; re-sold to South Yorkshire Railway Preservation Society, Meadowhall, Sheffield, and moved 2nd April 1996; used for spares; noted in blue livery with number 08216, no engine, Meadowhall, 27th June 1999; to Barrow Hill Engine Shed, Staveley, 10th April 2001; scrapped by Harry Needle staff, late April 2001.

D3290   Derby   1956   5A   3/86   P   D3290

08220   new (order 9248) as number 13290, 14th December 1956; withdrawn, 30th March 1986; moved to Chester Depot; sold to William Smith Ltd; despatched from Chester Depot, by road, 11th July 1988; to William Smith Ltd, Wakefield; re-sold to Steamtown, Carnforth, Lancashire, and moved by road, 12th July 1990; seen in black livery with no number (with 'Thomas' face), Steamtown, 2nd October 1993; seen in black livery with number 08220, Steamtown, 13th September 1997; re-sold and moved to Wrenbury Station, Cheshire, April 2006; re-sold to Great Central Railway (Nottingham), Ruddington, Nottingham, and moved October 2008; to Railway Support Services, Wishaw, on hire, for use in the RSS hire fleet; to Electro-Motive Diesel, Longport, Staffordshire, on sub-hire, week-ending 9th August 2014; noted in blue livery with number 08220, Longport, 14th July 2016; returned to Great Central Railway (Nottingham), Ruddington, 25th September 2019; noted at Railway Support Services, Wishaw, 3rd June 2022; to Great Central Railway (Nottingham), Ruddington, 2022; noted with number D3290, Great Central Railway (Nottingham), Ruddington, September 2023; repainted in BR blue livery with number D3290, June 2024.

D3308   Darlington   1956   85B   3/84   P   08238 / CHARLIE

08238   new as number 13308, 20th May 1956; withdrawn, 18th March 1984; sold to Forest Free Mining, Tetbury, August 1984; the loco was intended for use at Parkend Mine, Forest of Dean, but the mining licence application was turned down; loco stored at BR Gloucester Depot; noted at Gloucester Depot, 8th October 1986; re-sold to Swindon Railway Workshop Ltd, and moved on 11th November 1988; used as works pilot; re-sold to Dean Forest Railway, Lydney, and moved January 1993; named CHARLIE; to RFS (Engineering) Ltd, Doncaster, for repairs, December 1997; to Dean Forest Railway, Lydney, 1998; noted in black livery with number 13308 and name CHARLIE, Dean Forest Railway, May 2004; suffered generator failure, 5th November 2008; underwent major overhaul and repaint; returned to service, 7th March 2010; noted in blue livery with number 08238 and name CHARLIE, Dean Forest Railway, 4th August 2017; seen in green livery with number 08238 and name CHARLIE, Dean Forest Railway, 14th July 2021; to North Yorkshire Moors Railway, Grosmont, 11th January 2024.

D3309   Darlington   1956   55H   7/84   F   3309
08239   new as number 13309, 30th June 1956; withdrawn, 8th July 1984; sold to C.F. Booth Ltd, Rotherham, and moved 20th June 1995; seen in blue livery with numbers 08239 and 3309, C.F. Booth Ltd, 8th July 1995 and 19th January 1997; re-sold to Harry Needle Railroad Company; moved to South Yorkshire Railway Preservation Society, Meadowhall, Sheffield, 12th November 1998; used for spares; to European Metal Recycling, Kingsbury, about June 2001; noted in blue livery with number 3309, Kingsbury, 29th June 2004 and 18th January 2005; scrapped, October 2005.

3336   Darlington   1957   41J   3/85   P   08266
08266   new as number 13336, 20th February 1957; withdrawn, 17th March 1985; to Swindon Works, for storage, 14th June 1985; sold to KWVR; despatched from BREL Swindon Works, by road, 21st November 1985; to Keighley & Worth Valley Railway, Haworth; repainted as D3336; seen in black livery with number 13336, 12th May 1995; seen in green livery with number 08266 and 55F shedplate, Haworth, 19th March 2006; seen in green livery with number D3336, Haworth, 14th May 1990 and 12th February 1994; seen repainted in black livery with number 13336, Haworth, 27th October 1994; seen in green livery with number 08266, Haworth, 19th April 2003; noted in green livery, 30th September 2013; noted in grey livery with number 08266, Haworth, 18th June 2016 and 12th June 2022.

D3342   Derby   1957   41A   7/87   F   08272
08272   new (order D252) as number 13342, 12th April 1957; withdrawn, 6th July 1987; to Doncaster Works, for storage; noted at Doncaster Works, 8th November 1987; sold to RFS (Engineering) Ltd, Doncaster, and moved January 1988; noted in BR blue livery with number 08272, RFS (Engineering) Ltd, Doncaster, 7th February 1988; used for spares; noted with engine removed, RFS (Engineering) Ltd, Doncaster, 6th November 1988; remains scrapped, January 1991.

D3358   Derby   1957   82C   1/83   P   08288 / PHOENIX
08288   new (order D252) 28th June 1957; withdrawn, 21st January 1983; to Swindon Works, for storage; noted in blue livery with number 08288, Swindon Works, 14th February 1983; sold to Mid Hants Railway, June 1984; despatched from BREL Swindon Works, 13th September 1984; noted at Farnham, 14th and 16th September 1984; damaged by vandals at Farnham; repaired and arrived at Mid Hants Railway, Ropley, 1st November 1984; repainted as D3358; seen at Mid Hants Railway, Ropley, 11th September 1992; to Wabtec, Doncaster, for repairs, 20th August 2002; returned to Mid Hants Railway, Ropley, October 2002; repainted in BR blue with number 08288, with name PHOENIX, Mid Hants Railway, by March 2021.

D3362   Derby   1957   65A   5/84   F   08292
08292   new (order D252) as number 13362, 17th May 1957; withdrawn, 1st May 1984; sold to Deanside Transit Ltd, Glasgow, and moved in July 1984; noted in grey livery with number 08292, Deanside Transit Ltd, 9th June 1986; noted repainted light green, September 1986; later used for spares for the company's other 08s; remains scrapped on site, November 1994.

D3366   Derby   1957   55B   10/88   F   08296 / 08787
08296   new (order D252) as number 13366, 31st May 1957; withdrawn, 30th October 1988; to Goole Depot, for storage; despatched from Goole Depot to Doncaster, February 1989; moved from Doncaster to Bescot, 21st February 1989; to ABB Transportation,

Crewe, March 1989; locomotive included in sale when BREL works privatised in April 1989; at unknown date between July 1991 and October 1992 it was given the identity 08787; to C. F. Booth Ltd, Rotherham, for scrap, 16th February 1994; scrapped 13th May 1994.

D3378   Derby   1957   41A   2/92   P   NPT
08308   new (order D252) 30th August 1957; latterly ran on BR with unofficial painted name LANGWITH; withdrawn, 21st February 1992; sold to South Yorkshire Railway Preservation Society, Meadowhall, Sheffield, and moved on 15th July 1993; noted in blue livery with number 08308, Meadowhall, February 1994; re-sold to RT Rail, Crewe, for use in their hire fleet, 1st May 1998; to RFS, Doncaster, for repairs, 21st January 1999; to ScotRail, Inverness, on hire, 7th September 1999; to Wabtec, Doncaster, for repairs, 16th November 2005; to ScotRail, Inverness, on hire, 4th May 2006; RT Rail was acquired by RMS Locotec Ltd, 8th November 2007; noted in blue livery with white cab and number 08308, Inverness, May 2009; to Castle Cement Works, Ketton, on hire, 30th August 2013; to RMS Locotec Ltd, Wolsingham Depot, 16th January 2014; to PD Ports, Tees Dock, on hire, 1st August 2014; to RMS Locotec Ltd, Wolsingham Depot, 3rd October 2018; engine removed, 2019; to Positive Traction Ltd, Chesterfield, for conversion to battery power, June 2021; noted in stripped condition, with no number, September 2022; to Barrow Hill Engine Shed, by road, for trials, 16th October 2023; noted in green livery with number 08308, BHES, 16th October 2023.

D3390   Derby   1957   16A   12/82   F   P400D / SUSAN
08320   new (order D252) 16th October 1957; withdrawn, 19th December 1982; sold to Forest Free Mining, Tetbury, August 1984; despatched from 16A Toton Depot, 2nd October 1984; noted at Bescot, 4th October 1984; noted at Gloucester, 5th October 1984; the locomotive had been intended for use at Parkend Mine, Forest of Dean, but the mining licence application was turned down; locomotive stored at BR Gloucester Depot; noted at Gloucester Depot, 8th October 1986; re-sold to English China Clays PLC, Blackpool Dryers, Burngullow, Cornwall, and arrived by road on 2nd December 1988; seen in blue livery with plates P400D and SUSAN on cabside, Blackpool Dryers, 13th September 1995; noted at Blackpool Dryers, 5th May 2008; to Imerys Minerals Ltd, Rocks Dryers, Bugle, December 2008; noted in blue livery with numberplate P400D, on a Reid Freight low-loader, Rocks Dryers, Bugle, 28th July 2010; to European Metal Recycling, Kingsbury, for scrap, 9th September 2010; scrapped October 2010.

D3401   Derby   1957   36A   3/88   P   08331 / TERENCE
08331   new (order D252) 23rd November 1957; withdrawn, 14th March 1988; sold to RFS (Engineering) Ltd, Doncaster, and arrived in April 1988; given number 001 and name TERENCE; noted in blue and grey livery, RFS Doncaster, 2nd June 1989; to RFS (Engineering) Ltd, Kilnhurst, about April 1990; seen at RFS Kilnhurst, 6th May 1990; to RFS (Engineering) Ltd, Doncaster; seen in grey livery with number 001 and nameplate TERENCE, RFS (Engineering) Ltd, Doncaster, 29th October 1990 and 23rd March 1991; to RFS (Engineering) Ltd, Kilnhurst, 5th August 1991; to Inco Europe, Clydach, Swansea, on hire, 17th September 1991; returned to RFS (Engineering) Ltd, Kilnhurst, 6th November 1991; seen at RFS (Engineering) Ltd, Kilnhurst, 19th January 1992 and 24th February 1992; to Trans-Manche Link, Channel Tunnel contract, on hire, 26th September 1992; given number 95; to RFS (Engineering) Ltd, Kilnhurst, 6th February 1993; to Flixborough Wharf Ltd, Flixborough, Scunthorpe, on hire, 8th March 1993; to Allied Steel & Wire Ltd, Cardiff, on hire, 24th September 1994; returned to RFS (Engineering) Ltd, Doncaster, 27th November 1994; to Great North Eastern Railway, Craigentinny Depot, Edinburgh, on hire, 30th January 1997; returned to RFS (Engineering) Ltd, Doncaster, by February 1998; to

Hays Chemicals, Sandbach, on hire, August 1998; returned to RFS (Engineering) Ltd, September 1998; to GNER, Craigentinny Depot, Edinburgh, on hire, 12th October 1998; to Wabtec, Doncaster, for repairs, December 2000; returned to Craigentinny Depot, Edinburgh, on hire; to Wabtec, Doncaster, 11th August 2005; re-sold to RT Rail, Crewe, 7th May 2006; noted in blue livery with numbers 08331 and H001, Wabtec, January 2007; to Embsay & Bolton Abbey Railway, for testing, May 2007; RT Rail acquired by RMS Locotec Ltd, 8th November 2007; to Wabtec, Doncaster, by 5th December 2007; to Lafarge Aggregates, Barrow upon Soar, on hire, 13th December 2007; re-sold to Class 20189 Ltd, July 2008; to Midland Railway, Butterley, for storage, by 5th July 2008; noted in black livery with number 08331, Midland Railway, 21st March 2009; to Cemex Rail Products, Washwood Heath, on hire, August 2010; to Boden Rail Engineering, Washwood Heath, on hire, August 2010; to Wabtec, Doncaster, for repairs, early August 2011; to Cemex Rail Products, Washwood Heath, on hire, September 2011; to Midland Railway, Butterley, Derbyshire, early February 2012; noted in black livery with number 08331, Butterley, 10th June 2023.

D3405   Derby   1957   41A   1/87   F   08335
08335   new (order D252) 13th December 1957; withdrawn (due to collision damage), 16th January 1987; noted in blue livery with number 08335, stored at Tinsley Depot, 6th June 1987; sold to Thomas Hill (Rotherham) Ltd, Kilnhurst, and moved on 20th November 1987; seen at Kilnhurst on 22nd May 1988, stated to have been purchased for assessment as regards the feasibility of Hill's tendering for BR Class 08 contract overhauls; noted with engine removed, Kilnhurst, July 1989; scrapped on site by C.F. Booth Ltd, Rotherham, 26th July 1989.

D3407   Derby   1957   36A   2/87   F   08337
08337   new (order D252) 27th December 1957; withdrawn, 6th February 1987; to Doncaster Works, December 1987; sold to RFS (Engineering) Ltd, Doncaster, and moved in January 1988; noted in BR blue livery with number 08337, RFS (Engineering) Ltd, Doncaster, 7th February 1988; used for spares; noted with engine removed, RFS (Engineering) Ltd, Doncaster, 14th August 1988; remains scrapped, January 1989.

D3415   Derby   1958   67C   11/83   F   RUSSELL
08345   new (order D252) 14th February 1958; withdrawn, 1st November 1983; stored at Ayr Depot; sold to Deanside Transit Ltd, Glasgow, and moved in May 1985; noted in grey livery with number 08345, Deanside Transit Ltd, 9th June 1986; repainted as number 3 with no BR number, 1991; repainted and named RUSSELL, May 1998; noted in yellow livery, Deanside Transit, 25th February 2006; re-sold to Harry Needle Railroad Company, and moved to Long Marston on 21st June 2007; noted in yellow livery with no number and name RUSSELL, Long Marston, 1st June 2008; to C.F. Booth Ltd, Rotherham, for scrap, 4th November 2009; scrapped on 13th November 2009.

D3420   Crewe   1957   86A   1/84   F   D3420
08350   new (order E495) 7th December 1957; withdrawn, 17th January 1984; to Swindon Works, for storage, 2nd March 1984; sold to Churnet Valley Railway; despatched from BREL Swindon Works, by road, 17th September 1984; to Churnet Valley Railway, Cheddleton; arrived at Cheddleton, 18th September 1984; repainted as D3420 in 1986; seen in blue livery with number 08350, Cheddleton, 8th May 1991; to L&NWR Ltd, Carriage Works, Crewe, July 2004; to Midland Railway, Butterley, by road, for storage, 5th September 2007; noted in green livery with number D3420, Butterley, 30th September

2007; to Heanor Haulage, Langley Mill, for storage, early September 2008; to Ron Hull Ltd, Rotherham, for scrap, week-ending 19th September 2008; scrapped, March 2010.

D3429   Crewe   1958   86A   1/84   P   08359
08359   new (order E495) 28th March 1958; withdrawn, 22nd January 1984; to Swindon Works, for storage, 7th March 1984; sold to Churnet Valley Railway, Cheddleton; despatched from BREL Swindon Works, by road, 19th September 1984; arrived at Cheddleton, 21st September 1984; given number 3429 and name SIGNAL RADIO, Cheddleton, September 1985; re-sold to a private individual and moved to Peak Rail, Buxton, 9th January 1987; noted in green livery, Buxton, 27th September 1987; to Peak Rail, Darley Dale, December 1989; noted in green livery, Darley Dale, 25th December 1989; seen in green livery with number D3429 (but no name) and 17D shedplate, Darley Dale, 5th March 1994; to Battlefield Line, Shackerstone, 16th October 1996; to Tyseley Steam Depot, Birmingham, 29th June 1999; to Northampton & Lamport Railway, Chapel Brampton, 11th August 2005; noted in green livery with number D3429, Northampton & Lamport Railway, November 2005; to Telford Steam Railway, Shropshire, 20th January 2007; to Chasewater Railway, Staffordshire, on loan, 19th July 2010; noted in green livery with number D3429, Chasewater, 20th April 2013; to Telford Steam Railway, Shropshire, 25th November 2016; to Chasewater Railway, Staffordshire, 17th January 2017; to Telford Steam Railway, Shropshire, 12th November 2018; to Chasewater Railway, Staffordshire, 4th April 2019; noted in blue livery with number 08359, Chasewater, 4th September 2021; to Avon Valley Railway, Bitton, 22nd November 2021; to Bodmin & Wenford Railway, on hire, by road, 11th March 2022; to Avon Valley Railway, Bitton, 26th March 2024.

D3452   Darlington   1957   16A   6/68   P   D3452
new 22nd November 1957; withdrawn, 22nd June 1968; sold to E.C.C. Ports Ltd, Fowey, Cornwall, and moved in September 1968; noted at Fowey, 24th October 1970; noted at Fowey, 7th April 1979; seen in green livery with number D3452, Fowey, 14th September 1982; re-sold to Bodmin & Wenford Railway, Cornwall, and moved on 5th March 1989; (one of four preserved Class 10s); seen in black livery with number D3452, Bodmin & Wenford Railway, 13th September 1995; noted in black livery with number D3452, Bodmin & Wenford Railway, September 2007.

D3460   Darlington   1957   86A   11/91   P   08375 / PB357
08375   new 21st June 1957; withdrawn, 8th November 1991; sold to Railway Age, Crewe, and moved 13th August 1993; noted in black livery with number 08375, Railway Age, Crewe, 30th March 1996; displayed at open day, Crewe Electric Depot, 3rd May 1997; returned to Railway Age; noted in blue livery with number 08375, Railway Age, May 1997; re-sold to RT Rail, Crewe, for use as a hire loco, June 1997; to L&NWR Ltd, Carriage Works, Crewe, about January 1998; to RMS Locotec Ltd, Dewsbury, for air-brake equipping, 25th March 1998; to Direct Rail Services, Sellafield, on hire, about August 1998; to RFS (Engineering) Ltd, Doncaster, for repairs, about January 1999; to Port of Felixstowe, on hire, 2nd November 1999; to Freightliner, Ipswich, on hire, by 21st November 1999; to Freightliner, Port of Felixstowe, on hire, by 10th July 2001; to Wabtec, Doncaster, for repairs, by 7th October 2001; to Flixborough Wharf Ltd, Flixborough, Scunthorpe, on hire, January 2002; to Freightliner, Port of Felixstowe, on hire, by 16th February 2002; to Wabtec, Doncaster, May 2002; to Hays Chemicals, Sandbach, on hire, 5th July 2002; returned to Wabtec, Doncaster, for repairs, 2nd December 2002; to Hays Chemicals, Sandbach, on hire, about January 2003; to Wabtec, Doncaster, March 2003; to Port of Felixstowe, on hire, 4th June 2003; to Wabtec, Doncaster, for repairs, by 6th October 2003; to Alstom Transport, Eastleigh Works, on hire, 20th November 2003; to Hays Chemicals,

Sandbach, on hire, September 2004; to Wabtec, Doncaster, for repairs, 2005; returned to Hays, on hire, 15th September 2005; to PD Ports, Tees Dock, on hire, 15th October 2005; to Wabtec, Doncaster, for repairs, 15th February 2006; to Manchester Ship Canal, Trafford Park, Manchester, on hire, about May 2006; to Wabtec, Doncaster, December 2006; to DHL, ProLogis Park Industrial Estate, Coventry, on hire, 5th February 2007; RT Rail was acquired by RMS Locotec Ltd, 8th November 2007; to Corus, Trostre Works, Llanelli, on hire, 18th December 2009; to Castle Cement Works, Ketton, on hire, November 2010; to PD Ports, Tees Dock, on hire, 28th February 2011; to Tata Steel, Shotton, on hire, week commencing 25th March 2013; to Castle Cement Works, Ketton, on hire, 15th January 2014; noted in black livery with number 21, Ketton, June 2015; to RMS Locotec Ltd, Wolsingham Depot, for overhaul and repaint, 11th May 2019; re-sold to Port of Boston, June 2019; to Port of Boston, 11th July 2019; noted in blue livery with grey cab and numbers 08375, Port of Boston, 4th June 2020; later given extra number PB357.

D3462   Darlington   1957   84A   6/83   P   D3462
08377   new 27th June 1957; withdrawn, 26th June 1983; to Swindon Works, for storage, 7th July 1983; sold to Dean Forest Railway; despatched from BREL Swindon Works, by rail, 12th March 1986; arrived at Dean Forest Railway, Lydney, 19th March 1986; noted in blue livery, Dean Forest Railway, Lydney, 29th May 1986; seen in green livery with number D3462, Dean Forest Railway, 29th May 1990; to Rail & Marine Engineering, Thingley Junction, Chippenham, for storage, 1995; re-sold to West Somerset Railway, Minehead, and arrived there on 23rd April 1996; noted in green livery with number D3462, West Somerset Railway, June 2008; re-sold to Mid Hants Railway, Ropley, and moved on 26th March 2013; to South Devon Railway, for tyre turning, 21st March 2022.

D3476   Darlington   1957   16A   6/68   F   D3476
new 24th December 1957; withdrawn, 1st June 1968; sold to E.C.C. Ports Ltd; arrived at E.C.C. Ports Ltd, Fowey, Cornwall, 6th September 1968; noted at Fowey, 7th April 1979; seen in original BR green livery, with number D3476, E.C.C. Ports Ltd, Fowey, 14th September 1982; became disused when BR took over the shunting, early 1988; re-sold to Harry Needle Railroad Company; moved by road to storage in the Midlands, 5th March 1989; to South Yorkshire Railway Preservation Society, Meadowhall, Sheffield, 2nd October 1989; noted in green livery with number D3476, poor condition, no rods, Meadowhall, 9th March 1991 and 27th April 1996; re-sold to a private individual, August 2000; moved to Colne Valley Railway, Castle Hedingham, 1st December 2000; noted in repainted blue livery, no number, no rods, Colne Valley Railway, 19th August 2004 and 26th May 2007; planned restoration never took place; to hauliers yard (possibly Wishaw), to be used for spares, 26th February 2009; remains to T.J. Thomson & Son Ltd, Stockton-on-Tees, 3rd March 2009; scrapped March 2009.

D3489   Darlington   1958   16A   4/68   P   D3489
new 3rd May 1958; withdrawn, 6th April 1968; sold to Felixstowe Dock & Railway Co Ltd, Suffolk, and moved in August 1968; seen repainted dark blue with hand-painted BR number, and fitted with air brakes, Felixstowe, 27th September 1972; noted named COLONEL TOMLINE, Felixstowe, 11th September 1979; to BREL Swindon Works, for tyre turning and repairs, 30th January 1980; noted at BREL Swindon Works, 1st March 1980; to Port of Felixstowe, by road low-loader, 19th May 1980; noted in light green livery with number D3489, Port of Felixstowe, 11th May 1981; to BR Stratford Depot, for repairs to its traction motor, 12th October 1984; to Port of Felixstowe, December 1984; noted in lime green livery, 5th August 1987; to Wilmott Bros, Ilkeston, for repairs, 27th July 1990; to Port of Felixstowe, 18th September 1990; withdrawn in 1999; re-sold to Spa Valley Railway,

Tunbridge Wells, and moved on 17th August 2001; (one of four preserved Class 10s); noted in black livery with number D3489 and name COLONEL TOMLINE, Spa Valley Railway, 10th August 2003 and 1st July 2014; to Bodmin & Wenford Railway, for storage, 31st October 2023; to Helston Railway, 12th December 2023; to Bodmin & Wenford Railway, 26th March 2024.

D3497　　Doncaster　　　　　　1957　16B　　4/68　　F　D3497 / 73602
new (order EO5) 4th December 1957; withdrawn, 13th April 1968; sold to E.C.C. Ports Ltd; moved to E.C.C. Ports Ltd, Fowey, Cornwall, 21st August 1968; out of use and used for spares, July 1978; noted at Fowey, 7th April 1979; seen dumped beside estuary (used as a source of spares for D3452 and D3476) in original BR green livery with number D3497, E.C.C. Ports Ltd, Fowey, 14th September 1982; noted with small plate 73602 on cabside, E.C.C. Ports Ltd, Fowey, 15th June 1983; noted out of use, 18th June 1984; some parts purchased by Bodmin & Wenford Railway; remains scrapped on site at Fowey, February 1990.

D3503　　Derby　　　　　　　1958　40B　　6/96　　F　08388
08388　　new (order D658) 14th April 1958; withdrawn, 11th June 1996; noted in two-tone grey livery with number 08388, Immingham Depot, 20th September 1996; to RFS (Engineering) Ltd, Doncaster, for storage, 24th July 1998; sold to Mike Darnall, Newton Heath, Manchester, 1999; to Wabtec, Doncaster, for repairs, 1999; to Mike Darnall, Newton Heath, by road low-loader, August 2000; stored at a non-rail location on an industrial estate at Newton Heath; the preservation attempt failed; to European Metal Recycling, Kingsbury, for scrap, 15th July 2010; scrapped, early October 2010.

D3504　　Derby　　　　　　　1958　2F　　　9/09　　P　08389
08389　　new (order D658) 17th April 1958; unofficially named SAXON, 4th December 1993; withdrawn, 4th September 2009; to Toton Depot, for storage; sold to Harry Needle Railroad Company, July 2011; despatched from EWS Toton Depot, 5th October 2011; to Nemesis Rail, Burton upon Trent; noted in EWS red and yellow livery with number 08389, Nemesis Rail, 28th February 2012; to Barrow Hill Engine Shed, Staveley, 28th February 2014; seen in EWS red and yellow livery with number 08389, Barrow Hill Engine Shed, 10th July 2014; to Celsa, Cardiff, on hire, 28th January 2016; noted in EWS red and yellow livery at Celsa, Cardiff, 14th February 2016; to Railway Support Services, Wishaw, for repairs, 28th May 2016; noted at Wishaw, 9th July 2016; to Celsa, Cardiff, on hire, by 27th July 2016; to Barrow Hill Engine Shed, Staveley, for repairs, 26th November 2016; to Midland Road Depot, Leeds, for tyre turning, 24th January 2017; returned to Barrow Hill Engine Shed, Staveley, 27th January 2017; to Celsa, Cardiff, on hire, 10th February 2017; returned to Barrow Hill Engine Shed, Staveley, 27th June 2017; to Celsa, Cardiff, on hire, 7th July 2017; to Barrow Hill Engine Shed, Staveley, for repairs, 6th June 2019; to St Philip's Marsh Depot, Bristol, for tyre turning, 27th November 2019; to Celsa, Cardiff, on hire, 2nd December 2019; to Midland Road Depot, Leeds, for tyre turning, 19th January 2023; in faded EWS red and yellow livery at Midland Road, 20th January 2023; to Nemesis Rail, Burton upon Trent, 30th January 2023; to Celsa, Cardiff, on hire, 13th March 2023.

D3505　　Derby　　　　　　　1958　87E　　3/93　　F　08390
08390　　new (order D658) 18th April 1958; unofficially named BERNI, January 1987; withdrawn, 6th November 1993; to Margam, 20th November 1993; to Adtranz, Crewe, by road, for storage; sold to Harry Needle Railroad Company; despatched from Adtranz, Crewe, by road, 4th April 1997; to South Yorkshire Railway Preservation Society, Meadowhall, Sheffield; used for spares; noted in BR blue livery with number 08390, no

centre wheels, Meadowhall, 23rd August 1997; remains to Barrow Hill Engine Shed, Staveley, 6th March 2001; scrapped on site by Harry Needle Railroad Company staff, March 2004.

D3508   Derby   1958   16A   4/10   F   08393
08393   new (order D658) 25th April 1958; withdrawn, 18th April 2010; sold to EMR, July 2011; moved to European Metal Recycling, Kingsbury, 4th August 2011; re-sold to LH Group, Barton under Needwood, and moved on 11th January 2012; noted in EWS red and yellow livery with number 08393, LH Group, 6th May 2012; used for spares; noted with engine removed, LH Group, August 2015; scrapped on site, early February 2016.

D3513   Derby   1958   82A   7/85   F   402D / ANNABEL
08398   new (order D658) 16th May 1958; withdrawn, 15th July 1985; sold to E.C.C. Ports Ltd; moved from BR Bath Road Depot, Bristol to Taunton, 13th December 1985; moved to Exeter on 23rd January 1986; moved to Lostwithiel on 1st February 1986; moved to E.C.C. Ports Ltd, Fowey, Cornwall, 3rd February 1986; given number P402D; to BR Laira Depot, Plymouth, for repairs, summer 1988; to E.C.C. Marsh Mills, Devon, July 1988; to E.C.C. Blackpool Dryers, Burngullow, by August 1992; noted with identity 402D ANNABEL, Blackpool Dryers, 17th August 1992; to BR Laira Depot, Plymouth, for tyre turning, 29th September 1992; returned to E.C.C. Blackpool Dryers; to E.C.C. Rocks Dryers, Bugle, November 1992; seen in blue livery with numberplate 402D and nameplate ANNABEL, Rocks Dryers, Bugle, 13th September 1995; noted in blue livery with no number, Rocks Dryers, Bugle, 28th July 2010; re-sold to European Metal Recycling, Kingsbury, for scrap, and moved on 9th September 2010; scrapped, January 2011.

D3516   Derby   1958   40B   2/04   P   08401
08401   new (order D658) 23rd May 1958; withdrawn, February 2004; sold to Hunslet Engine Company for use as a hire loco, January 2011; moved to LH Group, Barton under Needwood, for repairs, 27th January 2011; to GBRf, Whitemoor Yard, March, on hire, 18th March 2011; to GBRf, Cardiff, on hire, 6th April 2011; to GBRf, Whitemoor Yard, March, on hire, 26th May 2011; noted in green livery with number 08401, Whitemoor, August 2011; to Celsa, Cardiff, on hire, October 2011; to GBRf, Whitemoor Yard, March, on hire, 8th December 2011; to LH Group, Barton under Needwood, 16th January 2013; to Cleveland Potash, Boulby Mine, on hire, 31st January 2013; to Celsa, Cardiff, on hire, 24th October 2013; to LH Group, Barton under Needwood, 5th March 2015; to Wabtec, Doncaster, 19th March 2015; to LH Group, Barton under Needwood, by 18th August 2015; noted in black livery with number 08401, LH Group, August 2015; to Bounds Green Depot, London, on hire, 24th August 2015; to LH Group, Barton under Needwood, by 21st August 2015; to Bounds Green Depot, London, on hire, 24th August 2015; to LH Group, Barton under Needwood, by 17th October 2015; to Midland Road Depot, Leeds, for tyre turning, 2nd August 2016; noted in green livery with number 08401, Midland Road, 18th August 2016; to LH Group, Barton under Needwood, late August 2016; to Hams Hall Rail Freight Terminal, Coleshill, on hire, 22nd August 2016; noted in green livery with number 08401, Hams Hall, March 2018; to LH Group, Barton under Needwood, for repairs, 14th May 2020; returned to Hams Hall, on hire; re-sold to Hunslet Ltd (part of the Ed Murray Group), 1st May 2021; suffered fire damage at Hams Hall; to Hunslet Ltd, Barton under Needwood, for repairs, March 2022; noted on a low-loader, M6 at Farleton, 20th July 2022; to LNER, Craigentinny Depot, Edinburgh, on hire, 20th July 2022; to LH Group, Barton under Needwood, for repairs, 8th June 2023; noted at LH Group, with number 401 on both buffer beams, 2nd July 2023; to Tata Steel Europe, Shotton, on hire, by 8th August 2023; to Hunslet Ltd, Barton under Needwood, 17th January 2024.

D3520   Derby   1958   5A   9/14   P   08405
08405   new (order D658) 6th June 1958; withdrawn, September 2014; noted in EWS red and yellow livery with number 08405, Crewe Electric Depot, 3rd December 2015; put on sale by tender by DB Cargo, with bids due by 5th October 2016; sold to Railway Support Services; despatched from DB Cargo, Crewe Electric Depot, January 2017; to Railway Support Services, Wishaw; to East Midlands Trains, Neville Hill Depot, Leeds, on hire, 23rd January 2017; noted at Neville Hill, in EWS livery, 10th December 2018 and 10th January 2024.

D3525   Derby   1958   PZ   ?   P   08410
08410   new (order D658) 20th June 1958; noted in green livery with number 08410, Long Rock Depot, Penzance, 19th August 2017; withdrawn, date not known; sold at auction by Great Western Railway; despatched from Long Rock Depot, 26th February 2020; arrived at A.V. Dawson Ltd, Middlesbrough, 27th February 2020; noted in green livery with number 08410, A.V. Dawson Ltd, 20th June 2020 and 19th September 2023.

D3526   Derby   1958   66B   2/07   P   08411
08411   new (order D658) 20th June 1958; noted in blue livery with number 08411, Allerton Depot, 30th October 2005; withdrawn, 28th February 2007; sold to Traditional Traction, Wishaw, March 2007; to Colne Valley Railway, Castle Hedingham, for storage, 29th March 2007; noted in blue livery with number 08411, Colne Valley Railway, April 2009; to Railway Support Services, Wishaw, 12th June 2015; used for spares; seen in blue livery with number 08411 (no engine), Railway Support Services, Wishaw, 23rd September 2023.

D3528   Derby   1958   41A   12/96   F   08413
08413   new (order D658) 27th June 1958; withdrawn, December 1996; to RFS (Engineering) Ltd, Doncaster, June 1998; sold to C.F. Booth Ltd; despatched from Wabtec, Doncaster, 13th June 2000; to C.F. Booth Ltd, Rotherham; seen in blue livery with number 08413, no rods, C.F. Booth Ltd, 17th May 2004; re-sold to RMS Locotec Ltd, Dewsbury, and moved 3rd June 2004; noted in grey livery with number 08413, RMS Locotec Ltd, 1st August 2004; used for spares; remains to Morley Waste Traders Ltd, Leeds, for scrap, February 2007; scrapped, February 2007.

D3529   Derby   1958   81A   2/07   F   08414
08414   new (order D658) 27th June 1958; noted in grey livery with numbers 08414 and D3529, EWS Toton Depot, 16th January 2004 and 11th March 2006; withdrawn, 28th February 2007; sold to Traditional Traction; despatched from Toton Depot, 9th March 2007; to Traditional Traction, Wishaw; noted at Wishaw, 17th April 2007; used for spares; remains to European Metal Recycling, Kingsbury, for scrap, 16th August 2007; scrapped, late August 2007.

D3530   Derby   1958   8F   9/96   F   08415
08415   new (order D658) 4th July 1958; withdrawn, 30th September 1996; sold to RFS (Engineering) Ltd, Doncaster, and moved October 1996; noted in BR blue livery with number 08415, Doncaster, November 1996; used for spares; remains to European Metal Recycling, Attercliffe, Sheffield, for scrap, 5th September 1997; scrapped, 1997.

D3531   Derby   1958   16A   2/92   F   08416
08416   new (order D658) 4th July 1958; withdrawn, 14th February 1992; sold to RFS (Engineering) Ltd, Kilnhurst, and moved on 10th April 1992; used for spares; seen in BR blue livery with number 08416, Kilnhurst, 23rd May 1992 and 21st May 1993; works closed,

May 1993; seen at Kilnhurst, 7th August 1993; remains scrapped on site by C.F. Booth Ltd of Rotherham, August 1993.

D3532   Derby   1958   DY   12/97 P   08417
08417   new (order D658) 4th July 1958; withdrawn, December 1997; stored at Etches Park; sold to Serco Railtest and worked at Derby RTC, 1998/99; to Foster Yeoman Quarries Ltd, Merehead Stone Terminal, on hire, 17th June 1999; to Whatley Quarry, on hire, by 5th December 1999; returned to Serco Railtest, Derby, for repairs, 30th March 2000; to Merehead Stone Terminal, on hire, by 5th April 2000; noted on low-loader on Bristol ring road, 8th April 2002; to Railway Technical Centre, Derby, 9th April 2002; noted at Derby, 12th April 2002; noted in blue livery with number 08417, Derby RTC, September 2006; noted in yellow livery with number 08417, Derby RTC, 20th February 2012 and 18th March 2016; worked Branch Line Society trip round Etches Park, 16th July 2017; to Peak Rail, Rowsley, 31st March 2021; put on sale, July 2021; seen in yellow livery with number 08417, Rowsley, 24th July 2021; re-sold to Harry Needle Railroad Company; to Barrow Hill Engine Shed, Staveley, 24th November 2021; seen in yellow livery with number 08417, HNRC, Barrow Hill Engine Shed, 12th March 2022; to Tyseley Locomotive Works, Birmingham, for repairs, 4th May 2023.

D3533   Derby   1958   16A   2/04   P   08418
08418   new (order D658) 10th July 1958; withdrawn, February 2004; noted in EWS livery, Bescot Yard, 21st September 2004 and 16th February 2006; sold to West Coast Railway Company; despatched from Bescot Depot, 4th August 2010; to West Coast Railway Company, Carnforth; noted in EWS red/yellow livery with number 08418, being used for spares, Carnforth, 21st January 2023.

D3534   Derby   1958   12B   4/93   F   08419
08419   new (order D658) 11th July 1958; withdrawn, 23rd April 1993; sold to ABB Transportation, Crewe, and moved about April 1994; noted in BR blue livery with number 08419, no middle wheels, Crewe, 4th November 1994; sold to Mike Darnall, Newton Heath, Manchester, and moved to Bombardier Transportation, Doncaster Works, for storage, 23rd November 2000; seen in blue livery with number 08419, Bombardier, Doncaster, 30th August 2004; to C.F. Booth Ltd, Rotherham, for scrap, 24th November 2004; noted at C.F. Booth Ltd, 25th November 2004; scrapped, November 2004.

D3536   Derby   1958   55G   8/92   P   09201
09201   withdrawn, August 1992; former number 08421 to 22nd September 1992; to RFS, Kilnhurst, 22nd September 1992; rebuilt and renumbered 09201; last allocation was to Immingham Depot in August 1999; sold to Harry Needle Railroad Company, 2015; to Hope Cement Works, Derbyshire, for storage, 14th September 2015; noted disused at rear of works, 15th May 2023; to Harry Needle Railroad Company, Worksop Depot, 5th June 2023.

D3538   Derby   1958   8F   11/88 P   2 / HO11
08423   new (order D658) 15th August 1958; withdrawn, 24th November 1988; sold to Trafford Park Estates Ltd, Manchester, and moved on 21st November 1988; repainted green, summer 1989; to Longsight Depot, for tyre turning, early 1990; seen in green livery with number 08423, Trafford Park, 17th November 1990 and 7th May 1994; to MoD Kineton, for trials, 1st August 1994; re-sold to RMS Locotec Ltd, Dewsbury, and moved 8th August 1994; to Mobil Oil Co Ltd, Coryton Bulk Terminal, Essex, on hire, about July 1995; to Flixborough Wharf Ltd, Flixborough, Scunthorpe, on hire, mid-June 1998; returned to

RMS Locotec Ltd, 3rd July 2003; to Flixborough Wharf Ltd, Flixborough, Scunthorpe, on hire, 18th February 2004; suffered fire damage, 11th September 2006; to RMS Locotec Ltd, Wakefield, for repairs, 10th October 2006; to PD Ports, Tees Dock, on hire, week-ending 23rd February 2007; noted in RMS blue livery with grey cab, PD Ports, February 2012; to RMS Locotec Ltd, Wolsingham Depot, 1st August 2014; to PD Ports, Tees Dock, on hire, 27th August 2014; noted in RMS blue/grey with numbers 2 and HO11, PD Ports, 19th February 2023; to RMS Locotec Ltd, Wolsingham Depot, 2nd March 2023; noted at Wolsingham, 5th January 2024; to Independent Railway Engineering, Chesterfield, 16th January 2024.

D3543   Derby   1958   40B   2016   P   08428
08428   new (order D658) 29th August 1958; withdrawn, 2016; put on sale by tender by DB Cargo, with bids due by 5th October 2016; sold to Harry Needle Railroad Company; noted in EWS red and yellow livery with number 08428, Warrington Depot, 25th November 2016; despatched from DB Cargo, Warrington Depot, 17th January 2017; to Harry Needle Railroad Company, Barrow Hill Engine Shed, Staveley; to Celsa, Cardiff, on hire, 24th January 2017; noted in EWS red and yellow livery with number 08428, Celsa, 4th February 2017; to Moveright International, Wishaw, Warwickshire, 4th March 2017; to Barrow Hill Engine Shed, Staveley, 14th March 2017; seen in EWS red and yellow livery with number 08428, Barrow Hill Engine Shed, 28th December 2017; noted at BHES, 16th December 2023.

D3551   Derby   1958   36A   1/92   P   08436 / D3551
08436   new (order D658) 3rd October 1958; latterly ran on BR with painted name BEIGHTON; withdrawn, 10th January 1992; sold to South Yorkshire Railway Preservation Society, Meadowhall, Sheffield, and moved late May 1993; re-sold to RT Rail, Crewe, April 1998; to RMS Locotec Ltd, Dewsbury, for overhaul, 1st May 1998; to Keighley & Worth Valley Railway, Haworth, for painting, 1998; to Hays Chemicals, Sandbach, on hire, 8th January 1999; to Railway Age, Crewe, for repairs, 17th December 2000; to Hays Chemicals, Sandbach, on hire, 22nd December 2000; to Wabtec, Doncaster, for repairs, 26th September 2001; to Hays Chemicals, Sandbach, on hire, 9th February 2002; to RT Rail, Crewe, November 2002; to Hays Chemicals, Sandbach, on hire, March 2003; to Wabtec, Doncaster, by July 2003; noted in black livery with number 08436, Wabtec, July 2003; re-sold to Swanage Railway, Dorset, and moved in August 2004; noted in black livery with number 08436, Swanage, 30th May 2007; noted in lined black livery with 'BRML Eastleigh Works' on side, numbers D3551 and 08436, and name BEIGHTON, Swanage Railway, 9th June 2024.

D3556   Derby   1958   66B   2/04   P   08441
08441   new (order D658) 30th October 1958; withdrawn, February 2004; noted in EWS red and yellow livery, Motherwell Depot, 9th June 2005; sold to Colne Valley Railway; despatched from Motherwell Depot, 14th May 2007; to Colne Valley Railway, Castle Hedingham; noted in red and yellow livery with number 08441, Colne Valley Railway, 15th May 2007; to LH Group, Barton under Needwood, for repairs, 16th February 2010; to Colne Valley Railway, Castle Hedingham, November 2010; to Port of Felixstowe, on hire, 7th February 2011; re-sold to Railway Support Services, Wishaw, and moved on 24th April 2014; noted on a low-loader leaving RSS, 1st December 2015; to Chasewater Railway, Staffordshire, for testing, 1st December 2015; noted at Chasewater Railway, January 2016; to Bounds Green Depot, London, on hire, 3rd February 2016; noted in RSS black livery with number 08441, Bounds Green, 3rd March 2016; noted at Bowes Park piloting a failed VTEC train into Bounds Green Depot, 1st June 2018; to Railway Support Services,

Wishaw, for repairs, 1st February 2020; to Bounds Green Depot, London, on hire, 29th June 2020; departed after depot passed to Hitachi ownership, March 2021; to Crown Point Depot, Norwich, on hire, March 2021; noted in RSS black, Crown Point, 23rd April 2023.

D3557   **Derby**   **1958**   **81A**   **1/04**   **P**   **0042**
08442   new (order D658) 3rd November 1958; withdrawn, January 2004; to London & North Western Railway Company, Traction and Rolling Stock Depot, Eastleigh, Hampshire (ex DB Schenker, with site), 21st April 2011; noted in LNWR grey livery with no number, Eastleigh, 15th May 2019; noted in blue livery with number 0042, Eastleigh, 23rd October 2020; dismantled for spares by early 2023; to Tyseley Locomotive Works, Birmingham, in dismantled condition, 3rd July 2023; to Nemesis Rail, Burton upon Trent, by 26th September 2023.

D3558   **Derby**   **1958**   **62A**   **7/85**   **P**   **D3558**
08443   new (order D658) to 64E Polmont, 7th November 1958; withdrawn, 7th July 1985; to Grangemouth Depot, for storage, November 1985; sold to Scottish Grain Distillers; despatched from BR Grangemouth Depot, 14th January 1986; to Scottish Grain Distillers, Cambus Distillery, Alloa; noted at Cambus Distillery, 4th May 1986; noted in black and red livery, September 1989; donated to Scottish Railway Preservation Society, Bo'ness, and moved 28th June 1993; seen in BR blue livery with number 08443, Bo'ness, 13th September 1998; noted in BR green livery with number D3558, Bo'ness, October 2012 and 5th July 2024.

D3559   **Derby**   **1958**   **86A**   **11/86**   **P**   **08444**
08444   new (order D658) 7th November 1958; withdrawn, 2nd November 1986; sold to Bodmin & Wenford Railway; despatched from Canton Depot; arrived at Bodmin & Wenford Railway, 27th March 1987; seen in green livery with number D3559, Bodmin & Wenford Railway, 13th September 1995; noted in green livery with number 08444, Bodmin & Wenford Railway, 7th September 2019.

D3560   **Derby**   **1958**   **40B**   **11/95**   **P**   **08445**
08445   new (order D658) 14th November 1958; withdrawn, November 1995; to RFS (Engineering) Ltd, Doncaster, by road, for storage, August 1998; sold to Mike Darnall, Newton Heath, Manchester, 2000; to Wabtec, Doncaster, for repairs, 2000; moved to Mike Darnall, non-rail location at Newton Heath, 17th July 2000; to East Lancashire Railway, Bury, on hire, by 14th September 2001; noted in light grey livery with number 08445, East Lancashire Railway, Bury, July 2002; to Carillion Construction Ltd, Manchester Metrolink upgrade, on hire, 26th May 2007; to East Lancashire Railway, Bury, by 17th June 2007; to former Corus Works, Castleton, for storage, from about 31st August 2007; to Castle Cement Works, Ketton, on hire, 18th February 2009; to Corus, Shotton Steelworks, on hire, 22nd July 2009; re-sold to LH Group, Barton under Needwood, March 2011; to LH Group, Barton under Needwood, for repairs, 27th March 2011; to GBRf, Trafford Park, Manchester, on hire, April 2011; to Daventry International Rail Freight Terminal, on hire, week-ending 24th June 2011; to LH Group, Barton under Needwood, for repairs, 14th August 2011; to Daventry International Rail Freight Terminal, on hire, September 2011; noted in yellow, blue and green livery with number 08445, Daventry International Rail Freight Terminal, January 2012 and July 2018; re-sold to Hunslet Ltd (part of the Ed Murray Group), 1st May 2021.

D3562   Derby   1958   12B   11/94 P   08447
08447   new (order D658) 28th November 1958; withdrawn, 11th November 1994; sold to Deanside Transit Ltd, Glasgow, about May 1995; noted in yellow livery, Deanside Transit, 25th February 2006; to Harry Needle Railroad Company, Long Marston, for overhaul, 12th June 2007; noted in lilac livery with black roof and number 08447, Long Marston, 7th June 2008; to Deanside Transit Ltd, Glasgow, 13th June 2008; to John G. Russell (Transport) Ltd, Hillington, Glasgow, for storage, by 2019; noted at Hillington, 8th June 2024.

D3566   Derby   1958   KD   2007   P   08451
08451   new (order D658) 5th December 1958; withdrawn, 2007; sold to Alstom Transport, Longsight Depot, Manchester, and moved in 2007; to Alstom Transport, Wembley Depot, London, 24th January 2012; to Alstom Transport, Longsight Depot, Manchester, 14th March 2012; noted in blue livery with number 08451, Longsight Depot, June 2012; to Alstom Transport, Polmadie Depot, Glasgow, by 22nd April 2016; noted in green livery with number 08451, Polmadie Depot, 6th June 2016; to Alstom Transport, Longsight Depot, Manchester, 21st June 2017; noted at Longsight, 22nd June 2017; noted at Longsight, September 2017; to Alstom Transport, Polmadie Depot, Glasgow, mid-November 2021; to Arlington Fleet Services Ltd, Eastleigh Works, for repairs, 16th February 2022; to Alstom Transport, Polmadie Depot, Glasgow, 27th October 2022.

D3569   Derby   1958   WN   ?   P   08454
08454   new (order D658) 19th December 1958; withdrawn, date not known; sold to Alstom Transport, Wembley Depot, London, and moved in 2007; to Alstom Transport, Longsight Depot, Manchester, 5th March 2012; to Alstom Transport, Oxley Depot, Wolverhampton, by 28th June 2014; to Alstom Transport, Edge Hill Depot, Liverpool, by 23rd February 2015; to Arlington Fleet Services Ltd, Eastleigh Works, for overhaul and repaint, by 27th September 2016; noted in fresh blue livery with number 08454, Eastleigh Works, 8th December 2016; to Alstom Transport, Edge Hill Depot, Liverpool, by 14th December 2016; to Alstom Transport, Technology Centre, Widnes (new Pendolino facility), 23rd June 2017; noted in blue livery with number 08454, Alstom Transport, Widnes, 30th June 2017 and 24th February 2024.

D3575   Crewe   1958   8J   4/04   P   08460
08460   new (order E496) 13th August 1958; withdrawn, April 2004; sold to Railway Support Services, Wishaw; despatched from EWS Allerton Depot, 29th March 2007; to Colne Valley Railway, Castle Hedingham, for storage; noted in grey livery with number 08460, Colne Valley Railway, October 2008; to Port of Felixstowe, on hire, 5th November 2009; to Colne Valley Railway, Castle Hedingham, for repairs, 13th January 2011; to Railway Support Services, Wishaw, June 2014; to Axiom Rail, Wagon Works, Stoke on Trent, on hire, about September 2014; to Railway Support Services, Wishaw, about September 2016; noted repainted in RSS black livery with name SPIRIT OF THE OAK, Wishaw, 15th December 2016; to DB Cargo Maintenance Ltd, Wheildon Road Works, Stoke on Trent, on hire, 5th January 2017; noted in RSS black livery with number 08460, Stoke on Trent, 16th December 2017 and 4th January 2019; to Railway Support Services, Wishaw, 8th February 2019; to GBRf, East Yard, Eastleigh, on hire, 28th March 2019; noted at Eastleigh, 29th March 2019; to Railway Support Services, Wishaw, 7th May 2021; to Felixstowe Docks, on hire, early 2022; to Railway Support Services, Wishaw, for repairs, 17th September 2022; noted in RSS grey livery with number 08460 and name SPIRIT OF THE OAK, at RSS, 20th September 2022; to Hams Hall Rail Freight Terminal, Coleshill, on

hire, 3rd November 2022; to Railway Support Services, Wishaw, 25th February 2023; to GBRf, Bescot Yard, on hire, 18th April 2023.

D3577   Crewe   1958   16A   3/15   P   08994
08994   new (order E496) 28th August 1958; moved to Landore, numbered 08462, 18th February 1987; rebuilt with cut down cab for working Burry Port & Gwendraeth Valley line; reinstated with new number 08994, 4th September 1987; worked at Deanside Transit, Glasgow, on hire, from 2nd June 2005 to 30th August 2006; withdrawn, March 2015; put on sale by DB Schenker, 5th May 2015; sold to Harry Needle Railroad Company, August 2015; despatched from DB Schenker, Toton Depot, 14th September 2015; to Nemesis Rail, Burton upon Trent, for storage; noted in red and yellow livery with number 08994, Nemesis Rail, March 2016 and February 2019; noted on a low-loader on the M50, 19th January 2022; to Gwendraeth Valley Railway, near Kidwelly, 19th January 2022.

D3585   Crewe   1958   5A   3/86   F   08470
08470   new (order E496) 7th November 1958; withdrawn, 23rd March 1986; to Basford Hall, Crewe, for storage, March 1987; noted in Basford Hall Yard, 27th March 1987 and 28th July 1987; moved to BREL Crewe Works, September 1987; noted at Crewe Works, 6th December 1987; acquired by ABB Transportation, Crewe, April 1989; locomotive included in sale when BREL works privatised in April 1989; overhauled and used as yard shunter; to TML, Cheriton, for tunnel use, 23rd November 1992; returned to Crewe; scrapped on site by M.R.J. Phillips (Metals) Ltd, August 1996.

D3586   Crewe   1958   15A   9/85   P   D3586
08471   new (order E496) 13th November 1958; to Swindon Works, for assessment, June 1985; withdrawn, 9th September 1985; sold to SVR; despatched from BREL Swindon Works, 12th March 1986; arrived at BR Kidderminster yard, 9th April 1986; moved by low-loader to Severn Valley Railway, Bridgnorth, 14th April 1986; entered service in original BR blue livery; seen repainted in blue livery with number 3586, Severn Valley Railway, 26th September 1986; was regular Bridgnorth pilot; re-engined, January 1994; repainted in BR green, summer 1994; noted in green livery with number D3586, Severn Valley Railway, October 2003 and 19th March 2016; to St Philip's Marsh Depot, Bristol, for tyre turning, January 2021; returned to SVR; noted in green livery with number D3586, 5th May 2018.

D3587   Crewe   1958   BD   11/97   P   08472
08472   new (order E496) 20th November 1958; withdrawn, November 1997; sold to Great North Eastern Railway and allocated to Bounds Green Depot, London, December 1997; re-sold to RFS (Engineering) Ltd, Doncaster, August 1998, but with the locomotive remaining at Bounds Green Depot, London, now on hire; to Craigentinny Depot, Edinburgh, on hire, 1999; to Wabtec, Doncaster, December 2000; to Craigentinny Depot, on hire, by August 2003; to Wabtec, Doncaster, by 1st April 2004; to Craigentinny Depot, on hire, 10th August 2005; noted in black livery with number 08472, Craigentinny Depot, February 2007; to Wabtec, Doncaster, 26th November 2010; noted in Wabtec black livery with number 08472, Wabtec, May 2011; to LNER, Craigentinny Depot, on hire, 12th September 2011; re-sold to Hunslet Ltd (Ed Murray Group), 1st May 2021; to Reid Freight Services Ltd, Cinderhill Industrial Estate, Longton, Stoke-on-Trent, for storage, 11th June 2023; to Aggregate Industries Ltd, Bardon Hill, on hire, 4th March 2024; to Chasewater Railway, 2nd August 2024.

D3588   Crewe   1958   86A   3/86   P   08473

08473   new (order E496) 28th November 1958; withdrawn, 17th March 1986; to Basford Hall, Crewe, for storage, 28th March 1987; to Leicester, for storage, 9th July 1987; sold to T.J. Thomson & Son Ltd; despatched from BR Leicester, 28th November 2000; arrived at T.J. Thomson & Son Ltd, Stockton-on-Tees, the following day; partly scrapped, February 2001; remains re-sold to Dean Forest Railway, Lydney, for spares (including radiator recovered from 08515) and moved 5th March 2001; noted in blue livery in heavily dismantled condition, Dean Forest Railway, 5th February 2006; seen dismantled, described as 'a rolling chassis for spares', Dean Forest Railway, 14th July 2021; to Nemesis Rail, Burton upon Trent, 21st November 2023.

D3591   Crewe   1958   62A   9/85   P   D3591

08476   new (order E496) 17th December 1958; withdrawn, 9th September 1985; to Swindon Works, for storage; sold to Swanage Railway; despatched from BREL Swindon Works, by road, 21st March 1986; delivered to Swanage Railway, Dorset; returned to working order and repainted as D3591, 1986; seen in green livery with number D3591, Swanage, 13th September 1992; noted in black livery with number D3591, Swanage, June 2001 and May 2009; to Battlefield Line, Shackerstone, 24th July 2009; to Swanage Railway, Dorset, 10th March 2012; noted in green livery with number D3591, Swanage Railway, May 2014; re-sold to a group of Swanage Railway members, about June 2024.

D3594   Horwich   1958   86A   11/91   P   08479 / 13594

08479   new (LOT 259) 1st September 1958; withdrawn, 8th November 1991; sold to East Lancashire Railway, Bury, and moved 30th April 1993; noted in BR blue livery with number 08479, East Lancashire Railway, 2nd April 2011; noted in black livery with number 13594, East Lancashire Railway, July 2014; serviceable and used as Baron Street pilot, July 2024.

D3595   Horwich   1958   16A   ?   P   08480

08480   new (LOT 259) 12th September 1958; noted in mustard yellow livery, with horizontal red stripe and TOTON No.1 along bonnet side, Toton Depot, 6th February 2006; withdrawn, date not known; put on sale by tender by DB Cargo, with bids due by 5th October 2016; sold to Railway Support Services; despatched from DB Cargo, Toton Depot, 1st February 2017; to Railway Support Services, Wishaw; noted in yellow livery with red stripe and with number 08480, RSS Wishaw, 2nd February 2017; repainted in RSS black livery, March 2017; to Great Central Railway, Loughborough, 17th March 2017; noted in RSS black livery with number 08480, Great Central Railway, 18th March 2017; to Crown Point Depot, Norwich, on hire, 18th July 2017; noted at Norwich, 18th July 2017; to Railway Support Services, Wishaw, 18th March 2020; to Arriva Traincare, Barton Hill Depot, Bristol, on hire, 16th June 2020; to Gemini Rail, Wolverton Works, on hire, 11th December 2020; to Railway Support Services, Wishaw, about March 2021; to Port of Felixstowe, on hire, May 2021; to Phillips Ltd, Killingholme Oil Refinery, on hire, early 2023; to LNER, Craigentinny Depot, Edinburgh, on hire, 25th May 2023.

D3596   Horwich   1958   86A   4/02   F   08481

08481   new (LOT 259) 19th September 1958; withdrawn, April 2002; noted in blue livery with number 08481, Wigan Depot, 7th September 2003; sold to Barry Rail Centre; despatched from Wigan Depot, 7th October 2005; to Barry Rail Centre, Barry Island; noted in blue livery with no number, Barry, 6th May 2007; to European Metal Recycling, Kingsbury, for scrap, May 2011; scrapped early June 2011.

D3598   **Horwich**   **1958**   **84A**   **?**   **P** **08483**
08483   new (LOT 259) 3rd October 1958; withdrawn, date not known; put up for auction by Great Western Railway; sold to Locomotive Services Ltd; despatched from Laira Depot, 25th February 2020; to Locomotive Services Ltd, Crewe, 26th February 2020; seen in green livery with number 08483, Locomotive Services Ltd, 7th June 2024; to PD Ports, Tees Dock, on hire, 10th July 2024.

D3599   **Horwich**   **1958**   **BY**   **9/88**   **P** **08484**
08484   new (LOT 259) 10th October 1958; moved from Bletchley to Wolverton Works, 2nd October 1988; locomotive included in sale when BRML works privatised; to LH Group, Barton under Needwood, for repair, 16th November 2003; returned to Wolverton Works, 10th February 2004; noted in grey livery with number 08484, Wolverton, July 2004; re-sold to Railway Support Services, Wishaw, April 2006; to Port of Felixstowe, on hire, 25th April 2006; to Railway Support Services, Wishaw, for storage, about 10th November 2006; to Nene Valley Railway, Wansford, for gala, February 2007; returned to Railway Support Services, February 2007; to St Philip's Marsh Depot, Bristol, for tyre turning, 19th April 2007; to Port of Felixstowe, on hire, week-ending 4th May 2007; to Tyseley Diesel Depot, Birmingham, for tyre turning, 13th July 2012; to Gloucestershire Warwickshire Railway, Toddington, on hire, 14th July 2012; noted in blue livery with number 08484 and name CAPTAIN NATHANIEL DARELL, Gloucestershire Warwickshire Railway, Toddington, July 2013; to Port of Felixstowe, on hire, 8th November 2013; to Railway Support Services, Wishaw, for repairs, 9th September 2014; to National Railway Museum, Shildon, on hire, 16th September 2015; noted in Railway Support Services black and red livery with number 08484, Shildon, 25th February 2016; to Network Rail, Rail Innovation Centre, Asfordby, on hire, 25th February 2016; to Bombardier Transportation, Litchurch Lane Works, Derby, on hire, 5th August 2016; to Railway Support Services, Wishaw, for repairs, 17th October 2016; to Cemex Rail Products, Washwood Heath, on hire, 25th October 2016; noted in RSS black livery with number 08484, Washwood Heath, 25th October 2016; to Hitachi, Newton Aycliffe, on hire, 4th April 2018; seen in RSS black livery with number 08484, Newton Aycliffe, 21st June 2018; to Crown Point Depot, Norwich, on hire, 17th March 2020; noted in RSS black, Crown Point, 23rd April 2023.

D3600   **Horwich**   **1958**   **12A**   **11/07** **P** **08485**
08485   new (LOT 259) 20th October 1958; withdrawn, November 2007; sold to West Coast Railway Company, Carnforth, and moved 1st September 2010; noted in blue livery with number 08485, Carnforth, 2nd September 2010 and 17th August 2023.

D3605   **Horwich**   **1958**   **62A**   **12/85** **P** **D3605**
08490   new (LOT 259) 14th November 1958; withdrawn, 18th December 1985; moved to Perth, for storage, January 1986; sold to Strathspey Railway, Aviemore; despatched from BR Perth, and arrived at Aviemore on 18th June 1987; noted in black livery with number D3605, Strathspey Railway, June 2010; noted in green livery with number D3605, Strathspey Railway, 17th November 2022.

D3607   **Horwich**   **1958**   **66B**   **12/00** **F** **08492**
08492   new (LOT 259) 26th November 1958; latterly ran on BR with painted name BARNSLEY; withdrawn, 15th December 2000; sold to Harry Needle Railroad Company; despatched from EWS Motherwell Depot, 2nd June 2006; to Barrow Hill Engine Shed, Staveley, arrived 5th June 2006; seen in blue livery with number 08492, Barrow Hill Engine Shed, Staveley, 21st June 2011; used for spares; to European Metal Recycling, Kingsbury, for scrap, 14th February 2012; scrapped, 15th February 2012.

D3608   Doncaster   1958   86A   7/99   F   08493
08493   new (order EO1) 28th February 1958; withdrawn, 2nd July 1999; sold to RT Rail of Crewe; despatched from Canton Depot, Cardiff, and moved to Wabtec, Doncaster, for storage, September 2003; noted in faded blue livery with number 08493, Wabtec, 29th August 2005 and 27th March 2006; used for spares; remains to C.F. Booth Ltd, Rotherham, for scrap, 19th May 2008; scrapped, August 2008.

D3610   Doncaster   1958   5A   ?   P   08495
08495   new (order EO1) 8th March 1958; withdrawn, date not known; put on sale by tender by DB Cargo, with bids due by 5th October 2016; sold to NYMR; noted in EWS red and yellow livery, Crewe Electric Depot, 17th November 2016; despatched from DB Cargo, Crewe Electric Depot, 22nd December 2016; to North Yorkshire Moors Railway, Grosmont; noted in blue livery with number 08495, North Yorkshire Moors Railway, August 2018; re-sold and moved to Nemesis Rail, Burton upon Trent, about 17th November 2023.

D3613   Darlington   1958   40A   3/69   F   DAVID / D14
new 27th June 1958; withdrawn, 15th March 1969; sold to George Cohen Ltd, Kettering, and moved early August 1969; quickly re-sold to NCB; to NCB Bestwood Colliery, Nottinghamshire, August 1969; noted with number D3613, Bestwood Colliery, 3rd September 1969; noted at Bestwood Colliery, 28th June 1970 and 27th March 1971; to Linby Colliery, September 1971; to Moor Green Colliery, Newthorpe, November 1971; named DAVID; noted at Moor Green Colliery, 4th June 1972; to 16A Toton Depot, for repairs, by 26th October 1975; returned to Moor Green Colliery, Newthorpe, by 22nd December 1975; numbered D14 about 1980; seen in green livery with name DAVID, Moor Green Colliery, 7th March 1981 and 6th May 1983; noted being scrapped on site by The Vic Berry Company of Leicester, 8th April 1985; Moor Green Colliery closed on 19th July 1985.

D3618   Darlington   1958   40A   4/69   F   ROBIN / D16
new 21st August 1958; withdrawn, 19th April 1969; sold to George Cohen Ltd, Kettering, and moved 16th August 1969; quickly re-sold to NCB; to NCB Bestwood Colliery, Nottinghamshire, August 1969; noted with number D3618, Bestwood Colliery, 3rd September 1969; to Annesley Colliery, March 1970; seen named ROBIN at Annesley Colliery, 30th May 1970; to BR Toton Depot for repairs, June 1974; returned to Annesley Colliery; to Cotgrave Colliery, 24th July 1980; numbered D16 about 1980; to Moor Green Colliery, Newthorpe, 30th March 1981; seen in red livery with ROBIN and D16 on cabside, Moor Green Colliery, 6th May 1983; scrapped on site by The Vic Berry Company of Leicester, 26th March 1985; Moor Green Colliery closed on 19th July 1985.

D3619   Darlington   1958   40A   3/69   F   SIMON / D15
new 26th August 1958; withdrawn, 15th March 1969; sold to NCB; to NCB Gedling Colliery, Nottinghamshire, September 1969; to Bestwood Colliery, November 1969; named SIMON; noted at Bestwood Colliery, 28th June 1970 and 27th March 1971; to Linby Colliery, July 1971; noted at Linby Colliery, 12th April 1972; to Moor Green Colliery, Newthorpe, 24th November 1975; numbered D15 about 1980; seen in green livery with SIMON and D15 on cabside, Moor Green Colliery, 7th March 1981 and 6th May 1983; noted being scrapped on site by The Vic Berry Company of Leicester, 8th April 1985; Moor Green Colliery closed on 19th July 1985.

D3638　　Darlington　　　　　1958　　52A　　　11/70　F　9185/61
new 24th November 1958; withdrawn, 7th November 1970; to NCB Bates Colliery, Blyth, on hire/trial, 19th November 1970; sold to NCB on 28th January 1971; to BR Gateshead Depot, for repairs, 4th February 1971; to BR Cambois Depot, Blyth, for repairs, 12th February 1971; to Bates Colliery, Blyth, 16th February 1971; to Ashington Colliery, 15th April 1971; to Bates Colliery, Blyth, 22nd April 1971; noted in BR blue livery, Bates Colliery, 25th October 1971; to Ashington Central Workshops, March 1975; noted with number 9185/61, Ashington Central Workshops, 20th July 1975; used for spares; remains scrapped on site by NCB staff, September 1975.

D3639　　Darlington　　　　　1958　　36A　　　7/69　F　?
new 25th November 1958; withdrawn, 13th July 1969; seen at 36A Doncaster Depot, 10th December 1969; sold by BR per tender reference 17/230/521T/107; sold to R.E. Trem Ltd of Finningley and moved to C.F. Booth Ltd, Doncaster, for storage, date not known; there are no known sightings at C.F. Booth Ltd, Doncaster; noted (per RO, July 1971, page 246) with number painted over and boards attached indicating export to Conakry, West Coast of Africa, at The Surrey Commercial Docks, London, 7th March 1970; exported from The Surrey Commercial Docks, London, March 1970 (see Appendix C).

D3642　　Darlington　　　　　1958　　36C　　　6/69　F　37
new 5th December 1958; withdrawn, 28th June 1969; sold to British Steel Corporation; to BSC Redbourn Works, Scunthorpe, September 1969; given number 37; to BSC Appleby-Frodingham Works, Scunthorpe, October 1975; scrapped on site by BSC staff, October 1978.

D3648　　Darlington　　　　　1959　　52A　　　1/71　F　9185/60
new 7th January 1959; withdrawn, 24th January 1971; to NCB Bates Colliery, Blyth, on hire/trial, 16th February 1971; to BR Cambois Depot, Blyth, for repairs, 27th February 1971; sold to NCB, 1st March 1971; to Bates Colliery, Blyth, 1st March 1971; to BR Cambois Depot, Blyth, for further repairs, 5th March 1971; to Bates Colliery, Blyth, 23rd March 1971; given number 9185/60; noted at Bates Colliery, 25th October 1971; scrapped on site by L. Marley & Co Ltd of Stanley, February 1977.

D3649　　Darlington　　　　　1959　　36A　　　7/69　F　?
new 9th January 1959; withdrawn, 20th July 1969; seen at 36A Doncaster Depot, 26th November 1969; sold by BR per tender reference 17/230/521T/107; sold to R.E. Trem Ltd of Finningley and moved to C.F. Booth Ltd, Doncaster, for storage, date not known; there are no known sightings at C.F. Booth Ltd, Doncaster; noted, (per RO, July 1971, page 246) with number painted over and boards attached indicating export to Conakry, West Coast of Africa, at The Surrey Commercial Docks, London, 7th March 1970; exported from The Surrey Commercial Docks, London, March 1970 (see Appendix C).

D3654　　Doncaster　　　　　1958　　86A　　　6/04　P　08499 / REDLIGHT
08499　　　new (order EO1) 29th March 1958; withdrawn, 1st June 2004; noted in EWS red and yellow livery with number painted over, Canton Depot, Cardiff, 25th July 2004; to Pullman Rail, Canton Depot, Cardiff (locomotive included in sale when depot sold), June 2005; noted in blue and black livery with no number, Canton Depot, 26th February 2006; noted in dark blue livery with black roof and small number 08499, Canton Depot, 5th November 2007 and 22nd May 2009; noted with REDLIGHT nameplates and number 08499, Canton Depot, 27th July 2017.

D3655   Doncaster   1958   16A   4/11   P   08500
08500   new (order EO6) 18th April 1958; withdrawn, April 2011; moved to Bescot Depot, for storage; put on sale by DB Schenker, 5th May 2015; sold to Harry Needle Railroad Company; despatched from DB Schenker, Bescot, 23rd October 2015; to Nemesis Rail, Burton upon Trent; noted in EWS red and yellow livery with number 08500, Nemesis Rail, 21st July 2017; to Harry Needle Railroad Company, Worksop Depot, 23rd April 2021.

D3657   Doncaster   1958   51L   10/88 P   08502 /
08502   LYBERT DICKINSON
new (order EO6) 6th May 1958; sold to ICI; despatched from 51L Thornaby Depot to ICI Wilton Works, Middlesbrough, 6th September 1988; officially withdrawn on 30th October 1988; repainted blue, 1995; noted in blue livery with number 08502 and name ANGIE, ICI Wilton Works, February 2007; re-sold to Harry Needle Railroad Company, February 2007; to Barrow Hill Engine Shed, Staveley, for servicing and repaint, 3rd August 2007; noted in pale blue livery with number 08502, Barrow Hill Engine Shed, 12th August 2007; to Northern Rail Ltd, Heaton Depot, Newcastle upon Tyne, on hire, 10th September 2007; to Barrow Hill Engine Shed, Staveley, 21st June 2012; noted in royal blue livery with number 08502, Barrow Hill Engine Shed, 28th July 2012; to Flixborough Wharf Ltd, Flixborough, Scunthorpe, on hire, October 2013; to Barrow Hill Engine Shed, Staveley, for repairs, 26th May 2014; seen shunting, in blue livery with number 08502, Barrow Hill Engine Shed, 24th September 2014; to Tyseley Diesel Depot, Birmingham, for tyre turning, 3rd November 2015; to Moveright International, Wishaw, Warwickshire, for storage, 20th December 2015; to GBRf, Speke, Liverpool, on hire, 20th December 2015; to Barrow Hill Engine Shed, Staveley, for repairs, 20th July 2017; to GBRf, c/o Ford Motor Co Ltd, Speke, Liverpool, on hire, late 2017; to Barrow Hill Engine Shed, Staveley, 10th April 2019; to East Kent Railway, Shepherdswell, 24th April 2019; noted in blue livery with number 08502, East Kent Railway, April 2019; to Harry Needle Railroad Company, Worksop Depot, 24th July 2023.

D3658   Doncaster   1958   55A   9/88   P   08503
08503   new (order EO6) 15th May 1958; sold to ICI; to BREL Doncaster Works, for overhaul, May 1988; officially withdrawn, 20th September 1988; despatched from BREL Doncaster Works, 26th September 1988; repainted blue, 1995; to ICI Wilton Works, Middlesbrough; re-sold to Harry Needle Railroad Company, February 2007; noted in blue livery with number 08503, ICI Wilton Works, 27th February 2007; to Heanor Heavy Haulage, Langley Mill, for storage, 7th December 2007; noted in blue livery with number 08503, Heanor, March 2008; to Moveright International, Wishaw, Warwickshire, for storage, late 2008; noted in pale blue livery with number 08503, Wishaw, 17th March 2012; re-sold to Railway Support Services, April 2013; to Colne Valley Railway, Castle Hedingham, 15th April 2013; noted in blue livery with number 08503, Colne Valley Railway, July 2013; to Barry Tourist Railway, Barry Island, on hire, 22nd October 2013; noted in blue livery with number 08503, Barry Tourist Railway, 31st August 2014 and 5th April 2015; to St Philip's Marsh Depot, Bristol, for tyre turning, 11th June 2019; returned to Barry, on hire, 12th June 2019; to Tyseley Locomotive Works, Birmingham, 7th October 2021; to Barry Tourist Railway, 21st April 2022.

D3662   Doncaster   1958   81D   8/93   P   08507
08507   new (order EO6) 20th June 1958; withdrawn, 13th August 1993; sold to S.M. McGregor & Sons, scrapyard, Bicester, and moved about April 1994; re-sold to South Yorkshire Railway Preservation Society, Meadowhall, Sheffield, and moved 23rd June 1995; to Barrow Hill Engine Shed, Staveley, for overhaul, 3rd November 1999; to Bombardier Transportation, Central Rivers Depot, Barton under Needwood, on hire, 7th

April 2001; noted in yellow and grey livery with number 08507, Central Rivers, October 2006; to Barrow Hill Engine Shed, Staveley, for repairs, 10th January 2011; to Port of Boston, on hire, 13th June 2011; re-sold to Riviera Trains Ltd of Crewe; to Crewe Diesel Depot, 19th September 2013; seen in blue livery with number 08507, Crewe Diesel Depot, 14th August 2014; to Nemesis Rail, Burton upon Trent, on hire, 12th January 2018; noted at Nemesis Rail, 12th January 2018; to Arlington Fleet Services Ltd, Eastleigh Works, Hampshire, for repairs, 23rd September 2019; to Cholsey & Wallingford Railway, 4th November 2021; to Railway Support Services, Wishaw, 5th July 2022; seen in blue livery with number 08507, Wishaw, 27th May 2023; to GBRf, Whitemoor Yard, March, on hire, 21st June 2023; to Railway Support Services, Wishaw, 11th April 2024; to ICL Cleveland Potash, Tees Dock, on hire, 21st April 2024.

D3673   Darlington   1958   66B   3/04   P   08511
08511   new 21st May 1958; withdrawn, March 2004; noted in EWS red and yellow livery with number 08511, Ayr Depot, 10th June 2005; sold to Railway Support Services, Wishaw; despatched from EWS Ayr Depot, 29th June 2007; to Railway Support Services, Wishaw, early July 2007; to Barry Rail Centre, Barry Island, 9th July 2007; to Port of Felixstowe, on hire, 8th July 2009; to Wembley Depot, London, for tyre turning, 2nd July 2010; to Colne Valley Railway, Castle Hedingham, for painting, 2nd August 2010; to Port of Felixstowe, on hire, 12th January 2011; to Tilbury Docks, on hire, 8th November 2013; to Railway Support Services, Wishaw, 2014; to Mid-Norfolk Railway, Dereham, 1st September 2014; to Port of Felixstowe, on hire, 8th September 2014; to Whatley Quarry, Somerset, on hire, 11th May 2015; noted at Whatley Quarry, 15th May 2015; to Railway Support Services, Wishaw, off-hire, 3rd July 2015; to Port of Felixstowe, on hire, 7th July 2015; to Railway Support Services, Wishaw, for repairs, December 2015; noted in red and yellow livery with number 08511, Wishaw, 16th December 2015; to Port of Boston, on hire, 7th February 2016; to Old Dalby Test Centre, on hire, 17th February 2016; to National Railway Museum, Shildon, 25th February 2016; to Railway Support Services, Wishaw, for repairs, 25th March 2016; to Bombardier Transportation, Litchurch Lane Works, Derby, on hire, 1st April 2016; noted in RSS black livery with number 08511, Litchurch Lane Works, 6th April 2016; to Railway Support Services, Wishaw, August 2016; to Chasewater Railway, Staffordshire, 12th September 2016; to Railway Support Services, Wishaw, 28th January 2017; to Arriva Traincare, Cambridge Depot, on hire (for moving Class 317 units), 23rd February 2017; noted with newly-fitted tightlock coupling, Cambridge Depot, 27th March 2017; noted on a low-loader leaving Cambridge, 8th December 2018; arrived at Arriva Traincare, Eastleigh, on hire, 10th December 2018; to GBRf, East Yard, Eastleigh, on hire, 28th March 2019; to Railway Support Services, Wishaw, 5th March 2021; to GBRf, East Yard, Eastleigh, on hire, about 14th April 2021; to Railway Support Services, Wishaw, 11th August 2022; to Felixstowe Docks, on hire, 17th September 2022; noted in RSS black livery with number 08511, Felixstowe Docks, 23rd September 2022; to Railway Support Services, Wishaw, for repairs, 3rd December 2022; to Telford Steam Railway, for working Santa trains, 19th December 2022; to Railway Support Services, Wishaw, about 10th January 2023; to GBRf, Whitemoor Yard, March, on hire, 4th March 2023; noted on low-loader on A141, 25th August 2023; to West Coast Railway Company, Carnforth, for tyre turning, 25th August 2023; to Railway Support Services, Wishaw, 29th August 2023; to GBRf, Eastleigh Yard, on hire, 22nd September 2023.

D3677   Darlington   1958   HT   1/92   F   08515
08515   new 18th June 1958; withdrawn, 20th January 1992; moved to Gateshead Depot, for storage; sold to Gwent Demolition of Margam, for scrap, 1995; firm ceased trading before locomotive could be despatched; remained in store at Gateshead to 2000;

re-sold to T.J. Thomson & Son Ltd, Stockton-on-Tees, for scrap, April 2000; parts removed at Gateshead Depot, February 2001; to T.J. Thomson & Son Ltd, Stockton-on-Tees, February 2001; to Foster Yeoman Quarries Ltd, Merehead Stone Terminal; used for spares, 2001; remains scrapped, February 2001.

D3678 Darlington   1958   16A   2/04   P   08516
08516   new 11th August 1958; withdrawn, February 2004; to DB Schenker, and noted in EWS red and yellow livery with number 08516, Barton Hill Depot, Bristol, 25th February 2005; to Arriva TrainCare, Barton Hill Depot, Bristol (ex DB Schenker, with site), April 2011; noted in LNWR grey livery with number 08516, Barton Hill Depot, 3rd April 2018; noted at Barton Hill Depot, 25th May 2020; to Railway Support Services, Wishaw, for repairs, by 8th August 2020; to Arriva TrainCare, Barton Hill Depot, 3rd December 2020; noted in Arriva TrainCare two-tone blue livery with number 08516, Barton Hill Depot, 12th January 2022.

D3679 Darlington   1958   30A   9/93   F   08517
08517   new 15th August 1958; withdrawn, 27th September 1993; sold to Harry Needle Railroad Company; despatched from 30A Stratford Depot, 4th June 2001; to HNRC, Barrow Hill Engine Shed, Staveley; to Wabtec, Doncaster, 27th June 2001; rebuilt using spares from 08668; noted in blue livery with yellow cab and no number, Wabtec, 1st May 2007; to C.F. Booth Ltd, Rotherham, for storage, 2007; to Harry Needle Railroad Company, Long Marston, 19th December 2007; noted in BR blue livery with number 08517, Long Marston, 30th June 2008 and 12th September 2010; used for spares, early 2011; remains to C.F. Booth Ltd, Rotherham, for scrap, 1st July 2011; seen at C.F. Booth Ltd, 2nd July 2011; scrapped, 21st July 2011.

D3685 Doncaster   1958   5A   11/03   P   08523 / HO61
08523   new (order EO2) 6th September 1958; noted in light blue livery with number 08523, in store, Crewe Depot, 3rd April 2003; withdrawn, November 2003; sold to RT Rail, Crewe, and remained stored at Crewe Diesel Depot, 2004; to L&NWR Ltd, Carriage Works, Crewe, for overhaul, 4th March 2004; to Heritage Centre, Crewe, by July 2005; to Hays Chemicals, Sandbach, on hire, about July 2005; to RMS Locotec Ltd, Wakefield, for repairs, February 2007; to Celtic Energy, Onllwyn Disposal Point, on hire, 30th March 2007; RT Rail was acquired by RMS Locotec Ltd, 8th November 2007; to RMS Locotec Ltd, Wakefield, November 2007; to Celtic Energy, Onllwyn Disposal Point, on hire, by 17th November 2007; to RMS Locotec Ltd, Wolsingham Depot, for repairs, 11th January 2012; to PD Ports, Tees Dock, on hire, 22nd August 2012; to RMS Locotec Ltd, Wolsingham Depot, for repairs, 4th December 2012; to ScotRail, Inverness, on hire, 28th August 2013; noted in blue livery with numbers 08523 and HO61, Inverness, June 2017; to RMS Locotec Ltd, Wolsingham Depot, for repairs, 19th August 2019; noted in dark blue livery with grey cab and number 08523, Wolsingham, 22nd October 2020; noted on low-loader, A50, 8th December 2022; to Electro-Motive Diesel Ltd, Longport, on hire, 8th December 2022; noted in 'Progress Rail' blue livery with grey cab and number 08523, Longport, 10th November 2023.

D3689 Darlington   1959   ZI   6/95   P   08527
08527   new 4th March 1959; noted in livery of light grey with blue upper body stripe and black roof, with LEVEL 5 on the side and number 08527, Stratford Depot, 10th September 1994; to ABB (Customer Support) Ltd, Ilford Level 5 Depot, June 1995; had ILFORD DEPOT PILOT and ILFORD No.1 in blue on battery box; locomotive included in sale when BRML works privatised; re-sold to Harry Needle Railroad Company, August 2006; to Barrow Hill Engine Shed, Staveley, September 2006; noted in grey livery with number

08527, BHES, March 2007; to Roberts Road Depot, Doncaster, on hire, 7th June 2007; to Flixborough Wharf Ltd, Flixborough, Scunthorpe, on hire, 31st August 2010; seen with number 08527, Flixborough, 15th March 2011; to Barrow Hill Engine Shed, Staveley, October 2013; noted in grey and black livery with number 08527, Barrow Hill Engine Shed, 19th October 2013; to GBRf, Trafford Park, Manchester, on hire, 10th February 2014; to Northern Rail, Allerton Depot, Liverpool, on hire, 28th May 2014; to Barrow Hill Engine Shed, Staveley, 11th August 2016; seen with number 08527, Barrow Hill Engine Shed, 19th November 2016; to GBRf, Immingham, on hire, December 2016; to Attero Recycling Ltd, Rossington, Doncaster, on hire, 19th January 2019; seen in two-tone grey livery with number 08527, Attero Recycling Ltd, 27th January 2019 and 11th April 2022; suffered bent side rod and shifted crank; to Nemesis Rail, Burton upon Trent, 12th April 2022; noted with no centre wheels and no rods, Nemesis Rail, 19th August 2022; to HNRC, Barrow Hill Engine Shed, 9th August 2023; seen in two-tone grey livery with number 08527, BHES, 14th October 2023; to HNRC, Worksop Depot, early April 2024.

D3690 **Darlington** **1959** **2F** **7/05** **P** **08528**
08528 new 9th March 1959; withdrawn, July 2005; sold to Battlefield Line; despatched from Bescot Yard; to Battlefield Line, Shackerstone, 31st July 2010; noted in grey and black livery with number 08528, Battlefield Line, September 2010; noted in green livery with number D3690, Battlefield Line, 17th November 2012; re-sold to Great Central Railway, Loughborough, and moved 3rd July 2014; noted in green livery with number 08528, Great Central Railway, 12th October 2020; to Derwent Valley Light Railway, Murton, 12th October 2020; noted at DVLR, 20th April 2024.

D3691 **Darlington** **1959** **40B** **12/99** **F** **08529**
08529 new 12th March 1959; withdrawn, 24th December 1999; to Doncaster Depot, for storage; sold to RT Rail, Crewe; despatched from Doncaster Depot to Wabtec, Doncaster, for repairs, July 2005; to RMS Locotec Ltd, Dewsbury, August 2005; to Wabtec, Doncaster, for storage, by November 2005; used for spares; noted in blue livery with number 08529, in semi-dismantled condition, Wabtec, 11th October 2007; remains to C.F. Booth Ltd, Rotherham, for scrap, 21st May 2008; scrapped, August 2008.

D3692 **Darlington** **1959** **5A** **?** **P** **08530**
08530 new 19th March 1959; withdrawn, date not known; sold to Traditional Traction, Wishaw, about August 2006; to Port of Felixstowe, on hire, about January 2007; to LH Group, Barton under Needwood, for repairs, 24th December 2007; re-sold to Freightliner, March 2009, and based initially at Southampton Docks.

D3699 **Darlington** **1959** **2F** **10/00** **F** **08535**
08535 new 7th May 1959; withdrawn, 1st October 2000; moved to Crewe Depot, for storage, May 2001; sold to RT Rail, Crewe; despatched from Crewe Diesel Depot, 4th March 2004; stored at L&NWR Ltd, Carriage Works, Crewe; noted in grey livery with number 08535, Crewe, March 2007; to RMS Locotec Ltd, Wakefield, for repairs, 5th September 2007; RT Rail acquired by RMS Locotec Ltd, 8th November 2007; to Corus, Shotton Steelworks, 21st January 2009; used for spares; remains to C.F. Booth Ltd, Rotherham, for scrap, 15th September 2009; scrapped, 17th September 2009.

D3700 **Darlington** **1959** **16C** **?** **P** **08536**
08536 new 13th May 1959; withdrawn, date not known; stored at Etches Park, Derby, from 1997; noted in faded blue livery with number 08536, Etches Park, 5th October 2009; despatched from Etches Park, July 2010; to Railway Vehicle Engineering Ltd, Railway

Technical Centre, Derby; noted (long disused) in blue livery with number 08536, Derby RTC, 23rd May 2018; re-sold to Railway Support Services, Wishaw; moved to RSS Wishaw, 9th October 2018; noted in blue livery with number 08536, RSS Wishaw, 16th October 2018 and 29th April 2024.

D3723   Darlington   1959   1A   7/90   P   08556
08556   new 30th October 1959; withdrawn, 3rd July 1990; stored at Willesden Depot; sold to North Yorkshire Moors Railway, Grosmont, and arrived on 23rd October 1993; seen in green livery with number 08556, North Yorkshire Moors Railway, 3rd April 1994; mostly used as pilot at New Bridge Permanent Way Depot; seen in green livery with number 08556, NBPWD, 11th September 2020; under repair in Grosmont shed, 21st June 2024.

D3734   Crewe   1959   5A   ?   P   08567
08567   new (order E498) 17th April 1959; withdrawn, date not known; put on sale by tender by DB Schenker, with bids due by 22nd September 2015; sold to Arlington Fleet Services Ltd, Eastleigh Works; noted in EWS red and yellow livery with number 08567, on a low-loader at DB Schenker, Crewe Electric Depot, 2nd December 2015; despatched from Crewe Electric Depot, to Arlington Fleet Services Ltd, Eastleigh Works, Hampshire, 3rd December 2015; noted at Eastleigh Works, 28th January 2016; noted in EWS red and yellow livery, on-test after repairs, Eastleigh Works, 25th June 2018; repainted in Arlington green livery, June 2023; received nameplates 'John Arlington Stephens, 20th May 1925 – 19th July 1984', July 2023.

D3735   Crewe   1959   ZH   6/95   P   08568 / St Rollox
08568   new (order E498) 24th April 1959; to Railcare Ltd, Springburn Works, Glasgow, June 1995; the locomotive was included in sale when the BRML works was privatised; noted in two-tone grey livery with number 08568 and name ST ROLLOX, Springburn Works, 28th June 2013 and 16th February 2016; it had spent several years in open storage when sold to Railway Support Services, Wishaw, May 2017; moved to Railway Support Services, 8th December 2017; noted at Wishaw, 21st December 2017; to Tyseley Diesel Depot, Birmingham, for repairs and tyre turning, 6th June 2019; returned to Railway Support Services, Wishaw, early August 2020; put up for sale by Railway Support Services, January 2022; seen in grey livery with number 08568 and name 'St Rollox' (lower case), and with no centre wheels, Wishaw, 27th May 2023.

D3738   Crewe   1959   ?   ?   P   08571
08571   new (order E498) 5th June 1959; withdrawn, date not known; to Great North Eastern Railway and allocated to Craigentinny Depot, Edinburgh; sold to RFS (Engineering) Ltd, Doncaster, about March 1997, and hired to GNER, Craigentinny Depot, Edinburgh; to RFS (Engineering) Ltd, Doncaster, 1998; to Whatley Quarry, on hire, January 1999; to RFS (Engineering) Ltd, Doncaster, by 17th February 1999; to Hanson, Whatley Quarry, on hire, by 19th September 1999; noted in blue livery with number 08571, Whatley Quarry, 1st July 2000; noted at Whatley Quarry, 24th June 2001; to Merehead Stone Terminal, on hire, September 2001; returned to Wabtec, Doncaster, off-hire, by 26th January 2002; to GNER, Bounds Green Depot, London, on hire, about March 2005; to LH Group, Barton under Needwood, for repairs, 24th August 2015; to Bounds Green Depot, London, on hire, 17th October 2015; to Wabtec, Doncaster, for repairs, 3rd February 2016; to Bombardier Transportation, Litchurch Lane Works, Derby, on hire, March 2016; to LH Group, Barton under Needwood, 1st August 2016; to Freightliner, Port of Felixstowe, on hire, 26th August 2016; to Wabtec, Doncaster, for repairs, 29th August 2016; to LH Group, Barton under Needwood, 29th September 2016; to Freightliner, Port

of Felixstowe, on hire, 9th November 2016; noted at Felixstowe, 23rd November 2016 and 6th February 2017; to LH Group, Barton under Needwood; to Freightliner, Southampton Docks, on hire, 8th May 2017; to LH Group, Barton under Needwood, 14th August 2017; to Daventry International Rail Freight Terminal, on hire, 19th October 2017; noted in Wabtec black livery with number 08571, Daventry International Rail Freight Terminal, March 2018; to LH Group, Barton under Needwood, 2019; to Daventry International Rail Freight Terminal, on hire, 6th February 2020; re-sold to Hunslet Ltd (Ed Murray Group), 1st May 2021; to Hunslet Ltd, Barton under Needwood, about 27th July 2022; noted in Wabtec black livery with number 08571, Barton under Needwood, 2nd November 2022; re-sold to Harry Needle Railroad Company; to DIRFT, Daventry, on hire, 6th June 2023; to HNRC, Barrow Hill Engine Shed, 17th June 2023; noted at Barrow Hill, 29th April 2024.

D3740   Crewe   1959   ZI   6/95   P   08573
08573   new (order E498) 5th June 1959; to ABB (Customer Support) Ltd, Ilford, June 1995; locomotive was included in sale when BRML works privatised; sold to RT Rail, Crewe, February 2001; to Wabtec, Doncaster, for repairs, 25th October 2001; to Channel Tunnel Rail Link, Beechbrook Farm, near Ashford, on hire, 9th November 2001; to Wabtec, Doncaster, for repairs, by 10th January 2003; to Bombardier Transportation, Ilford, on hire, by September 2003; to Wabtec, Doncaster, for repairs, August 2004; to Freightliner, Coatbridge, on hire, 12th March 2005; to ScotRail, Inverness, on hire, April 2005; to Wabtec, Doncaster, 8th May 2006; to Tubelines, Ruislip, London, on hire, 17th May 2006; to Wabtec, Doncaster, for repairs, about May 2007; RT Rail acquired by RMS Locotec Ltd, 8th November 2007; to Bombardier Transportation, Ilford Depot, on hire, 12th July 2007; noted in black livery with number 08573, Ilford, 28th February 2008 and 20th February 2017; to hauliers yard, 27th February 2018; to RMS Locotec Ltd, Wolsingham Depot, 6th March 2018; noted in black livery with number 08573, Wolsingham, 5th January 2024; to Independent Railway Engineering, Chesterfield, 26th January 2024.

D3743   Crewe   1959   86A   6/00   F   08576
08576   new (order E498) 20th June 1959; withdrawn, 28th June 2000; sold to Battlefield Line; despatched from Canton Depot, Cardiff, 12th February 2004; to Battlefield Line, Shackerstone; used for spares; remains to T.J. Thomson & Son Ltd, Stockton-on-Tees, for scrap, 8th June 2006; scrapped, May 2007.

D3745   Crewe   1959   16A   ?   P   08578
08578   new (order E498) 26th June 1959; withdrawn, date not known; noted in EWS red and yellow livery with number 08578, Toton Depot, 19th June 2013; put on sale by tender by DB Schenker, with bids due by 22nd September 2015; sold to Harry Needle Railroad Company; despatched from DB Schenker, Toton Depot, 18th January 2016; to Harry Needle Railroad Company, Long Marston; noted at Long Marston, 23rd July 2021; noted on a low-loader, M42 northbound, 10th August 2021; to Harry Needle Railroad Company, Worksop Depot, 12th August 2021.

D3747   Crewe   1959   2F   4/11   P   08580
08580   new (order E498) 4th July 1959; withdrawn, April 2011; put on sale by tender by DB Schenker, with bids due by 22nd September 2015; sold to Railway Support Services; despatched from Bescot Depot, late September 2015; noted in EWS red and yellow livery with number 08580, Railway Support Services, Wishaw, October 2015; to Colne Valley Railway, Castle Hedingham, 25th February 2016; to Railway Support Services, Wishaw, 25th May 2017; noted in EWS red and yellow livery with number 08580, Wishaw, 23rd June

2017; noted in workshops with no engine, Wishaw, 15th February 2018; noted repainted in RSS black livery with number 08580, Wishaw, 16th May 2018; to Castle Cement Works, Ketton, for open weekend, 29th June 2018; noted at Ketton, 1st July 2018; to Bounds Green Depot, London, on hire, 4th July 2018; noted at Bounds Green, 10th February 2019; to Marcroft, Stoke-on-Trent, on hire, about 18th March 2021; to Serco, Old Dalby Test Centre, on hire, 31st March 2021; to Direct Rail Services, Garston, Liverpool, on hire, 7th October 2021; to Ribble Steam Railway, to use repair facilities, 29th March 2022; to Direct Rail Services, Garston, Liverpool, 30th March 2022; to Railway Support Services, Wishaw, 1st September 2023; to GBRf, Whitemoor Yard, March, on hire, about 4th September 2023.

D3755   Crewe   1959   55H   ?   P   08588 / HO47
08588   new (order E498) 22nd September 1959; withdrawn, date not known; sold to RT Rail, Crewe, April 2005; despatched from Neville Hill Depot direct to Wabtec, Doncaster, for repairs, 16th April 2005; to Network Rail, Whitemoor Yard, March, on hire, 15th September 2005; to Wabtec, Doncaster, for repairs, 31st March 2006; to RMC Aggregates, Dove Holes, on hire, 7th April 2006; noted in black livery with number 08588, Dove Holes, 26th September 2006; to Channel Tunnel Rail Link, Dagenham, London, on hire, 28th September 2006; to Wabtec, Doncaster, for repairs, 6th November 2006; to Bombardier Transportation, Ilford Depot, on hire, 20th November 2006; RT Rail was acquired by RMS Locotec Ltd, 8th November 2007; to Network Rail, Whitemoor Yard, March, on hire, December 2007; to PD Ports, Tees Dock, on hire, July 2008; to RMS Locotec Ltd, Wolsingham Depot, 1st March 2011; to Cemex Rail Products, Washwood Heath, on hire, 10th April 2012; to Tyseley Diesel Depot, Birmingham, for tyre turning, 18th January 2017; to RMS Locotec Ltd, Wolsingham Depot, for repaint, 20th January 2017; to Northern Rail, Heaton Depot, Newcastle upon Tyne, on hire, 16th March 2017; noted in RMS blue livery with number 08588, Heaton Depot, 27th March 2017; to RMS Locotec Ltd, Wolsingham Depot, 11th October 2017; to Loram UK Ltd, Railway Technical Centre, Derby, on hire, 6th February 2018; noted in RMS Locotec blue livery with grey cab and numbers 08588 and HO47, Railway Technical Centre, Derby, 6th February 2018; to Eastern Rail Services, carriage sidings, Runham, Great Yarmouth, on hire, 31st March 2021; to Alstom Transport, Ilford Depot, on hire, 10th June 2021; noted at Ilford Depot, 20th January 2024.

D3757   Crewe   1959   52B   9/93   P   08590 / RED LION
08590   new (order E498) 5th October 1959; withdrawn, 27th September 1993; sold to Midland Railway, Butterley, Derbyshire, and arrived July 1994; seen in black livery with number 08590, Butterley, 26th July 1999; noted in pink livery with number 08590, Butterley, 16th June 2006 and 2nd April 2007; seen in black livery with number 08590 and name RED LION, Butterley, 30th April 2011; noted in blue livery with number 08590, Butterley, 27th February 2014 and 10th July 2022.

D3759   Crewe   1959   ?   3/15   P   08993 / ASHBURNHAM
08993   new (order E498) 27th October 1959; originally numbered 08592; renumbered 08993 on 13th April 1985 after rebuilding with cut down cab for working Burry Port & Gwendraeth Valley line; named ASHBURNHAM, January 1986; withdrawn, March 2015; put on sale by DB Schenker, 5th May 2015; sold to KWVR; despatched from DB Schenker, Stoke on Trent, 1st September 2015; arrived at Keighley & Worth Valley Railway, Haworth, 5th September 2015; noted in EWS red and yellow livery, with cast plate '08993 ASHBURNHAM', Haworth, 5th September 2015; overhauled; entered service on 18th June 2016; noted in EWS livery, 1st September 2019.

D3760   Crewe   1959   5A   4/11   P   08593
08593   new (order E498) 12th November 1959; withdrawn, April 2011; put on sale by tender by DB Schenker, with bids due by 22nd September 2015; noted in EWS red and yellow livery with number 08593, Crewe Electric Depot, 3rd December 2015; sold to Railway Support Services; despatched from DB Schenker, Crewe Electric Depot, 20th January 2016; to Railway Support Services, Wishaw; used for spares; put up for sale by Railway Support Services, January 2022; seen in EWS red and yellow livery with number 08593, no engine, Railway Support Services, Wishaw, 23rd September 2023.

D3761   Crewe   1959   16A   2/97   F   08594
08594   new (order E498) 13th November 1959; withdrawn, February 1997; sold to Mike Darnall, Newton Heath, Manchester, November 2000; to Wabtec, Doncaster, by April 2001; noted in BR blue livery with number 08594, Wabtec, 26th August 2005; to Mike Darnall, 9th January 2008; to Railway Support Services, Wishaw, by road low-loader, 15th July 2010; noted in BR blue livery with number 08594, Wishaw, 17th July 2010; to European Metal Recycling, Kingsbury, for scrap, 21st July 2010; scrapped, September 2010.

D3763   Derby   1959   81A   3/77   P   08596
08596   new (order D1400) 18th April 1959; withdrawn, 12th March 1977; sold to Bowaters; despatched from 81A Old Oak Common Depot, 16th May 1977; to Bowaters UK Paper Co Ltd, Sittingbourne, Kent; noted at Bowaters, 1st September 1977; noted with no number, Bowaters, 28th April 1979; to BREL Swindon Works, for overhaul, 20th November 1981; returned to Bowaters UK Paper Co Ltd, 4th January 1982; re-sold to RFS (Engineering) Ltd, Kilnhurst, and moved 5th June 1991; to E.C.C. Calcium Carbonates Ltd, Quidhampton, on hire, and left Kilnhurst by road on 29th July 1991; offloaded at Salisbury Station and tripped to E.C.C. Calcium Carbonates Ltd, 30th July 1991; tripped to Salisbury Station and loaded, 5th September 1991; returned to RFS (Engineering) Ltd, Kilnhurst, 6th to 7th September 1991; serviced, repainted grey and numbered 006, Kilnhurst, September 1991; to Channel Tunnel, Folkestone, (Balfour Beatty Ltd), on hire, 23rd September 1991; given number CTTG 19; seen with numbers 006 and No.19 and RFS logo, Cheriton, 8th March 1992; to RFS (Engineering) Ltd, Doncaster, 6th July 1993; to Sheerness Steel Co Ltd, Sheerness, Kent, on hire, July 1993; returned to RFS (Engineering) Ltd, Doncaster, by 1st November 1996; reinstated for main line use, 30th April 1999; to EWS, Decoy Yard, Doncaster, on hire, 1st May 1999; returned to RFS (Engineering) Ltd, Doncaster; to Balfour Beatty, Leeds Station contract, on hire, about March 2000; returned to RFS (Engineering) Ltd, Doncaster, by 14th May 2001; to Blue Circle, Hope Cement Works, Derbyshire, on hire, about February 2002; returned to Wabtec, Doncaster, November 2002; to GNER, Bounds Green Depot, London, on hire, by May 2003; to Wabtec, Doncaster, by 27th July 2003; to Mendip Rail, Whatley Quarry, on hire, 5th January 2004; returned to Wabtec, Doncaster, by 7th March 2004; seen in black livery with number 08596, Wabtec, 26th August 2004; to Daventry International Rail Freight Terminal, on hire, by 5th October 2004; to Wabtec, Doncaster, for repairs, 12th November 2004; to Daventry International Rail Freight Terminal, on hire, by 8th December 2004; to Wabtec, Doncaster, for repairs, by 12th March 2005; to Hanson Quarry Products, Whatley Quarry, on hire, by 22nd January 2006; to Wabtec, Doncaster, by 5th February 2006; noted in Wabtec black livery with number 08596, Wabtec, December 2007; to Bounds Green Depot, London, on hire, 16th February 2008; to LH Group, Barton under Needwood, for repairs, 5th December 2014; to Bounds Green Depot, London, on hire, 1st July 2015; to Railway Support Services, Wishaw, 4th February 2016; to Craigentinny Depot, Edinburgh, on hire; noted in Wabtec black livery, being offloaded from a low-loader, Craigentinny, 11th February 2016; re-sold to Hunslet Ltd (Ed Murray Group), 1st May 2021; to Reid Freight Services Ltd, Cinderhill

Industrial Estate, Longton, Stoke-on-Trent, for storage, 22nd July 2022; re-sold to Harry Needle Railroad Company, June 2023.

D3765   Derby   1959   5A   11/86   P   08598 / HO16
08598   new (order D1400) 2nd May 1959; withdrawn, 10th November 1986; sold to PD Fuels; despatched from BREL Crewe Works; to Powell Duffryn Fuels Ltd, NCBOE Gwaun-cae-Gurwen, 16th January 1987; repainted in blue and white, about September 1991; seen in blue and white livery, with number 08598, Gwaun-cae-Gurwen, 23rd April 1992; seen in blue and white livery, with number 08598, Gwaun-cae-Gurwen, 17th August 1994; re-sold to RMS Locotec Ltd, Dewsbury, September 1995; given number HO16 in 1995; remained at Gwaun-cae-Gurwen, now on hire; noted at Gwaun-cae-Gurwen, 5th October 1995 and 12th March 1998; to RMS Locotec Ltd, Dewsbury, September 1998; to Cleveland Potash Ltd, Boulby Mine, on hire, 21st June 1999; returned to RMS Locotec Ltd, Dewsbury, by 14th July 2000; re-sold to Potter Group, Knowsley, September 2001; to Potter Group, Selby, 24th May 2002; to Potter Group, Knowsley, by 12th April 2003; to Gloucestershire Warwickshire Railway, Toddington, for gala, 7th July 2010; noted in yellow livery with number 08598, GWR, July 2010; to Potter Group, Ely, 1st September 2010; re-sold to Ed Murray & Sons Ltd, Hartlepool, mid-2013; to Chasewater Railway, Staffordshire, 10th September 2013; noted in yellow livery with number 08598, Chasewater Railway, 28th September 2013 and 16th April 2016; re-sold to A.V. Dawson Ltd, Middlesbrough, and moved 4th May 2017; noted in yellow livery with number 08598, A.V. Dawson Ltd, 5th May 2017; to Railway Support Services, Wishaw, for overhaul, 19th July 2019; to Tyseley Diesel Depot, Birmingham, for tyre turning, 30th September 2019; to A.V. Dawson Ltd, Middlesbrough, 17th January 2020; noted in yellow livery with number 08598, A.V. Dawson Ltd, Middlesbrough, 11th February 2023.

D3767   Derby   1959   70D   7/99   P   08600
08600   new (order D1400) 16th May 1959; at one time ran as 97800 IVOR; despatched from Eastleigh Depot, 17th November 1997; to A.V. Dawson Ltd, Middlesbrough, on hire, November 1997; off-hire, September 1998; to RFS (Engineering) Ltd, Doncaster, for storage; withdrawn, 30th July 1999; sold to A.V. Dawson Ltd, Middlesbrough; to EWS Thornaby Depot, for tyre turning, 14th June 2004; returned to A.V. Dawson Ltd, Middlesbrough, 1st July 2004; to LH Group, Barton under Needwood, for repairs, 14th October 2008; returned to A.V. Dawson Ltd, Middlesbrough, 23rd March 2009; noted in red livery with number 08600, in store, A.V. Dawson Ltd, 11th February 2023.

D3769   Derby   1959   12A   3/86   F   004
08602   new (order D1400) 30th May 1959; withdrawn, 30th March 1986; seen with numbers painted over, with BUXTON nameplates, BR Carlisle Depot, 19th May 1988; moved to Tyne Yard, for storage, 24th June 1988; sold to RFS (Engineering) Ltd, Doncaster, and moved 30th June 1988; given number 004 and named CLARENCE; noted at RFS (Engineering) Ltd, Doncaster, 14th August 1988; to Foster Yeoman Quarries Ltd, Isle of Grain Stone Terminal, on hire, December 1988; returned to RFS (Engineering) Ltd, Doncaster, off-hire, by 12th March 1989; to Sheerness Steel Co Ltd, Sheerness, Kent, on hire, December 1989; to ABB Transportation, Litchurch Lane Works, Derby, on hire, April 1990; to ABB Transportation, York, on hire, November 1990; to ABB Transportation, Litchurch Lane Works, Derby, on hire, November 1990; re-sold to ABB Transportation, Litchurch Lane Works, Derby, April 1991; repainted in green livery with number D3769, Litchurch Lane Works, Derby, summer 1991; to Fragonset, Derby, for overhaul, about September 2003; to Bombardier Transportation, Litchurch Lane Works, Derby, about April 2004; noted in blue livery with number 004, Litchurch Lane Works, 30th April 2009 and 31st

July 2019; re-sold to Harry Needle Railroad Company, about January 2020; noted at Litchurch Lane Works, June 2020; to Harry Needle Railroad Company, Worksop Depot, 21st December 2020; used for spares; noted in blue livery, no engine, Worksop, 3rd May 2023; remains to European Metal Recycling, Attercliffe, Sheffield, for scrap, 10th May 2023; scrapped (wheelsets returned to HNRC), 6th November 2023.

D3771   Derby   1959   36A   7/93   P   604 / PHANTOM
08604   new (order D1400) 13th June 1959; whilst allocated to Longsight was unofficially named ARDWICK; withdrawn, 16th July 1993; to Etches Park, for storage; sold to Didcot Railway Centre, Oxfordshire, and moved by road on 28th September 1994; given number WD40 and name PHANTOM, autumn 1995; repainted in lined green; repainted plain green, August 2010; noted in green livery with cast GWR-style 604 numberplate and name PHANTOM, Didcot, 31st July 2011; repainted in BR blue, early August 2017; noted in blue livery with number 604 and name PHANTOM, Didcot, 19th August 2017; repainted (still blue), August 2022.

D3772   Derby   1959   8F   ?   P   G.R. WALKER
08605   new (order D1400) 13th June 1959; noted in red livery with no number but named WIGAN 2, Springs Branch Depot, Wigan, 1st February 2016; withdrawn, date not known; put on sale by tender by DB Cargo, with bids due by 5th October 2016; re-sold to Riviera Trains Ltd of Crewe; noted still at Springs Branch Depot, 5th August 2017; noted on a low-loader, leaving Springs Branch Depot, 7th June 2018; to Ecclesbourne Valley Railway, Wirksworth, on loan, 7th June 2018; noted in red livery, with nameplate G.R. WALKER, Wirksworth, 9th June 2018 and 5th December 2020; to Riviera Trains, Knottingley Depot, to shunt coaching stock, 7th January 2021; purchased by Railway Support Services, Wishaw, June 2022; to W.H. Davies Ltd, Ferrybridge Depot, 12th October 2022; to Railway Support Services, Wishaw, 14th October 2022; to Balfour Beatty Ltd, Willesden Junction, on hire, 25th November 2022; to Railway Support Services, Wishaw, for repairs, 10th May 2023; to Balfour Beatty Ltd, Willesden Junction, on hire, 18th May 2023; to Railway Support Services, Wishaw, 18th June 2024.

D3778   Derby   1959   9A   2006   P   08611
08611   new (order D1400) 8th August 1959; withdrawn, early 2006; acquired by Alstom Transport; to Longsight Depot, Manchester, 1st July 2006; to Edge Hill Depot, Liverpool, 9th January 2007; noted in red and black livery with number 08611, Edge Hill, 11th April 2008; to Alstom Transport, Longsight Depot, Manchester, by 27th March 2011; to Arlington Fleet Services Ltd, Eastleigh Works, for overhaul, 13th January 2016; noted in red and black livery with number 08611, Eastleigh Works, 13th January 2016; to Alstom Transport, Stonebridge Park Carriage Maintenance Depot, London, 20th July 2016; noted in blue livery with number 08611, Stonebridge Park, 14th October 2016 and 29th March 2024.

D3780   Derby   1959   8J   12/93   P   08613 / HO64 / 3
08613   new (order D1400) 22nd August 1959; withdrawn, 1st December 1993; sold to Trafford Park Estates Ltd, Manchester, and moved 2nd February 1994; re-sold to Wabtec, Doncaster, and moved by road, November 2000; overhauled; re-sold to RT Rail, Crewe, January 2001; to Bombardier Transportation, Ilford Depot, on hire, March 2001; to Wabtec, Doncaster, for overhaul, November 2006; returned to Ilford Depot, on hire; to RMS Locotec Ltd, Wakefield, for repairs, 8th May 2007; to Barrow Hill Engine Shed, Staveley, 5th October 2007; noted in blue livery with number 08613, Barrow Hill Engine Shed, 7th October 2007; RT Rail was acquired by RMS Locotec Ltd, 8th November 2007; to Corus, Shotton, on hire, November 2007; to Castle Cement Works, Ketton, on hire, 8th April 2009;

to Boden Rail Engineering, Washwood Heath, for repairs, November 2011; to Celtic Energy, Onllwyn Disposal Point, on hire, 11th January 2012; noted in blue livery with number 08613, Onllwyn, April 2016; to RMS Locotec Ltd, Wolsingham Depot, 4th January 2017; to PD Ports, Tees Dock, on hire, 3rd October 2018; to RMS Locotec Ltd, Wolsingham Depot, for overhaul, 11th July 2019; noted in dark blue livery with grey cab and numbers 08613 and HO64, Wolsingham, February 2020; to Tyseley Diesel Depot, Birmingham, for tyre turning, about 12th November 2022; to PD Ports, Tees Dock, on hire, 22nd December 2022; noted in RMS blue/grey livery with numbers 08613, HO64, and LOCO 3, PD Ports, 31st December 2023.

D3782   Derby   1959   8J   12/93 P 08615 / UNCLE DAI
08615   new (order D1400) 5th September 1959; withdrawn, 1st December 1993; sold to Trafford Park Estates Ltd, Manchester, and moved 2nd February 1994; seen in BR blue livery with no number, Trafford Park Estates, 17th February 1995; re-sold to Wabtec, Doncaster, and moved in November 2000; to Hanson Quarry Products, Whatley Quarry, on hire, by 5th May 2002; to Merehead Stone Terminal, on hire, October 2002; noted at Merehead, 28th June 2003; returned to Wabtec, Doncaster, 12th August 2003; noted at Wabtec, 5th October 2003; to GNER, Craigentinny Depot, Edinburgh, on hire, by March 2004; noted in Wabtec black livery with number 08615, Craigentinny, February 2007; to LH Group, Barton under Needwood, for repairs, 11th February 2016; noted in black livery with number 08615, LH Group, 31st March 2018; noted in blue and orange livery with number 08615, LH Group, 30th December 2018; to Tata Steel, Shotton, on hire, 31st January 2019; re-sold to Hunslet Ltd (Ed Murray Group), 1st May 2021; noted in Hunslet royal blue and orange livery with number 08615 and nameplate UNCLE DAI, Shotton Steelworks, 1st November 2022; to Hunslet Ltd, Barton under Needwood, July 2023; to Tata Steel, Shotton, on hire, 17th January 2024.

D3784   Derby   1959   ?   2007   P 08617/STEVE PURSER
08617   new (order D1400) 12th September 1959; withdrawn, 2007; sold to Alstom Transport, Wembley Depot, London, 2007; noted in black livery with number 08617, Wembley Depot, March 2007 and December 2015; to Arlington Fleet Services Ltd, Eastleigh Works, for overhaul and repaint, 29th March 2016; noted in blue livery with number 08617 and nameplates STEVE PURSER, Eastleigh Works, 29th April 2016; to Alstom Transport, Oxley Depot, Wolverhampton, 23rd September 2017; noted at Oxley, 1st May 2023.

D3785   Derby   1959   52A   9/90   F 08618
08618   new (order D1400) 19th September 1959; withdrawn, 14th September 1990; sold to Gwent Demolition, Margam, 1995; firm went out of business before locomotive was moved and sale cancelled; re-sold to T.J. Thomson & Son Ltd, Stockton-on-Tees, for scrap, April 2000; dismantled for spares by Thomson staff, Gateshead Depot, April 2001; remains to T.J. Thomson & Son Ltd, Stockton-on-Tees, April 2001; to Freightliner, Southampton Docks, for spares, April 2001; scrapped on site (by Southampton Steel Ltd), September 2003.

D3789   Derby   1959   66B   12/95 P 08622 / HO28 / 19
08622   new (order D1400) 17th October 1959; withdrawn, 10th December 1995; sold to RMS Locotec Ltd, Dewsbury, and moved 4th May 2002; to Flixborough Wharf Ltd, Flixborough, Scunthorpe, on hire, 6th March 2003; to RMS Locotec Ltd, Dewsbury, 19th April 2004; to PD Ports, Tees Dock, on hire, 6th August 2004; to RMS Locotec Ltd, Dewsbury, 14th February 2006; to RMS Locotec Ltd, Wakefield, 29th June 2006; to Flixborough Wharf Ltd, Flixborough, Scunthorpe, on hire, 9th October 2006; to Corus,

Trostre Works, Llanelli, on hire, 3rd January 2008; to PD Ports, Tees Dock, on hire, week commencing 9th November 2009; to RMS Locotec Ltd, Wolsingham Depot, for repairs, 18th January 2011; to Castle Cement Works, Ketton, on hire, 1st March 2011; to Nene Valley Railway, Wansford, for gala, September 2011; to Castle Cement Works, Ketton, on hire, October 2011; to RMS Locotec Ltd, Wolsingham Depot, for repairs, 20th November 2012; to Castle Cement Works, Ketton, on hire, 31st January 2013; noted in black livery with numbers 08622 and 19, Castle Cement Works, Ketton, June 2015; to RMS Locotec Ltd, Wolsingham Depot, for repairs, 7th November 2016; to Ketton Cement Works, on hire, 6th July 2018; noted in RMS black livery with numbers 08622 and HO28, Ketton, 6th December 2018; to RMS Locotec Ltd, Wolsingham Depot, (where repainted in RMS blue/grey livery), 1st March 2023; to Ketton Cement Works, by road, on hire, 13th October 2023.

D3790   Derby   1959   2F   ?   P   08623
08623   new (order D1400) 24th October 1959; noted in DB Cargo red livery, Bescot Depot, 2nd February 2016; withdrawn, date not known; put up for sale by tender by DB Cargo, with bids due by 5th October 2016; sold to Harry Needle Railroad Company; despatched from Bescot Depot, 31st August 2017; noted on a low-loader leaving Bescot, and moved direct to Hope Cement Works, Derbyshire, for storage, 31st August 2017; noted in DB Cargo red livery with number 08623, Hope Cement Works, 1st November 2017; noted disused at rear of works, 15th May 2023; to Harry Needle Railroad Company, Worksop Depot, 7th June 2023; to Tyseley Depot, for tyre turning, 8th July 2024.

D3792   Derby   1959   2F   3/97   F   08625
08625   new (order D1400) 31st October 1959; withdrawn, 20th March 1997; moved to Canton Depot, Cardiff, for storage; noted in BR blue livery, Canton Depot, 20th August 1998; sold to Dean Forest Railway; despatched by road from Canton Depot, 17th June 2000; delivered to Dean Forest Railway, Lydney; to Cotswold Rail, Moreton in Marsh, about June 2001; to European Metal Recycling, Kingsbury, for scrap, February 2004; noted in BR blue livery with no number and no wheels, Kingsbury, 18th February 2004; scrapped, 26th February 2004.

D3795   Derby   1959   2F   8/05   F   08628
08628   new (order D1400) 21st November 1959; withdrawn, 31st August 2005; despatched from Saltley Depot, 28th September 2005; to Moveright International, Wishaw, Warwickshire; to European Metal Recycling, Kingsbury, about October 2005; to Moveright International, Wishaw, 13th April 2006; to Bryn Engineering, and stored at Redrock Plant & Truck Services, Blackrod, Bolton, April 2006; to Ribble Steam Railway, Preston, 5th August 2011; noted in blue livery, in dismantled condition after spares recovery, Ribble, 13th September 2014; remains to Railway Support Services, Wishaw, 19th November 2015; used for spares; remains to European Metal Recycling, Kingsbury, for scrap, 19th May 2016; scrapped.

D3796   Derby   1959   ZN   6/95   P   08629 / WOLVERTON
08629   new (order D1400) 21st November 1959; latterly worked at Wolverton Works, where repainted in claret livery, August 1993; locomotive included in sale when BRML works privatised; site became Railcare Ltd; noted in red livery with number 08629, Wolverton Works, July 2004 and March 2008; to Great Central Railway, for gala, 11th February 2011; to Alstom Transport, Wolverton Works, 14th February 2011; to Chinnor & Princes Risborough Railway, Oxfordshire, for gala, 1st October 2013; to Knorr Bremse Rail Services Ltd, Wolverton Works, 8th October 2013; noted in green, white and blue livery,

with number 08629, Wolverton Works, 6th April 2018; the works and locomotive acquired by Gemini Rail, November 2018; re-sold to Meteor Power Ltd, and moved to Chinnor & Princes Risborough Railway, Oxfordshire, 14th February 2020; to UK Rail Leasing, Leicester, 9th June 2021; noted in green, white and blue livery, with number 08629, Leicester, 11th June 2021; re-sold to Railway Support Services, Wishaw, about February 2022; tripped to Knighton Junction, 21st February 2022; to Electro-Motive Diesel Ltd, Longport, by road, on hire, 22nd February 2022; to Railway Support Services, Wishaw, 4th July 2022; seen in green, white and blue livery, with number 08629, RSS, 9th July 2022; to GBRf, East Yard, Eastleigh, on hire, 10th August 2022; to Railway Support Services, Wishaw, 28th July 2023; repainted in grey RSS livery, December 2023; to Daventry International Rail Freight Terminal, on hire, about 12th January 2024.

D3797   Derby   1959   ?   ?   P   3
08630   new (order D1400) 28th November 1959; withdrawn, date not known; sold to Harry Needle Railroad Company; to Barrow Hill Engine Shed, Staveley, 17th February 2016; overhauled, repainted black with orange cab, and numbered CELSA 3; to Celsa, Cardiff, on hire, 23rd August 2016; noted at Celsa, 4th February 2017; left Celsa, 19th April 2018; stored in haulier's yard; to Railway Support Services, Wishaw, for repairs, 26th April 2018; noted in black livery with red cab and large number 3, Wishaw, 18th May 2018; to Celsa, Cardiff, on hire, 11th June 2018; to HNRC, Barrow Hill Engine Shed, Staveley, for repairs, 25th February 2019; to Celsa, Cardiff, on hire, week-ending 23rd March 2019; to HNRC, Barrow Hill Engine Shed, Staveley, for repairs, 14th December 2020; to Celsa, Cardiff, on hire, 18th June 2021; to Nemesis Rail, Burton upon Trent, for repairs, 17th March 2022; to Celsa, Cardiff, on hire, May 2022; to Nemesis Rail, Burton upon Trent, for repairs, late 2022; to Celsa, Cardiff, on hire, 18th January 2023; to HNRC, Barrow Hill Engine Shed, for repairs, 13th February 2023; to Celsa, Cardiff, on hire, 18th April 2023.

D3798   Derby   1959   31B   12/92   P   08631 / EAGLE
08631   new (order D1400) 5th December 1959; withdrawn, 11th December 1992; sold to Mid-Norfolk Railway, Dereham, Norfolk, spring 1994; to Fragonset, Derby, on hire, 13th December 1997; to Anglian Railways, Crown Point Depot, Norwich, on hire, November 1999; to Fragonset, Derby, on hire, May 2001; noted in blue, white and red livery with number 08631, Fragonset, September 2002; to Bombardier Transportation, Litchurch Lane Works, Derby, on hire, December 2003; to Fragonset (FM Rail), Derby, on hire, about June 2004; to Mid-Norfolk Railway, Dereham, 22nd March 2007; to FM Rail, Derby, on hire; to Gwili Railway, Bronwydd Arms, on hire, 25th May 2010; to Mid-Norfolk Railway, Dereham, 29th June 2011; to Nemesis Rail, Burton upon Trent, 27th May 2014; to Bombardier Transportation, Litchurch Lane Works, Derby, on hire, 2nd June 2014; to Nemesis Rail, Burton upon Trent, 22nd April 2015; re-sold to Locomotive Services Ltd and moved to Crewe Diesel Depot, 1st June 2015; noted in blue livery with number 08631, Crewe Diesel Depot, July 2015 and 10th March 2017; to RMS Locotec Ltd, Wolsingham Depot, for repairs, 27th February 2019; to Locomotive Services Ltd, Crewe Diesel Depot, 6th January 2021; noted in blue livery with number 08631, Locomotive Services Ltd, 6th January 2023.

D3799   Derby   1959   16A   ?   P   08632
08632   new (order D1400) 5th December 1959; noted in EWS red and yellow livery and number 08632, Toton, 1st September 2013; withdrawn, date not known; put on sale by tender by DB Cargo, with bids due by 5th October 2016; sold to Railway Support Services, Wishaw; to Knorr Bremse Rail Services Ltd, Springburn Works, Glasgow, on hire, 19th January 2017; to RSS Wishaw, for repairs, late January 2017; to Chasewater Railway, Staffordshire, for testing, 21st February 2017; to Railway Support Services, Wishaw, for

repairs, March 2017; noted repainted in RSS/Runtech black livery with number 08632, Railway Support Services, 22nd May 2017; to Tata Steel, Trostre Works, Llanelli, on hire (via Runtech Ltd of Port Talbot), 24th May 2017; noted with remote control fitted, Trostre Works, 9th June 2017; to Railway Support Services, Wishaw, 28th July 2017; noted at Railway Support Services, Wishaw, 1st August 2017; to East Midlands Trains, Neville Hill Depot, Leeds, on hire, August 2017; to Railway Support Services, Wishaw, 22nd May 2019; noted at Wishaw, 28th May 2019; to GBRf, East Yard, Eastleigh, on hire, 3rd June 2019; noted at Eastleigh, 7th August 2019; to GBRf, Bescot Yard, on hire, 30th October 2019; noted in RSS black livery with number 08632, Bescot, 23rd June 2020; to Chasewater Railway, 1st July 2020; to Railway Technical Centre, Derby, on hire, 13th July 2020; noted repainted in Loram white livery with red cab, Derby, 25th October 2020; to Railway Support Services, Wishaw, for repairs, 21st April 2021; noted in new black livery, under repair, 22nd June 2023; to GBRf, Peterborough Depot, on hire, 7th July 2023.

D3800   Derby   1959   ?   ?   P   D3800
08633   new (order D1400) 12th December 1959; withdrawn, date not known; put on sale by tender by DB Cargo, with bids due by 5th October 2016; sold to Churnet Valley Railway; despatched from DB Wagon Works, Stoke on Trent, 10th December 2016; to Churnet Valley Railway, Cheddleton; noted in EWS red and yellow livery with number 08633, Churnet Valley Railway, 11th December 2016; noted in dark green livery with number D3800, Churnet Valley Railway, 3rd January 2019.

D3801   Derby   1959   30A   2/93   F   08634
08634   new (order D1400) 19th December 1959; withdrawn, 1st February 1993; noted in BR blue livery with numbers 90 (to signify 90-volt electrical systems) and 08634, Stratford Depot, 10th September 1994; sold to Harry Needle Railroad Company; to Barrow Hill Engine Shed, Staveley, 2nd July 2001; to West Coast Railway Company, Carnforth, 11th July 2002; noted in blue livery, being used for spares, Carnforth, 15th January 2005; scrapped, February 2005.

D3802   Derby   1959   16A   2/07   P   H3802
08635   new (order D1400) 26th December 1959; noted in blue livery with number 08635, Toton Depot, September 2006; withdrawn, February 2007; sold to T.J. Thomson & Son Ltd; despatched from EWS Toton Depot, 28th February 2007; to T.J. Thomson & Son Ltd, Stockton-on-Tees; re-sold to Severn Valley Railway, Bridgnorth, and moved 26th April 2007; placed in store and used as a source of spares; re-evaluated for possible overhaul, February 2017; assessment deemed unsuccessful, June 2017; returned to store at Kidderminster Carriage Shed, June 2017; chosen for a joint SVR/University of Birmingham Vanguard Sustainable Transport Solutions project, known as the 'Harrier Hydro Shunter Project', for developing a hydrogen/hybrid power pack; work undertaken at SVR Kidderminster Carriage Shed; given repaint in modified green livery and renumbered H3802, autumn 2023.

D3810   Horwich   1959   82B   5/01   P   08643
08643   new (LOT 267) 30th January 1959; withdrawn, May 2001; sold to Foster Yeoman Quarries Ltd, Merehead Stone Terminal, and moved 16th April 2003; noted in blue livery with number 08643, Merehead Stone Terminal, 28th June 2003; to Whatley Quarry, by 15th April 2004; to Merehead Stone Terminal, by 5th June 2005; to Isle of Grain Stone Terminal, Kent, by February 2007; to West Somerset Railway, Minehead, for gala held 12th to 14th June 2008; to Merehead Stone Terminal, 14th June 2008; noted in blue livery with number 08643, Merehead Stone Terminal, 21st June 2008; to Whatley Quarry, by 11th

June 2009; to Merehead Stone Terminal, by 9th June 2011; noted in blue livery with number 08643, Merehead Stone Terminal, 31st May 2019; to LH Group, Barton under Needwood, for repairs, June 2021; to Merehead Stone Terminal, 21st December 2021.

D3814   Horwich   1959   CW   4/93   F   08647
08647   new (LOT 267) 26th February 1959; latterly was a 'celebrity' shunter in LNER apple green livery, and named CRIMPSALL, at Doncaster Works; withdrawn, 22nd April 1993; stored at Doncaster Depot, then at Crewe Works, and noted there on 17th August 1996; sold to Harry Needle Railroad Company, early 1997; despatched from Adtranz, Crewe, 12th March 1997; to South Yorkshire Railway Preservation Society, Meadowhall, Sheffield; used for spares; remains to Mayer Parry Ltd, Snailwell, Cambridgeshire, for scrap, 20th November 1998; scrapped.

D3815   Horwich   1959   84A   4/02   P   08648 / HO65
08648   new (LOT 267) 4th March 1959; withdrawn, April 2002; sold to RT Rail, Crewe, July 2002; despatched from Laira Depot, Plymouth, and moved direct to Wabtec, Doncaster, for repairs, 24th July 2002; to Midland Road Depot, Leeds, for fitting with auto couplers, mid-2004; to Brunner Mond, Winnington Works, Northwich, on hire, about September 2004; to Wabtec, Doncaster, for repairs, about September 2005; to Brunner Mond, Northwich, on hire, 13th January 2006; to Wabtec, Doncaster, for repairs, August 2007; RT Rail was acquired by RMS Locotec Ltd, 8th November 2007; noted in yellow livery with number 08648 and name OLD GEOFF, Wabtec, February 2009; to PD Ports, Tees Dock, on hire, 17th January 2011; noted in yellow livery with numbers 08648 and 28, Tees Dock, February 2012; to RMS Locotec Ltd, Wolsingham Depot, for repairs, 27th August 2014; to Tyseley Diesel Depot, Birmingham, for tyre turning, 24th November 2014; to Northern Rail, Heaton Depot, Newcastle upon Tyne, on hire, 27th November 2014; to RMS Locotec Ltd, Wolsingham Depot, for repairs, 17th March 2017; to Grand Central Railway, Heaton Depot, Newcastle upon Tyne, on hire, 11th October 2017; to RMS Locotec Ltd, Wolsingham Depot, 7th March 2018; to ScotRail, Inverness, on hire, 11th April 2018; noted in RMS black livery with number 08648, Inverness, 7th June 2018.

D3816   Horwich   1959   ZG   6/95   P   08649 / BRADWELL
08649   new (LOT 267) 10th June 1959; to Wessex Traincare Ltd, Eastleigh Works, June 1995; locomotive included in sale when BREL works privatised; noted in green livery with number D3816 and name G.H. STRATTON, Eastleigh Works, 26th April 1996; to Wimbledon Depot, for tyre turning, December 2000; returned to Eastleigh Works, January 2001; to LH Group, Barton under Needwood, for repairs, 14th October 2003; returned to Eastleigh Works, late 2003; noted in grey, white and red livery with number D3816, Eastleigh Works, May 2004; to Alstom Transport, Wolverton Works, 4th April 2006; to Chinnor & Princes Risborough Railway, Oxfordshire, for gala, 1st October 2013; noted in blue, white and green livery with number 08649, Chinnor & Princes Risborough Railway, October 2013; re-sold to Knorr Bremse Rail Services Ltd, Wolverton Works, and moved 22nd October 2013; the works and locomotive acquired by Gemini Rail, November 2018; to Meteor Power Ltd, Silverstone, 8th January 2020; to Very Light Rail Innovation Centre, Dudley, 30th August 2021; to Gemini Rail, Wolverton Works, 17th February 2022; to Meteor Power, Silverstone, Northamptonshire, 4th March 2023.

D3817   Horwich   1959   70D   ?   P   08650
08650   new (LOT 267) 18th March 1959; withdrawn, date not known; sold to Foster Yeoman Quarries Ltd, Merehead Stone Terminal, Somerset, and moved 18th March 1989; given number 55; to Foster Yeoman Quarries Ltd, Isle of Grain Stone Terminal, Kent, 7th

May 1989; seen in blue livery with number 55, Isle of Grain, 8th March 1992; to Whatley Quarry, by 24th June 2001; to Foster Yeoman Quarries Ltd, Merehead Stone Terminal, 28th June 2002; to Foster Yeoman Quarries Ltd, Isle of Grain Stone Terminal, August 2002; to Wabtec, Doncaster, for repairs, about June 2003; to Foster Yeoman Quarries Ltd, Isle of Grain Stone Terminal, late 2003; noted in blue livery with number 08650, Isle of Grain, 31st August 2005; to Foster Yeoman Quarries Ltd, Merehead Stone Terminal, by 16th February 2007; to Foster Yeoman Quarries Ltd, Isle of Grain Stone Terminal, 2nd March 2007; to Merehead Stone Terminal, 8th June 2012; to Knights Rail Services, Eastleigh Works, for repairs, 3rd July 2012; noted at Eastleigh Works, 15th April 2013; to Isle of Grain Stone Terminal, 2nd May 2013; to Arlington Fleet Services Ltd, Eastleigh Works, 19th March 2015; to Aggregate Industries, Isle of Grain Stone Terminal, 27th March 2015; to Railway Support Services, Wishaw, for repairs to traction motors, 11th December 2015; noted in Mendip Rail blue livery with number 08650, Wishaw, 13th January 2016; to Whatley Quarry, 2nd March 2016; to Merehead Stone Terminal, by 27th May 2016; to Whatley Quarry, by 1st August 2016; noted at Whatley, 14th August 2016; noted on low-loader on A36, 20th March 2017; to Arlington Fleet Services Ltd, Eastleigh Works, for overhaul, 20th March 2017; to Hanson, Whatley Quarry, 23rd August 2017; to Railway Support Services, Wishaw, for repairs, 27th March 2019; noted in Mendip Rail blue livery with number 08650, Railway Support Services, Wishaw, 16th July 2019 and 14th August 2021; noted on a low-loader, M5 southbound, 24th November 2021; to Merehead Stone Terminal, 24th November 2021; to Whatley Quarry, early January 2022.

D3819 Horwich   1959   86A   6/92   P 08652
08652   new (LOT 267) 26th March 1959; withdrawn, 20th June 1992; sold to Foster Yeoman Quarries Ltd, Merehead Stone Terminal, Somerset, and moved 5th June 1993; seen in Yeoman blue livery with number 66, Merehead Stone Terminal, 7th September 1994; to ARC (Southern) Ltd, Whatley Quarry, Somerset, March 1995; to Merehead Stone Terminal, 2001; to Whatley Quarry, 22nd October 2002; to Foster Yeoman Quarries Ltd, Acton Rail Terminal, London, 10th May 2004; to Whatley Quarry, October 2004; to Acton Rail Terminal, November 2004; to Whatley Quarry, by 20th February 2005; to Acton Rail Terminal, by 27th May 2005; to Whatley Quarry, by 13th November 2005; to LH Group, Barton under Needwood, for overhaul, by 24th March 2008; to Whatley Quarry, about October 2008; noted in blue livery with number 08652, Whatley Quarry, February 2012; to Railway Support Services, Wishaw, for repairs, 14th May 2015; to Whatley Quarry, 2nd July 2015; to Merehead Stone Terminal, December 2015; to Whatley Quarry, February 2016; noted in blue livery with number 08652, Whatley Quarry, 14th August 2016; to Railway Support Services, Wishaw, on hire, 2nd December 2016; sub-hired by RSS to Axiom Rail, Wagon Works, Stoke on Trent, from 6th December 2016; to Merehead Stone Terminal, off hire, 4th January 2017; re-sold to Railway Support Services, Wishaw, May 2020; to Railway Support Services, Wishaw, 30th May 2020; noted in blue livery with number 08652, inside workshops with no engine, Wishaw, 25th May 2024.

D3820 Horwich   1959   16A   ?   P 08653 / VERNON
08653   new (LOT 267) 2nd April 1959; withdrawn, date not known; noted in EWS red and yellow livery, Toton Depot, 19th June 2013; put on sale by tender by DB Schenker, with bids due by 22nd September 2015; sold to Harry Needle Railroad Company; despatched from DB Schenker, Toton Depot, 18th January 2016; to Harry Needle Railroad Company, Long Marston; noted at Long Marston, 23rd July 2021; to Battlefield Line, Shackerstone (for overhaul by HTRS Ltd), 23rd March 2022.

D3822   Horwich   1959   51L   8/05   F   08655
08655   new (LOT 267) 15th April 1959; noted in blue livery with number 08655, Thornaby Depot, 6th March 2004; withdrawn, 31st August 2005; sold to T.J. Thomson & Son Ltd, Stockton-on-Tees; despatched from Thornaby Depot, 29th September 2005; noted at T.J. Thomson & Son Ltd, 8th October 2005; re-sold to LH Group, Barton under Needwood, and moved 16th June 2006; noted in blue livery with number 08655, LH Group, November 2006; used for spares; remains scrapped by Donald Ward of Burton upon Trent, April 2007.

D3830   Horwich   1959   82B   ?   P   08663
08663   new (LOT 267) 5th June 1959; withdrawn, date unknown; seen in blue livery with number 08663, St Philip's Marsh Depot, Bristol, 18th February 2016; named ST SILAS, in honour of the St Silas Royal British Legion Branch, Bristol, in ceremony at St Philip's Marsh, 11th April 2016; noted at Temple Meads, Bristol, 23rd November 2016; to Railway Technical Centre, Derby, 15th April 2017; noted in blue livery with number 08663 and ST SILAS nameplate, Derby RTC, 7th July 2017; to St Philip's Marsh, 7th July 2017; sold by Great Western Railway to Avon Valley Railway, June 2019; to Avon Valley Railway, Bitton, 8th July 2019; to Railway Support Services, Wishaw, for repairs, 23rd July 2019; noted in blue livery with number 08663 but no name, Railway Support Services, 29th July 2019; to GBRf, Dagenham, on hire, 13th September 2019; to Chrysalis Rail, Landore Depot, on hire, 12th March 2020; to GBRf, Felixstowe Docks, on hire, 16th October 2020; to Railway Support Services, Wishaw, for repairs, 17th April 2021; to Loram UK Ltd, Railway Technical Centre, Derby, on hire, 21st April 2021; to Railway Support Services, Wishaw, for repairs, early June 2022; seen in blue livery with number 08663, Railway Support Services, 9th July 2022; to Avon Valley Railway, 31st October 2022; to Hitachi, Newton Aycliffe, on hire, 6th September 2023.

D3832   Crewe   1960   40B   2/04   F   08665
08665   new (order E500) 1st January 1960; withdrawn, February 2004; sold to C.F. Booth Ltd, Rotherham, and moved 16th December 2009; to Harry Needle Railroad Company (swapped for 08695); to Barrow Hill Engine Shed, Staveley, 17th December 2009; used for spares; seen with number 08665, Barrow Hill Engine Shed, 21st June 2011; to European Metal Recycling, Kingsbury, for scrap, 4th October 2011; scrapped October 2011.

D3835   Crewe   1960   5A   11/95   F   08668
08668   new (order E500) 2nd January 1960; withdrawn, 8th November 1995; moved to Crewe Works, for storage; noted on scrap line at Crewe Works, 17th August 1996; sold to Harry Needle Railroad Company; despatched from Adtranz, Crewe, 21st March 1997; to South Yorkshire Railway Preservation Society, Meadowhall, Sheffield; to Barrow Hill Engine Shed, Staveley, 3rd November 1999; to Wabtec, Doncaster, 2nd July 2001; some of its parts used in the rebuild of 08517; noted in blue livery with number 08668, Wabtec, 27th July 2003; noted in blue livery with green cab and number 08668, Wabtec, 20th July 2007; to Harry Needle Railroad Company, Long Marston, 20th December 2007; noted at Long Marston, June 2008; used for spares, February 2011; remains to European Metal Recycling, Kingsbury, for scrap, 29th June 2011; scrapped on arrival.

D3836   Crewe   1960   9A   3/89   P   08669 / BOB MACHIN
08669   new (order E500) 2nd January 1960; withdrawn, March 1989; immediately sold to Trafford Park Estates Ltd, Manchester, and moved 28th March 1989; repainted green, summer 1989; seen in green livery with number 08669, Trafford Park, 17th November

1990; re-sold to Wabtec, Doncaster, November 2000; to First Great Western, Laira Depot, Plymouth, on hire, 27th April 2001; to Wabtec, Doncaster, 14th November 2002; to Freightliner, Port of Felixstowe, on hire, February 2003; to Wabtec, Doncaster, May 2003; to GNER, Bounds Green Depot, London, on hire, 28th May 2003; to Wabtec, Doncaster, about March 2005; seen in black livery with number 08669 and name BOB MACHIN, Wabtec, 1st December 2011; to Tyseley Diesel Depot, Birmingham, for tyre turning, 19th February 2014; returned to Wabtec, Doncaster; seen at Wabtec, 6th May 2014; to LH Group, Barton under Needwood, for repairs, 19th March 2015; to Wabtec, Doncaster, 1st May 2015; to Midland Road Depot, Leeds, for tyre turning, 18th May 2017; returned to Wabtec, Doncaster, 26th May 2017; to LH Group, Barton under Needwood, for repairs, 31st May 2019; returned to Wabtec, Doncaster, 7th August 2019; to Hams Hall Rail Freight Terminal, Coleshill, on hire, 14th May 2020; re-sold to Hunslet Ltd (Ed Murray Group), 1st May 2021; to Wabtec, Doncaster; seen in Wabtec black livery with number 08669 and name BOB MACHIN, Wabtec, 9th June 2024.

D3837   Crewe   1960   66B   9/08   P   08670
08670   new (order E500) 2nd January 1960; noted in EWS red and yellow livery with number 08670, Motherwell Depot, 9th June 2005; withdrawn, 30th September 2008; sold to T.J. Thomson & Son Ltd, Stockton-on-Tees, January 2009; re-sold to Railway Support Services, Wishaw, early 2009; re-sold to Colne Valley Railway, early 2009; despatched from DB Schenker, Motherwell Depot, 5th March 2009; to Colne Valley Railway, Castle Hedingham; noted in EWS red and yellow livery with number 08670, Colne Valley Railway, April 2009; re-sold to Railway Support Services, Wishaw, early 2013; to Pullman Rail, Canton Depot, Cardiff, on hire, 22nd April 2013; to Railway Support Services, Wishaw, 11th July 2014; to Port of Felixstowe, on hire, 11th May 2015; to Railway Support Services, Wishaw, for repairs, 7th July 2015; to Chasewater Railway, Staffordshire, for testing, 11th December 2015; to Railway Support Services, Wishaw, 16th December 2015; noted repainted in RSS black livery, 25th January 2016; to Bounds Green Depot, London, on hire, 4th February 2016; noted at Hornsey Depot fitted with emergency use Dellner coupler for recovery of broken-down EMU 700026 at Harringay, 2nd May 2018; to Tyseley Diesel Depot, Birmingham, for tyre turning, 1st March 2019; to GBRf, Bescot Yard, Walsall, on hire, 28th March 2019; noted in RSS black livery with number 08670, Bescot Yard, 29th March 2019; started work at Bescot, 10th April 2019; noted in Bescot Yard, 1st September 2019; to Railway Support Services, Wishaw, for repairs, early May 2022; to GBRf, Bescot Yard, on hire, mid-May 2022; noted in RSS black livery with number 08670, Bescot Yard, 27th November 2022; to Railway Support Services, Wishaw, for repairs, 9th January 2023; to GBRf, Bescot Yard, on hire, 16th January 2023; to Railway Support Services, Wishaw, for repairs, early April 2023; to GBRf, Bescot Yard, on hire, mid-April 2023.

D3843   Horwich   1959   16A   ?   P   08676 / DAVE 2
08676   new (LOT 270) 4th July 1959; withdrawn, date not known; sold to Harry Needle Railroad Company; despatched from DB Schenker, Toton Depot, 17th February 2016; to Barrow Hill Engine Shed, Staveley; noted in EWS red and yellow livery, Barrow Hill Engine Shed, 27th February 2016; to East Kent Railway, Shepherdswell, for storage, 12th October 2016; noted in EWS red and yellow livery with number 08676 and name DAVE 2, East Kent Railway, 6th June 2017 and 27th September 2022.

D3845   Horwich   1959   41A   10/88   P   555
08678   new (LOT 270) 1st August 1959; latterly ran on BR with painted name STAVELEY; withdrawn, 30th October 1988; noted at BR Tinsley Depot, 12th March 1989; moved to Doncaster Depot, mid-March 1989; moved to Tyne Yard, 22nd March 1989; sold

to Glaxochem Ltd; moved to Carlisle Upperby Depot, for inspection and repairs, and repaint in light blue and navy livery with number 678, 29th March 1989; moved by rail to Glaxochem Ltd, Ulverston, Cumbria; noted in Plumpton Loop, Penrith, 29th April 1989; arrived at Glaxochem, 4th May 1989; seen in two-tone light blue/navy blue livery and number 678, Ulverston, 11th August 1989; to Steamtown, Carnforth, Lancashire, for repairs, 3rd August 1990; returned to Glaxochem; to Steamtown, Carnforth, for repairs, 20th July 1992; returned to Glaxochem; re-sold to Steamtown, Carnforth, and moved by road on 8th November 1994; seen with number 678, Carnforth, 13th September 1997; to Fragonset, Derby, for repairs, May 2001; to Maintrain, Etches Park, on hire, by 9th June 2001; to Fragonset, Derby, by 3rd November 2001; to West Coast Railway Company, Carnforth, December 2001; noted in blue and navy livery with number 08678 and name ARTILLA, Carnforth, 27th July 2008; noted in WCRC maroon livery with number 08678, Carnforth, 21st February 2018; noted in WCRC maroon livery with number 555, Carnforth, 21st January 2023.

D3846 Horwich   1959   8J   6/76   F   08679
08679   new (LOT 270) 8th August 1959; withdrawn, 19th June 1976; sold to NCB; to North Gawber Colliery, Mapplewell, Barnsley, 25th June 1976; seen in BR blue livery with number 08679 and BR two-way arrow logo, North Gawber Colliery, 26th June 1976; to Royston Drift Mine, Barnsley, 7th September 1976; seen in BR blue livery with number 08679, Royston Drift Mine, 24th December 1977 and 25th November 1978; to North Gawber Colliery, Mapplewell, 5th April 1979; seen with 'Registered No. LDS.27' plate, North Gawber Colliery, 16th April 1979; rail traffic ceased and seen dumped, with no engine, North Gawber Colliery, 21st September 1985; seen on a CDR low-loader, arriving at C.F. Booth Ltd, Rotherham, for scrap, at 12:35 on 18th April 1986; allocated C.F. Booth Ltd plant number 91009; scrapped by 25th April 1986.

D3849 Horwich   1959   ZF   6/95   P   LIONHEART
08682   new (LOT 270) 29th August 1959; to ABB (Customer Support) Ltd, Doncaster Works (locomotive included in sale when BRML works privatised), 5th June 1995; noted in blue livery with number 08682 and name LIONHEART, Doncaster Works, July 1996; re-sold to Bombardier Transportation, Litchurch Lane Works, Derby, about January 2008; to Railway Vehicle Engineering Ltd, Railway Technical Centre, Derby, on hire, 31st July 2010; returned to Bombardier Transportation, Litchurch Lane Works, October 2010; to Railway Vehicle Engineering Ltd, Railway Technical Centre, for repairs, June 2013; to Bombardier Transportation, Litchurch Lane Works, 24th July 2013; to Railway Vehicle Engineering Ltd, for repairs, 22nd May 2014; to Bombardier Transportation, Litchurch Lane Works, 9th June 2014; noted in purple, blue, red and green livery with number 08682, Litchurch Lane Works, 5th March 2017; re-sold to Harry Needle Railroad Company, about January 2020; noted at Litchurch Lane Works, 30th May 2020; to Hope Cement Works, for storage, 22nd December 2020; to Battlefield Line, Shackerstone, for storage, 27th October 2022; re-sold by HNRC, August 2023; noted in multi-colour livery, Shackerstone, 28th October 2023.

D3850 Horwich   1959   16A   2/04   P   08683
08683   new (LOT 270) 5th September 1959; withdrawn, February 2004; noted at Toton Depot, 11th March 2006; sold to Railway Support Services, Wishaw, and moved on 8th March 2007; noted in EWS red and yellow livery with number 08683, Wishaw, 6th July 2007; to Gloucestershire Warwickshire Railway, Toddington, for storage, November 2009; to Freightliner, Port of Felixstowe, on hire, 10th February 2011; to Colne Valley Railway, Castle Hedingham, for storage, 17th August 2011; to Pullman Rail, Canton Depot, Cardiff, on hire, 16th April 2013; to Electro-Motive Diesel, Longport, Staffordshire, on hire, 3rd

October 2013; to Railway Support Services, Wishaw, late 2013; to Bombardier Transportation, Litchurch Lane Works, Derby, on hire, 23rd February 2015; to Railway Support Services, Wishaw, for overhaul, 1st April 2016; noted in EWS red and yellow livery with number 08683, Wishaw, 13th April 2016; noted in new Railway Support Services black livery, Wishaw, 21st May 2016; to Chasewater Railway, Staffordshire, for testing, 27th July 2016; to Epping Ongar Railway, Essex, for gala, 12th September 2016; to Bombardier Transportation, Litchurch Lane Works, Derby, on hire, 17th October 2016; to Railway Support Services, Wishaw, 27th January 2017; noted in RSS black livery with number 08683, Railway Support Services, 2nd April 2017; to Great Central Railway, Loughborough, 3rd May 2017; to Crown Point Depot, Norwich, on hire, 18th July 2017; noted at Norwich, 18th July 2017; to GBRf, Eastleigh Yard, on hire, March 2021.

D3852   Horwich   1959   40B   2/09   P   08685
08685   new (LOT 270) 19th September 1959; withdrawn, February 2009; sold to Harry Needle Railroad Company; despatched from Immingham Depot, 19th September 2011; to HNRC, Barrow Hill Engine Shed, Staveley; seen in EWS red and yellow livery with number 08685, Barrow Hill Engine Shed, 21st September 2011 and 15th April 2016; to Hope Cement Works, Derbyshire, on hire, 11th October 2016; noted at Wishaw, in transit, 4th December 2016; to East Kent Railway, Shepherdswell, for storage, 5th December 2016; noted in EWS red and yellow livery with number 08685, East Kent Railway, 24th June 2017; to HNRC, Barrow Hill Engine Shed, 4th May 2023.

D3854   Horwich   1959   5A   3/15   P   08995
08995   new (LOT 270) 3rd October 1959; to Landore Depot, 18th February 1987; rebuilt with cut down cab for working Burry Port & Gwendraeth Valley line and renumbered 08995; formerly numbered 08687 until 4th September 1987; named KIDWELLY, October 1987; withdrawn, March 2015; put on sale by DB Schenker, 5th May 2015; sold to Railway Support Services, Wishaw; despatched from Crewe Electric Depot, 7th September 2015; to Railway Support Services, Wishaw; noted in EWS red and yellow livery with number 08995, Railway Support Services, Wishaw, 23rd June 2017 and 16th July 2019; to Gwendraeth Valley Railway, Mudlescwm, Kidwelly, 16th December 2020.

D3859   Horwich   1959   5A   1/94   F   NFT
08692   new (LOT 270) 7th November 1959; withdrawn, 31st January 1994; sold to ABB Transportation, Crewe, and moved 5th May 1994; to ABB (Customer Support) Ltd, Doncaster, about August 1995; returned to ABB Transportation, Crewe, by 17th August 1996; noted on scrap line at Crewe Works, 17th August 1996; used as a spares donor; re-sold to Harry Needle Railroad Company, 2002; to West Coast Railway Company, Carnforth, July 2002; noted in blue livery with no number, Carnforth, 20th May 2003 and 10th March 2005; scrapped, May 2005.

D3861   Horwich   1959   81A   9/08   P   08694
08694   new (LOT 270) 21st November 1959; withdrawn, 30th September 2008; sold to C.F. Booth Ltd, Rotherham, and moved 8th January 2009; re-sold to Great Central Railway, Loughborough, and moved 22nd May 2009; noted in EWS red and yellow livery with number 08694, Great Central Railway, June 2012 and July 2019; to Great Central Railway, Ruddington, 24th September 2020; repainted in BR grey livery, and named ANNESLEY MPD, June 2024.

D3862   Horwich   1959   66B   2/04   F   08695
08695   new (LOT 270) 28th November 1959; withdrawn, 28th February 2007; noted in EWS red and yellow livery with number 08695, Ayr Depot, March 2007; sold to T.J.

Thomson & Son Ltd; despatched from Ayr Depot, 27th July 2007; to T.J. Thomson & Son Ltd, Stockton-on-Tees; re-sold to Harry Needle Railroad Company; moved to Barrow Hill Engine Shed, Staveley, 11th September 2007; noted in EWS red and yellow livery with number 08695, Barrow Hill Engine Shed, 3rd October 2007; to C.F. Booth Ltd, Rotherham (swapped for 08665); scrapped December 2009.

| D3863 | Horwich | 1959 | 9A | ? | P | 08696 |

08696 new (LOT 270) 5th December 1959; withdrawn, date not known; sold to Alstom Transport, Longsight Depot, Manchester, 2007; noted in green livery with number 08696, Longsight Depot, August 2007; to Alstom Transport, Wembley Depot, London, 5th March 2012; to Arlington Fleet Services Ltd, Eastleigh Works, for repairs, 3rd August 2012; to Alstom Transport, Wembley Depot, London, 23rd August 2012; to Alstom Transport, Polmadie Depot, Glasgow, 14th December 2012; to Arlington Fleet Services Ltd, Eastleigh Works, for overhaul, 19th October 2015; noted at Eastleigh Works, 20th October 2015; noted being repainted blue, Eastleigh Works, 28th January 2016; noted in blue livery with number 08696, Eastleigh Works, 26th February 2016; to Alstom Transport, Wembley Depot, London, 26th February 2016; noted in blue livery with number 08696, Wembley Depot, 3rd January 2019; to Alstom Transport, Stonebridge Park Carriage Maintenance Depot, London, by January 2022.

| D3864 | Horwich | 1959 | 16C | ? | F | 08697 |

08697 new (LOT 270) 19th December 1959; withdrawn, date not known; noted in BR blue livery with number 08697, Etches Park, Derby, 16th April 2005 and 5th October 2009; to Railway Vehicle Engineering Ltd, Railway Technical Centre, Derby, July 2010; to D. Ward Ltd, Ilkeston, for scrap, 28th October 2014; scrapped, October 2014.

| D3866 | Horwich | 1960 | 5A | 10/93 | F | 08699 |

08699 new (LOT 270) 14th January 1960; withdrawn, 22nd October 1993; sold to ABB Transportation, Crewe, and moved 5th May 1994; noted in blue livery with no number, Crewe Works, 21st May 2000 and 31st May 2003; re-sold to Cotswold Rail, Moreton in Marsh, early 2005; to Daventry International Rail Freight Terminal, on hire, about March 2005; to Cotswold Rail, May 2005; to Tyseley Steam Depot, Birmingham, for storage, 2005; to Allelys Ltd, Studley, Warwickshire, for storage, about 24th February 2006; noted in blue livery with number 08699, Allelys Ltd, 18th March 2006; re-sold to RMS Locotec Ltd, January 2007; noted in blue livery with number 08699, Studley, May 2007; to Corus, Shotton, for storage, 8th July 2009; to RMS Locotec Ltd, Wolsingham Depot, 28th March 2013; noted in blue livery with number 08699, no rods, Wolsingham, 17th February 2017; used for spares; scrapped on site by J. Denham Metals Ltd of Bishop Auckland, 23rd March 2018.

| D3867 | Horwich | 1960 | 30A | 9/97 | P | D3867 |

08700 new (LOT 270) 20th January 1960; withdrawn, 21st September 1997; stored at Stratford Depot; sold to Harry Needle Railroad Company; to Barrow Hill Engine Shed, Staveley, 27th June 2001; to West Coast Railway Company, Carnforth, 7th August 2002; noted in blue livery with number painted over, Carnforth, 20th May 2003; to Bryn Engineering Ltd, Wigan, 22nd June 2004; to Embsay & Bolton Abbey Railway, 23rd June 2004; to East Lancashire Railway, Bury, about April 2007; noted in blue livery with number 08700, East Lancashire Railway, Bury, June 2011; to Harry Needle Railroad Company, Barrow Hill Engine Shed, Staveley, 17th March 2014; to Bombardier Transportation, Ilford Depot, on hire, 10th June 2015; noted in black livery with grey cab and number 08700, Ilford Depot, 5th March 2018; noted in blue livery with number 08700, Ilford Depot, 6th June

2021 and 24th September 2023; to Harry Needle Railroad Company, Worksop Depot, 4th December 2023; to Freightliner Terminal, Garston, on hire, 9th July 2024.

D3868   **Horwich**   **1960**   **16A**   **?**   **P**   **08701**

08701   new (LOT 270) 5th February 1960; withdrawn, date not known; noted in red and black livery with number 08701, Toton Depot, 6th March 2012; put on sale by tender by DB Schenker, with bids due by 22nd September 2015; sold to Harry Needle Railroad Company; noted on a low-loader leaving DB Schenker, Toton Depot, 19th January 2016; to Harry Needle Railroad Company, Long Marston; noted at Long Marston, 23rd July 2021; to Battlefield Line, Shackerstone (for overhaul by HTRS Ltd), 4th March 2022; noted in red and black livery with number 08701, Shackerstone, 28th October 2023.

D3870   **Horwich**   **1960**   **2F**   **?**   **P**   **08703**

08703   new (LOT 270) 17th February 1960; arrived at Bescot from Immingham, 5th August 2016; noted in EWS red and yellow livery, working at Bescot, 12th August 2016; was the last serviceable shunting locomotive at Bescot; withdrawn, date not known; put on sale by tender by DB Cargo, with bids due by 5th October 2016; sold to Railway Support Services, Wishaw; despatched from DB Cargo, Bescot Depot, 19th January 2017; to Railway Support Services, Wishaw; to Network Rail, Springs Branch Depot, Wigan, on hire, 20th January 2017; noted in EWS red and yellow livery with number 08703, Springs Branch, 18th May 2017; to Electro-Motive Diesel, Longport, on hire, 20th November 2020; to Chasewater Railway, 25th November 2020; to Railway Support Services, Wishaw, for repaint, about 8th February 2021; noted in GBRf blue livery with orange cab, Wishaw, 8th June 2021; to Balfour Beatty, Willesden Junction, on hire, 20th July 2021; noted with JERMAINE nameplates, Willesden, 14th September 2021; to Chinnor & Princess Risborough Railway, 26th October 2021; to Tyseley Diesel Depot, Birmingham, for tyre turning, 7th November 2021; to Balfour Beatty Ltd, Willesden Junction, on hire, 18th November 2021; to Railway Support Services, Wishaw, for repairs, 28th November 2022; to GBRf, Whitemoor Yard, March, on hire, 23rd December 2022; to Railway Support Services, Wishaw, 26th January 2023; to Tyseley Locomotive Works, Birmingham, for repairs, 11th May 2023; to Wishaw, 15th May 2023; received nameplates 'Steve Blick (Concrete Bob) ShunterSpot', in a ceremony at Wishaw on 21st May 2023; seen in RSS blue and orange livery with number 08703, Railway Support Services, Wishaw, 23rd September 2023; to GBRf, Whitemoor Yard, March, on hire, 10th April 2024; to Railway Support Services, Wishaw, 17th June 2024.

D3871   **Horwich**   **1960**   **1E**   **12/89**   **P**   **08704**

08704   new (LOT 270) 5th March 1960; withdrawn, 18th December 1989; sold to Port of Boston; despatched from BR Bletchley, by road, 19th April 1990; arrived at Port of Boston, 8th May 1990; first working on 10th May 1990; to Doncaster Depot, for tyre turning, 20th January 1992; to Port of Boston, 29th January 1992; worked last train prior to docks line closure, 29th January 1993; to Nene Valley Railway, Wansford, by road, on loan, 2nd February 1993; seen in blue livery with number 08704, Wansford, 25th February 1993; to Port of Boston, by road, 11th September 1997; line to docks reopened, and 08704 broke ribbon across entrance, 30th September 1997; noted in black livery, 30th September 1997; to Wabtec, Doncaster, for repairs, 6th June 2001; to Port of Boston, 26th July 2001; noted working, Boston, 18th May 2004; re-sold to Harry Needle Railroad Company, June 2011; to East Lancashire Railway, Bury, 13th April 2012; re-sold to Riviera Trains Ltd of Crewe, 2013; to Crewe Diesel Depot, 18th September 2013; noted in green livery with number D3871, Crewe, October 2013; to Bombardier Transportation, Litchurch Lane Works, Derby, for repaint, 12th May 2014; to Riviera Trains Ltd, Crewe Diesel Depot, late 2014; to

Nemesis Rail, Burton upon Trent, 23rd October 2015; to Ecclesbourne Valley Railway, Wirksworth, 5th March 2018; noted in dark blue livery with number 08704, Wirksworth, 21st April 2018; to Midland Road Depot, Leeds, for tyre turning, 6th March 2020; to Ecclesbourne Valley Railway, Wirksworth, 13th March 2020; to Riviera Trains, Knottingley Depot, to shunt coaching stock, 12th January 2021; noted at Knottingley Depot, 4th July 2024.

D3873   Crewe   1960   5A   ?   P   08706
08706   new (order E503) 19th March 1960; withdrawn, date not known; put on sale by tender by DB Cargo, with bids due by 5th October 2016; sold to Harry Needle Railroad Company; despatched from Crewe Electric Depot; to Railway Support Services, Wishaw, for storage, 31st August 2017; noted in red and yellow livery with number 08706, Railway Support Services, Wishaw, September 2017 and 8th August 2020; to Battlefield Line, Shackerstone, 17th January 2022; re-sold to Colne Valley Railway, and moved 24th April 2023; to Railway Support Services, Wishaw, on hire, 25th August 2023; to Direct Rail Services, Garston, Liverpool, on sub-hire, 1st September 2023; noted at Garston, 2nd December 2023; to Railway Support Services, Wishaw, about 9th February 2024.

D3874   Crewe   1960   55G   7/93   F   08707
08707   new (order E503) 19th March 1960; hauled from Holbeck to Doncaster by 56067, 3rd June 1993; withdrawn, 23rd July 1993; to Crewe Works, for storage; sold to Harry Needle Railroad Company; despatched from Adtranz, Crewe, 4th April 1997; to South Yorkshire Railway Preservation Society, Meadowhall, Sheffield; noted in BR blue livery with number 08707, no centre wheels, Meadowhall, 23rd August 1997; to Barrow Hill Engine Shed, Staveley, 27th July 2001; to West Coast Railway Company, Carnforth, 11th July 2002; noted in BR blue livery, being used for spares, Carnforth, 15th January 2005; scrapped on site, by Harry Needle staff, February 2005.

D3876   Crewe   1960   2F   4/11   P   08709
08709   new (order E503) 1st April 1960; withdrawn, April 2011; put on sale by tender by DB Schenker, with bids due by 22nd September 2015; noted in EWS red and yellow livery with number 08709, Bescot Depot, 12th October 2015; sold to Railway Support Services, Wishaw; despatched from Bescot Depot, October 2015; to Railway Support Services; to Colne Valley Railway, Castle Hedingham, for storage, 3rd March 2016; to Railway Support Services, Wishaw, 26th May 2017; put up for sale by Railway Support Services, January 2022; seen in EWS livery with number 08709, RSS, 23rd September 2023.

D3878   Crewe   1960   51L   ?   P   08711
08711   new (order E503) 28th April 1960; withdrawn, date not known; put on sale by DB Schenker, 5th May 2015; sold to Harry Needle Railroad Company; noted in red and black livery with number 08711, Tees Yard, 24th January 2013 and 1st April 2017; despatched from Tees Yard, 1st June 2017; to Harry Needle Railroad Company and moved to Nemesis Rail, Burton upon Trent, for storage, 2nd June 2017.

D3881   Crewe   1960   5A   ?   P   08714
08714   new (order E503) 4th May 1960; withdrawn, date not known; put on sale by tender by DB Schenker, with bids due by 22nd September 2015; sold to Harry Needle Railroad Company; despatched from Crewe Electric Depot, 19th October 2016; to Harry Needle Railroad Company and moved direct to Hope Cement Works, Derbyshire, for storage, 19th October 2016; noted in EWS red and yellow livery with number 08714, Hope

Cement Works, 25th November 2016; noted disused at rear of works, 15th May 2023; to Harry Needle Railroad Company, Worksop Depot, 8th June 2023.

D3889   Crewe   1960   ?   ?   P   08721
08721   new (order E503) 15th June 1960; withdrawn, date not known; noted in Express Parcels blue livery with red stripe and number 08721 and name STARLET, Longsight Depot, 28th June 2006; acquired by Alstom Transport, Longsight Depot, Manchester, 2007; noted in blue livery with red stripe, with number 08721 and name M.A. SMITH, Longsight Depot, 15th September 2007; to Alstom Transport, Edge Hill Depot, Liverpool, December 2011; to Arlington Fleet Services Ltd, Eastleigh Works, for repairs and repaint, 11th February 2015; noted in BR blue livery with number 08721, on a low-loader leaving Eastleigh Works, 30th November 2015; to Alstom Transport, Wembley Depot, London, 30th November 2015; noted on a low-loader, arriving at Alstom Transport, Longsight Depot, Manchester, 11th January 2016; noted in blue livery with number 08721, Longsight, 12th January 2016; to East Lancashire Railway, Bury, for gala, 14th April 2016; to Alstom Transport, Longsight Depot, Manchester, by 22nd April 2016; to Alstom Transport, Technology Centre, Widnes, 1st August 2017; noted in blue livery with number 08721, Widnes, 24th August 2017 and 24th February 2024.

D3892   Crewe   1960   ?   ?   P   08724
08724   new (order E503) 1st July 1960; withdrawn, date not known; to Great North Eastern Railway and allocated to Craigentinny Depot, Edinburgh; sold to RFS (Engineering) Ltd, Doncaster, 1997; to Foster Yeoman Quarries Ltd, Isle of Grain Stone Terminal, Kent, on hire, 9th September 1998; to ARC, Whatley Quarry, on hire, October 1998; to RFS (Engineering) Ltd, Doncaster, for repairs, 11th January 1999; to Blue Circle, Hope Cement Works, Derbyshire, on hire, March 1999; to RFS (Engineering) Ltd, Doncaster, 1999; to ARC, Whatley Quarry, on hire, by 19th September 1999; noted at Whatley Quarry, 5th April 2000; to RFS (Engineering) Ltd, Doncaster, 10th June 2000; to Neville Hill Depot, Leeds, on hire, about September 2000; to Wabtec, Doncaster, for repairs, early 2003; returned to Neville Hill Depot, Leeds, on hire, by 19th June 2003; seen in black livery with number 08724, Neville Hill, 21st March 2004 and 7th December 2011; to Wabtec, Doncaster, 10th January 2013; seen in black livery with number 08724, Wabtec, 10th March 2013; to Midland Road Depot, Leeds, for tyre turning, 2nd August 2017; seen in Wabtec black livery with number 08724, Midland Road Depot, 8th August 2017; returned to Wabtec, Doncaster, August 2017; re-sold to Hunslet Ltd (Ed Murray Group), 1st May 2021; noted in Wabtec black livery with number 08724, Wabtec, Doncaster, 29th June 2023.

D3896   Crewe   1960   66B   10/87   F   08728
08728   new (order E503) 12th August 1960; withdrawn, 12th October 1987; sold to Deanside Transit Ltd, Glasgow, and moved November 1987; noted in BR blue livery with number 08728, Deanside Transit, 25th February 2006; re-sold to Harry Needle Railroad Company, 2007; noted on a low-loader, Stafford Services, M6 southbound, 14th June 2007; arrived at HNRC, Long Marston, 14th June 2007; noted in BR blue livery with number 08728 and 'Deanside Transit' on side, Long Marston, 6th June 2009; to C.F. Booth Ltd, Rotherham, for scrap, 4th November 2009; scrapped, November 2009.

D3898   Crewe   1960   ZH   6/95   P   08730 / THE CALEY
08730   new (order E503) 23rd August 1960; to Railcare Ltd, Springburn Works, Glasgow, June 1995; locomotive included in sale when BRML works privatised; to LH Group, Barton under Needwood, for repairs, 19th August 2009; noted in black and white

livery with number 08730, LH Group, October 2009; returned to Springburn Works, Glasgow, about March 2010; to Midland Road Depot, Leeds, for tyre turning, 20th January 2017; seen in blue, white and green livery with number 08730, Midland Road Depot, 1st February 2017; to Knorr Bremse Rail Services Ltd, Springburn Works, Glasgow, 16th February 2017; Springburn Works closed, 29th July 2019; re-sold to Railway Support Services, Wishaw, and moved 18th October 2019; to Gemini Rail, Wolverton Works, on hire, 21st October 2019; noted in blue, white and green livery with no number, Wolverton Works, January 2020; to Railway Support Services, Wishaw, 23rd April 2020; to Chasewater Railway, 21st May 2020; to Railway Support Services, Wishaw, for repaint, 1st July 2020; noted in blue livery with ABP logo, Wishaw, 14th August 2020; to Chasewater Railway, 18th August 2020; noted on a low-loader on M1 Motorway, 24th September 2020; to Gemini Rail, Wolverton Works, on hire, 24th September 2020; to Railway Support Services, Wishaw, 3rd October 2020; to Electro-Motive Diesel, Longport, on hire, 16th November 2020; to Railway Support Services, Wishaw, 20th November 2020; to Whitemoor Yard, March, on hire, 23rd November 2020; to Railway Support Services, Wishaw, March 2021; to Whatley Quarry, on hire, 2nd June 2021; noted at Whatley Quarry, 13th June 2021; to location near Aldershot, for filming, early October 2021; to Railway Support Services, Wishaw, 12th October 2021; to Tyseley Locomotive Works, Birmingham, for repairs, 17th October 2021; to European Metal Recycling, Kingsbury, on hire, about 22nd April 2023; to GBRf, Whitemoor Yard, on hire, March, 2024; to European Metal Recycling, Kingsbury, on hire, April 2024; noted in blue livery, European Metal Recycling, 27th May 2024.

D3899   Crewe   1960   66B   3/96   F   08731
08731   new (order E503) 1st September 1960; to BREL Swindon Works, for dual brake fitting, June 1983; found to have cracked frames and so, for unknown reasons, identity changed with 08572 which was in works at the time for scrap; although really 08572, it emerged from BREL Swindon Works with number 08731; withdrawn, 7th March 1996; noted at Motherwell Depot, December 2001; sold to T.J. Thomson & Son Ltd, Stockton-on-Tees; despatched from EWS Motherwell Depot, by road, 21st June 2002; delivered to T.J. Thomson, 21st June 2002; re-sold to Foster Yeoman Quarries Ltd, Merehead Stone Terminal, and moved 13th August 2003; to LH Group, Barton under Needwood, used for spares, early March 2004; noted at LH Group, 14th March 2004; remains returned to Merehead Stone Terminal, 16th April 2004; noted in blue livery with number 08731, with no engine and no middle wheels, Merehead Stone Terminal, 27th November 2005 and 21st June 2008; to Bodmin & Wenford Railway, Cornwall, for spares, August 2008; noted at Bodmin, 15th August 2008; noted on a road low-loader on A303, 29th August 2008; to Merehead Stone Terminal, 29th August 2008; remains to J.W. Ransome & Sons, Frome, for scrap, March 2009; scrapped March 2009.

D3902   Crewe   1960   86A   12/99   F   08734
08734   new (order E503) 20th September 1960; withdrawn, December 1999; sold to Dean Forest Railway; despatched from Canton Depot, Cardiff, by road, 17th June 2000; to Dean Forest Railway, Lydney; noted in BR blue livery with number 08734, no centre wheels, Lydney, 23rd July 2009 and 3rd July 2010; used for spares; remains to Sims Metals Ltd, Newport, for scrap, 19th August 2011; scrapped August 2011.

D3904   Crewe   1960   66B   10/87   F   08736 / 4
08736   new (order E503) 24th September 1960; withdrawn, 12th October 1987; sold to Deanside Transit Ltd, Glasgow, and moved November 1987; given number 4; noted in BR blue livery with number 08736, Loco 4 on cab door, Deanside Transit, 25th February 2006;

re-sold to Harry Needle Railroad Company; noted on a low-loader on the M6, 23rd May 2007; to Harry Needle Railroad Company, Long Marston, 23rd May 2007; noted in BR blue livery with number 08736, Long Marston, June 2009; used for spares; to C.F. Booth Ltd, Rotherham, for scrap, 5th November 2009; scrapped, November 2009.

D3905   Crewe   1960   5A   ?   P   08737
08737   new (order E503) 8th October 1960; withdrawn, date not known; noted in EWS red and yellow livery with number 08737, Crewe Electric Depot, January 2016; moved from Crewe Electric Depot to Locomotive Services Ltd, Crewe Diesel Depot, 21st January 2016; noted at Crewe Diesel Depot, 27th November 2018; to Locomotive Services, Southall Depot, London, 9th October 2019.

D3906   Crewe   1960   16A   ?   P   08738
08738   new (order E503) 14th October 1960; withdrawn, date not known; put on sale by DB Schenker, 5th May 2015; noted in grey livery with number 08738, Toton Depot, September 2015; sold to Railway Support Services, Wishaw; despatched from DB Schenker, Toton Depot, 17th September 2015; to Colne Valley Railway, Castle Hedingham, for storage; noted in Euro Cargo Rail grey livery with number 08738, Colne Valley Railway, 14th February 2016; to Railway Support Services, Wishaw, 24th April 2017; noted in Euro Cargo Rail grey livery with number 08738, Railway Support Services, 23rd June 2017; noted repainted in RSS black livery with number 08738, Railway Support Services, 25th February 2019; to GBRf, East Yard, Eastleigh, on hire, 30th March 2019; to GBRf, Felixstowe, on hire, 22nd October 2019; to GBRf, East Yard, Eastleigh, on hire, 1st November 2019; to GBRf, Felixstowe, on hire, 5th February 2020; to Railway Support Services, Wishaw, 17th October 2020; noted in grey livery with number 08738, Railway Support Services, 24th October 2020; to Chasewater Railway, 8th February 2021; to Railway Support Services, Wishaw, for repairs, January 2022; to Chasewater Railway, for gala, 3rd November 2022; to GBRf, Felixstowe Docks, on hire, 2nd December 2022.

D3908   Crewe   1960   ?   8/05   F   08740
08740   new (order E503) 20th October 1960; withdrawn, 31st August 2005; sold to T.J. Thomson & Son Ltd, Stockton-on-Tees; despatched from Ferrybridge, early October 2005; noted at T.J. Thomson & Son Ltd, 8th October 2005; re-sold to LH Group, Barton under Needwood, and moved 15th June 2006; noted in grey livery with number 08740, LH Group, September 2006; used for spares; remains scrapped by Donald Ward of Burton upon Trent, April 2007.

D3910   Crewe   1960   81F   ?   P   08742
08742   new (order E503) 1st November 1960; withdrawn, date not known; noted in red and black livery with number 08742, Didcot, 24th February 2016; put on sale by tender by DB Cargo, with bids due by 5th October 2016; sold to Harry Needle Railroad Company; despatched from Didcot; to East Kent Railway, Shepherdswell, for storage, 30th April 2018; to Barrow Hill Engine Shed, Staveley, 25th April 2019; seen in faded red and black livery with number 08742, Barrow Hill Engine Shed, Staveley, 21st September 2019 and 14th October 2023.

D3911   Crewe   1960   55H   3/93   P   BRYAN TURNER
08743   new (order E503) 9th November 1960; withdrawn, 3rd March 1993; sold to RFS (Engineering) Ltd, Doncaster, March 1993; given number 024; to Grovehurst UK Paper Ltd, Sittingbourne, on hire, 25th January 1993; returned to RFS (Engineering) Ltd, Doncaster, by 1st November 1996; re-sold to ICI Billingham Works, Stockton-on-Tees, and moved 27th January 1997; briefly unofficially named ANGIE; used on Billingham to Seal Sands

HCN Train trial, 10th October 1997; to Thornaby Depot, for tyre turning, 15th March 1999; to ICI Wilton Works, Middlesbrough, April 2004; to Cleveland Potash Ltd, Teesport, on hire, about July 2005; returned to ICI Wilton Works, off hire, October 2005; to Hunslet Engine Company, Barton under Needwood, for repairs, 13th October 2011; noted in blue livery with number 08743, Barton under Needwood, 14th January 2012; to ICI Wilton Works, 2nd May 2012; to Wensleydale Railway, Leeming Bar, for gala, early September 2016; noted in blue livery with name BRYAN TURNER, Wensleydale Railway, 13th September 2016; to Sembcorp Utilities Ltd, Wilton, 27th September 2016.

D3914   Crewe   1960   66B   4/99   F   08746
08746   new (order E503) 26th November 1960; withdrawn, 8th April 1999; moved to Doncaster Depot, July 1999; noted in grey livery with number 08746, semi-dismantled, Doncaster Depot, 11th August 2002; sold to Barrow Hill Engine Shed Society, Staveley; despatched from Doncaster Depot, 21st July 2003; to Barrow Hill Engine Shed, Staveley; used for spares to repair 08928; remains to C.F. Booth Ltd, Rotherham, for scrap, 31st August 2003; noted in grey livery with number 08746, with no engine and no rods, C.F. Booth Ltd, 16th December 2003; scrapped, January 2004.

D3918   Crewe   1960   30A   6/98   F   08750
08750   new (order E503) 20th December 1960; to store at Stratford, 21st September 1997; withdrawn, June 1998; sold to RT Rail, Crewe, 2000; despatched from Stratford Depot, 7th August 2000; later to Wabtec, Doncaster, for assessment; to Wessex Traincare Ltd, Eastleigh Works, on hire, 1st December 2000; returned to RT Rail, Crewe; to Ilford Depot, on hire, 29th January 2001; noted on a low-loader, northbound on A1, 27th March 2001; noted under repair, Wabtec, Doncaster, 18th November 2001; to Channel Tunnel Rail Link, Beechbrook Farm, near Ashford, on hire, by March 2002; to Wabtec, Doncaster, 1st August 2002; to Wensleydale Railway, Leeming Bar, 21st July 2003; noted in black livery with number 08750, Wensleydale, August 2003; to Wabtec, Doncaster, December 2003; to Imerys Minerals Ltd, Quidhampton, Salisbury, on hire, 26th March 2004; to Wabtec, Doncaster, for repairs, about April 2005; returned to Quidhampton, on hire, about August 2005; to Wabtec, Doncaster, for repairs, 13th October 2005; to Tubelines, Ruislip, London, on hire, 16th May 2006; to First Capital Connect, Hornsey Depot, London, on hire, 30th May 2007; RT Rail was acquired by RMS Locotec Ltd, 8th November 2007; to RMS Locotec Ltd, Wakefield, for repairs, about November 2007; returned to Hornsey Depot, London, on hire; to Bombardier Transportation, Ilford Depot, on hire, about April 2011; noted in black livery with number 08750, Ilford Depot, July 2011; to Castle Cement Works, Ketton, on hire, November 2011; to RMS Locotec Ltd, Wolsingham Depot, for repairs, 20th August 2012; to Castle Cement Works, Ketton, on hire, 20th November 2012; to RMS Locotec Ltd, Wolsingham Depot, 1st February 2013; noted in black livery with number 08750, Wolsingham, 30th April 2016 and 17th October 2017; scrapped on site by Wanted Metal Recycling Ltd of Shildon, 18th December 2018.

D3919   Crewe   1960   41A   5/97   F   08751
08751   new (order E503) 28th December 1960; withdrawn, May 1997; sold to RFS Engineering, Doncaster, and moved by June 1998; used for spares; seen in two-tone grey livery with number 08751, RFS Doncaster, 19th October 2000; remains to C.F. Booth Ltd, Rotherham, January 2004; noted in grey livery with number 08751, C.F. Booth Ltd, 14th February 2004; re-sold to RT Rail, Crewe; to Wabtec, Doncaster, for spares recovery, April 2004; remains to C.F. Booth Ltd, Rotherham, for scrap, 7th July 2004; noted in grey livery with number 08751, with no engine, C.F. Booth Ltd, 29th August 2004; scrapped, 4th September 2004.

D3920   Crewe   1960   2F   ?   P   08752

08752   new (order E503) 30th December 1960; withdrawn, date not known; despatched from Tyne Yard; arrived at Bescot Depot, 11th August 2016; put on sale by tender by DB Cargo, with bids due by 5th October 2016; sold to Railway Support Services, Wishaw; despatched from DB Cargo, Bescot Depot, 9th January 2017; to Railway Support Services, Wishaw; to Bombardier Transportation, Litchurch Lane Works, Derby, on hire, 13th January 2017; noted in EWS red and yellow livery with number 08752, Litchurch Lane, 15th January 2017; to Railway Support Services, Wishaw, for repairs to its generator, 22nd March 2017; noted at RSS Wishaw, 23rd March 2017; to Gemini Rail, Wolverton Works, on hire, 5th February 2019; noted in grey livery with number 08752, Wolverton Works, January 2020; to Railway Support Services, Wishaw, for repairs, 25th September 2020; to Gemini Rail, Wolverton Works, on hire, 3rd October 2020; to Chasewater Railway, 28th January 2021; to Railway Support Services, Wishaw, about 8th February 2021; to Heritage Centre, Crewe, on hire, February 2021; to GBRf, Eastleigh, on hire, 18th May 2022; damaged during unloading; to Railway Support Services, Wishaw, for repairs, 25th May 2022; to Tyseley Diesel Depot, Birmingham, for tyre turning, about 30th June 2022; to Railway Support Services, Wishaw, early July 2022; to Electro-Motive Diesel, Longport, 17th August 2022; noted in grey RSS livery, leaving Longport on a low-loader, 5th January 2023; to Railway Support Services, Wishaw; to European Metal Recycling, Kingsbury, on hire, 11th February 2023; to Railway Support Services, Wishaw, 22nd April 2023; seen in RSS dark blue livery, Wishaw, 27th May 2023; to Imerys Minerals Ltd, Rocks Works, Cornwall, on hire, 6th June 2023; noted at Rocks Works, 18th April 2024.

D3922   Horwich   1961   60A   8/99   P   08754 / HO41

08754   new 12th January 1961; withdrawn, August 1999; sold to RT Rail, Crewe, August 1999; to Wabtec, Doncaster, autumn 1999; to Port of Felixstowe, on hire, about April 2000; to Freightliner, Garston Railport, Liverpool, on hire, 2000; to Freightliner, Dagenham Dock, on hire, by 12th November 2001; to Wabtec, Doncaster, for repairs, by 9th August 2002; to Grant Rail, March, on hire, 22nd April 2003; to Silverlink, Bletchley, on hire, 31st March 2004; to Reading PW Yard, by 24th November 2004; re-sold to RMS Locotec Ltd, Dewsbury, early May 2005; to PD Ports, Tees Dock, on hire, by 14th May 2005; to RMS Locotec Ltd, Dewsbury, 1st September 2005; to PD Ports, Tees Dock, on hire, 16th September 2005; to RMS Locotec Ltd, 25th November 2005; to Network Rail, Whitemoor Yard, March, on hire, 31st March 2006; to Bombardier Transportation, Ilford Depot, on hire, April 2008; to Network Rail, Whitemoor Yard, March, on hire, by July 2008; to Wabtec, Doncaster, October 2008; seen in black livery with number 08754, Wabtec, 26th February 2012; to Wabtec, Kilmarnock, April 2012; to Tyseley Diesel Depot, Birmingham, for tyre turning, 21st September 2012; to Allelys, Studley, for storage, September 2012; noted in blue livery with number 08754, Allelys, 24th October 2012; to North Norfolk Railway, Sheringham, 19th December 2012; to Crown Point Depot, Norwich, on hire, 20th December 2012; noted in blue livery with number 08754, Crown Point Depot, 20th April 2016; to Mid-Norfolk Railway, Dereham, 31st August 2017; noted in blue livery with number 08754, Mid-Norfolk Railway, September 2018; to RMS Locotec Ltd, Wolsingham Depot, 4th October 2018; to ScotRail, Inverness, on hire, 20th August 2019; to RMS Locotec Ltd, Wolsingham Depot, 4th June 2020; noted in RMS dark blue livery with grey cab and number 08754, Wolsingham, 22nd October 2020; to Gemini Rail, Wolverton Works, on hire, 25th November 2020; to Eastern Rail Services, carriage sidings, Runham, Great Yarmouth, on hire, 28th November 2023; to Independent Rail Engineering Ltd, Chesterfield, for modifications, 5th April 2024; to ScotRail, Haymarket Depot, Edinburgh, for marshalling of HST sets, 30th May 2024.

D3924   Horwich   1961   86A   1/03   P   HO59
08756   new 27th January 1961; withdrawn, January 2003; sold to RT Rail, Crewe; despatched from Canton Depot, Cardiff, 9th September 2003; to RT Rail, Crewe; to Wabtec, Doncaster, for storage, September 2003; noted in grey livery with number 08756, Wabtec, May 2004; to West Yard, Doncaster, for storage, July 2005; to Wabtec, Doncaster, for overhaul, 2006; to Brunner Mond, Northwich, on hire, September 2006; to RMS Locotec Ltd, on hire, 1st November 2006; used on Stirling to Alloa line contract; to Elsecar Steam Railway, near Barnsley, 14th November 2006; noted in grey livery with number 08756, Elsecar, April 2007; RT Rail was acquired by RMS Locotec Ltd, 8th November 2007; to Network Rail, Whitemoor Yard, March, on hire, 2nd October 2008; to Corus, Shotton Steelworks, on hire, 6th April 2009; noted in grey livery with number 08756, Shotton Steelworks, May 2016; to Tyseley Diesel Depot, Birmingham, for tyre turning, 28th January 2017; to Tata Steel, Shotton, 31st January 2017; to RMS Locotec Ltd, Wolsingham Depot, 19th May 2017; to Independent Rail Engineering Ltd, Chesterfield, for repaint, 10th June 2022; to Loram UK Ltd, Derby RTC, on hire, 23rd June 2022; noted in blue livery with green cab, Loram UK Ltd, 8th October 2023.

D3925   Horwich   1961   5A   ?   P   08757 / EAGLE
08757   new 1st February 1961; whilst in capital stock was hired to Mendip Rail in the period 2004 to 2008 and used at Whatley Quarry, Merehead Stone Terminal and Acton Yard; withdrawn, date not known; put on sale by tender by DB Cargo, with bids due by 5th October 2016; sold to Telford Steam Railway; despatched from DB Cargo, Crewe Electric Depot, 16th January 2017; to Telford Steam Railway, Shropshire; noted in RES red and black livery with number 08757, Telford Steam Railway, March 2018; to Railway Support Services, Wishaw, on hire, 5th June 2019 and moved direct to GBRf, South Terminal, Port of Felixstowe, on contract sub-hire, 19th June 2019; to GBRf, Dagenham, on sub-hire, 15th July 2019; to Telford Steam Railway, 14th September 2019; to Railway Support Services, Wishaw, on hire, about 14th May 2022; noted in grey and red livery with number 08757, Railway Support Services, Wishaw, 16th May 2022; to Hams Hall Rail Freight Terminal, Coleshill, on sub-hire, late May 2022; noted on a low-loader on the A449, 3rd November 2022; to Telford Steam Railway, 3rd November 2022.

D3930   Horwich   1961   60A   8/99   P   08762 / HO67
08762   new 6th March 1961; withdrawn, August 1999; sold to RT Rail; to RT Rail, Crewe, August 1999; to Wabtec, Doncaster, September 1999; to Freightliner, Dagenham Dock, on hire, by 26th April 2000; to Port of Felixstowe, on hire, after 13th December 2001; to Channel Tunnel Rail Link, Beechbrook Farm, near Ashford, on hire, by February 2002; to Europort, Wakefield, on hire, May 2002; to Wabtec, Doncaster, for repairs, by 10th June 2002; to Freightliner, Dagenham, on hire, about September 2002; to Imerys Minerals Ltd, Quidhampton, on hire, 16th January 2003; to Wabtec, Doncaster, for repairs, 30th March 2004; to Midland Road Depot, Leeds, on hire, by 28th July 2004; fitted with auto couplers, 2004; to Brunner Mond, Winnington Works, Northwich, on hire, about September 2004; suffered collision damage, early 2006; to Wabtec, Doncaster, for repairs, 19th April 2006; returned to Brunner Mond, Northwich, on hire; returned to Wabtec, Doncaster; RT Rail was acquired by RMS Locotec Ltd, 8th November 2007; noted in black livery with number 08762, Wabtec, 14th December 2007; to Cemex Rail Products, Washwood Heath, on hire, early August 2011; noted in black livery with number 08762, Washwood Heath, February 2012; to Network Rail, Derby, on hire, 31st July 2015; noted in black livery with number 08762, Derby, 8th April 2016; to RMS Locotec Ltd, Wolsingham Depot, 17th May 2017; noted on a low-loader on the M6 (junction 23), 27th February 2019; to Locomotive Storage, Crewe, 28th February 2019; noted in RMS dark blue livery with grey cab and numbers

08762 and HO67, Crewe, 13th March 2019; to RMS Locotec Ltd, Wolsingham Depot, 9th September 2019; to Northern Rail, Heaton Depot, Newcastle upon Tyne, on hire, 16th March 2020; noted at Heaton Depot, 17th March 2020; to Eastern Rail Services, Vauxhall Carriage Sidings, Runham, Great Yarmouth, on hire, 16th June 2020; to Gemini Rail, Wolverton Works, on hire, 15th February 2021; left Wolverton Works, 7th September 2021; noted on a low-loader, M1 northbound, 7th September 2021; arrived at Eastern Rail Services, Vauxhall Carriage Sidings, Great Yarmouth, on hire, 9th September 2021; noted in RMS green livery, Great Yarmouth, 20th September 2021; noted on a low-loader, Blyth Services, 25th March 2022; to Eco-Power Environmental, Rossington, on hire, 25th March 2022; to Castle Cement Works, Ketton, on hire, 1st March 2023; noted in RMS blue livery with grey cab and number 08762, Ketton, 5th June 2023; to Light Rail Innovation Centre, Dudley, by road, about 12th October 2023; to Alstom Transport, Ilford Depot, on hire, 1st December 2023.

D3932   **Horwich**   **1961**   **64B**   **5/88**   **P**   **08764**
08764   new 17th March 1961; withdrawn due to collision damage at Haymarket, 6th May 1988; sold to RFS (Engineering) Ltd, Doncaster, and moved May 1988; noted at RFS (Engineering) Ltd, Doncaster, under repair, 17th July 1988 and 22nd January 1989; given number 003 and name FLORENCE; to ARC Ltd, Machen Quarry, near Newport, on hire, 26th September 1990; returned to RFS (Engineering) Ltd, 31st October 1990; to ABB, York, on hire, 1st November 1990; to Flixborough Wharf Ltd, Flixborough, Scunthorpe, on hire, 7th December 1990; to RFS (Engineering) Ltd, Doncaster, 8th February 1993; to Trans-Manche Link, Channel Tunnel (given number 96), on hire, 10th February 1993; to RFS (Engineering) Ltd, Doncaster, June 1993; to Sheerness Steel Co Ltd, Sheerness, Kent, on hire, by 26th June 1993; returned to RFS (Engineering) Ltd, 1993; to Hartlepool Power Station, on hire, October 1993; returned to RFS (Engineering) Ltd, Doncaster, by 10th April 1994; to Flixborough Wharf Ltd, Flixborough, Scunthorpe, on hire, June 1994; returned to RFS (Engineering) Ltd, Doncaster, by 5th December 1996; to Transfesa UK Ltd, Riverside Terminal, Tilbury, on hire, 16th August 1997; to RFS (Engineering) Ltd, Doncaster, for repairs, by February 1998; to Transfesa Rail Terminal, Tilbury, on hire, by 13th January 1999; sold to Transfesa, 1999; re-sold to Alstom Transport, late 2012; to Alstom Transport, Stonebridge Park Heavy Repair Depot, London, 30th January 2013; noted in blue livery with number 08764, Wembley Depot, March 2015; to Alstom Transport, Polmadie Depot, Glasgow, 15th October 2015; noted in blue livery with number 08764, Polmadie Depot, 16th February 2016; noted in green livery with number 08764, Polmadie Depot, 6th June 2016.

D3933   **Horwich**   **1961**   **81A**   **5/08**   **P**   **NPT**
08765   new 23rd March 1961; withdrawn, May 2008; sold to Harry Needle Railroad Company; despatched from DB Schenker, Eastleigh, date not known; to Harry Needle Railroad Company; to Boden Rail Engineering, Washwood Heath, 28th June 2011; to Nemesis Rail, Burton upon Trent, for storage, 24th November 2011; noted in red and yellow livery with number 08765, Nemesis Rail, February 2012; to Rail Restorations North East, Shildon, for repairs, 9th February 2015; to Harry Needle Railroad Company, Barrow Hill Engine Shed, Staveley, 8th May 2015; seen in orange livery with no number, Barrow Hill Engine Shed, 25th July 2015; noted at BHES, 16th December 2023.

D3935   **Horwich**   **1961**   **30A**   **1/94**   **P**   **D3935**
08767   new 6th April 1961; noted at Colchester Depot, 15th January 1994; withdrawn, 31st January 1994; sold to Bird, scrap merchant, and stored at Ipswich Wagon Repair Depot, March 1994; re-sold to North Norfolk Railway, Sheringham; despatched from

Ipswich, 16th August 1994; seen in blue livery with number D3935, Sheringham, 2nd March 1997; noted in green livery with number D3935, Sheringham, August 2019.

D3937   Derby   1960   87E   5/89   P   D3937/ GLADYS
08769   new (order D2277) 4th March 1960; withdrawn, 23rd May 1989; despatched from 87E Landore Depot, 15th April 1990; at Worcester Depot from 15th April 1990; moved to MoD Long Marston, Worcestershire (hauled by 47314), 20th April 1990; seen in BR blue livery with number 08769 and two-way arrow logo, Long Marston, 24th July 1990; to The Fire Service College, Moreton in Marsh, by road, 12th November 1991; used as fire training locomotive; re-sold to a private individual, May 1999; moved to Dean Forest Railway, Lydney, 2nd March 2000; underwent restoration including rewire and naming GLADYS; to Severn Valley Railway, Bridgnorth, on loan, 12th May 2003; to Dean Forest Railway, Lydney, March 2010; after extended repairs returned to service, February 2013; noted in green livery with number D3937 and name GLADYS, Dean Forest Railway, 30th May 2016.

D3940   Derby   1960   30A   1/94   P   D3940
08772   new (order D2277) 16th March 1960; withdrawn, 14th January 1994; sold to East Anglian Railway Museum; noted at BR Colchester Depot, 5th March 1994; despatched from Colchester Depot, March 1994; to East Anglian Railway Museum, Wakes Colne, Essex; noted at East Anglian Railway Museum, 27th March 1994; re-sold to North Norfolk Railway, Sheringham, 18th September 2001; noted in green livery with number D3940, North Norfolk Railway, 11th June 2016.

D3941   Derby   1960   16A   3/94   P   08773
08773   new (order D2277) 19th March 1960; withdrawn, 29th March 1994; sold to Mike Darnall, Newton Heath, Manchester, and moved 25th July 2000; re-sold to Embsay & Bolton Abbey Railway, and moved February 2006; noted in blue livery with number 08773, Embsay & Bolton Abbey Railway, 24th March 2007; seen in green livery with number D3941, Embsay & Bolton Abbey Railway, 21st August 2021; repainted in blue livery with number 08773, spring 2022.

D3942   Derby   1960   51L   9/88   P   08774
08774   new (order D2277) 26th March 1960; to A.V. Dawson Ltd, Middlesbrough, on hire, 8th July 1988; withdrawn, 20th September 1988; sold to A.V. Dawson Ltd, Middlesbrough, and stayed at their site; seen in red livery with no number, A.V. Dawson Ltd, 13th April 1989 and 30th November 1991; to Cobra Railfreight, Middlesbrough, on hire, 4th January 1994; returned to Dawson, 31st May 1994; to Cobra Railfreight, Middlesbrough, on hire, October 1998; seen at Cobra, 7th November 1998; returned to A.V. Dawson Ltd, Middlesbrough, late 1998; to Wabtec, Doncaster, for overhaul, late 2001; returned to A.V. Dawson Ltd, 22nd February 2002; noted in red livery with white roof and number 08774, A.V. Dawson Ltd, 25th June 2017; to Railway Support Services, Wishaw, for repairs, 26th October 2017; returned to A.V. Dawson Ltd, 13th July 2018; noted in red livery with number 08774, Dawson, 12th August 2023.

D3948   Derby   1960   87E   1/01   P   D3948 / ZIPPY
08780   new (order D2277) 18th April 1960; withdrawn, January 2001; sold to Cotswold Rail; despatched from Landore Depot, about June 2001; to Cotswold Rail, Moreton in Marsh; to East Lancashire Railway, Bury, November 2002; to Transplant Ltd (London Underground maintenance), West Ruislip, on hire, 26th April 2005; to Midland Road Depot, Leeds, 15th September 2005; to Wabtec, Doncaster, 27th March 2006; to L&NWR Ltd, Carriage Works, Crewe, 1st June 2006; noted in blue livery with number 08780, Crewe,

February 2007; re-sold to Locomotive Services Ltd; moved to Locomotive Services, Southall Depot, London, 21st December 2007; noted in blue livery with number 08780 and name FRED, Southall Depot, October 2012; to Locomotive Services, Crewe Diesel Depot, 11th June 2019; noted in blue livery with number 08780, Crewe, 7th August 2019; repainted in BR green livery with number D3948 and name ZIPPY, July 2020.

D3950   Derby   1960   36A   ?   P   08782
08782   new (order D2277) 8th April 1960; noted in grey livery with number 08782 and name CASTLETON WORKS, Doncaster, 20th January 2013; withdrawn, date not known; put on sale by tender by DB Cargo, with bids due by 5th October 2016; sold to Harry Needle Railroad Company; despatched from Doncaster Depot; to Barrow Hill Engine Shed, Staveley, June 2017; seen in grey livery with number 08782, Barrow Hill Engine Shed, 12th March 2022; noted at BHES, 16th December 2023.

D3951   Derby   1960   16A   4/11   P   08783
08783   new (order D2277) 9th April 1960; stored, May 2008; withdrawn, April 2011; sold to European Metal Recycling, Kingsbury, Warwickshire; despatched from DB Schenker, Toton Depot, 4th August 2011; noted in EWS red and yellow livery with number 08783, Kingsbury, February 2012; noted at Kingsbury, 2nd August 2016; to Railway Support Services, Wishaw, for repairs, about 11th February 2023; seen in faded EWS red and yellow livery with number 08783, Wishaw, 23rd September 2023.

D3952   Derby   1960   16A   ?   P   08784
08784   new (order D2277) 23rd April 1960; withdrawn, date not known; put on sale by tender by DB Cargo, with bids due by 5th October 2016; sold to Great Central Railway (Nottingham); despatched from DB Cargo, Toton Depot, 14th December 2016; to Great Central Railway (Nottingham), Ruddington, Nottingham; noted in EWS red and yellow livery with number 08784, Ruddington, 27th December 2016; to Railway Support Services, Wishaw, for repairs, 24th July 2019; to Chasewater Railway, for testing, 3rd March 2020; to Electro-Motive Diesel, Longport, on hire, 12th March 2020; to Great Central Railway (Nottingham), Ruddington, 22nd February 2022; noted in grey livery with number 08784, Ruddington, 28th May 2023.

D3953   Derby   1960   86A   3/89   P   08785
08785   new (order D2277) 23rd April 1960; suffered collision damage, Alexandra Dock Junction, Newport, early March 1989; withdrawn, 13th March 1989; sold to RFS (Engineering) Ltd, Kilnhurst, and moved on 25th September 1990; seen being overhauled prior to becoming a hire locomotive, RFS Kilnhurst, 23rd March 1991; to Grovehurst Paper Ltd, Sittingbourne, Kent, on hire, 4th June 1991; to Channel Tunnel Rail Link, on hire, 11th September 1992; to RFS (Engineering) Ltd, Doncaster, July 1993; seen at RFS Doncaster, 6th August 1994; to BASF Chemicals Ltd, Seal Sands, on hire, 9th August 1994; to RFS (Engineering) Ltd, Doncaster, 25th November 1994; seen in grey livery with number 004 and CLARENCE nameplate, RFS Doncaster, 13th May 1995; re-sold to Freightliner, and initially allocated to Crewe, 14th July 1997.

D3954   Derby   1960   16A   12/08 P   08786
08786   new (order D2277) 29th April 1960; withdrawn, December 2008; moved to Doncaster Depot, for storage; sold to Harry Needle Railroad Company; despatched from DB Schenker, Doncaster Depot, 24th January 2011; to Barrow Hill Engine Shed, Staveley; seen in grey livery with number 08786, Barrow Hill Engine Shed, 4th May 2011; noted at BHES, 16th December 2023.

D3955   Derby   1960   86A   2/91   P   08296

08787   new (order D2277) 23rd April 1960; withdrawn, 14th February 1991; moved to BRML Crewe Works, for storage; acquired by ABB Transportation, Crewe Works, 1992; overhauled, 1992; given number 001, by October 1992: repainted light grey, white and dark grey, with number 001, January 1995; to ABB British Wheelset Ltd, Trafford Park, Manchester, January 1995; to Adtranz, Crewe, August 1996; re-sold to Foster Yeoman Quarries Ltd, February 2000; to East Somerset Railway, Cranmore, for storage, February 2000; to Foster Yeoman Quarries Ltd, Merehead Stone Terminal, about March 2000; to Hanson Aggregates, Whatley Quarry, by 1st October 2000; given identity 08296 in late 2000; to Foster Yeoman Quarries Ltd, Isle of Grain Stone Terminal, 28th June 2002; to Foster Yeoman Quarries Ltd, Whatley Quarry, 29th July 2002; to Merehead Stone Terminal, by 28th June 2003; to Whatley Quarry, by 17th April 2004; to Machen Quarry, near Newport, by 27th May 2004; to Whatley Quarry, by 12th June 2004; to Acton Rail Terminal, late 2004; to Isle of Grain Stone Terminal, by February 2005; to Merehead Stone Terminal, by 11th July 2005; to Hanson Quarry Products, Machen Quarry, near Newport, 6th April 2006; noted in blue livery with number 08296, Machen Quarry, February 2016; noted covered in graffiti, Machen Quarry, 22nd August 2017 and 24th January 2019; to Whatley Quarry, 27th March 2019; noted at Whatley Quarry, 3rd April 2019; noted partially repainted in blue livery, Whatley Quarry, 20th April 2019; moved by low-loader to Merehead Stone Terminal, 6th October 2021; to LH Group, Barton under Needwood, 23rd December 2021; noted in MRL blue livery with number 08296, Barton under Needwood, 2nd July 2023.

D3956   Derby   1960   16C   1/94   P   08788 / HO68

08788   new (order D2277) 7th May 1960; withdrawn, 14th January 1994; noted still at Derby Depot, September 1995; sold to RT Rail, Crewe; despatched from Derby Depot to Great Central Railway, Loughborough, for storage, 1996; to RT Rail, Crewe, 24th February 1999; to ScotRail, Inverness, on hire, 1st September 1999; noted in black livery with number 08788, Inverness, November 2004; to Manchester Ship Canal Company, Barton Dock, on hire, 21st April 2005; to Wabtec, Doncaster, for repairs, 2nd June 2005; to ScotRail, Inverness, on hire, 16th November 2005; RT Rail was acquired by RMS Locotec Ltd, 8th November 2007; to Craigentinny Depot, Edinburgh, for tyre turning, 19th December 2012; to ScotRail, Inverness, on hire, 20th December 2012; to RMS Locotec Ltd, Wolsingham Depot, 17th November 2015; to Tata Steel, Shotton, on hire, 17th August 2016; noted on a low-loader, on M62 near Huddersfield, 10th July 2019; to PD Ports, Tees Dock, on hire, 10th July 2019; noted in RMS blue/grey livery with numbers HO68 and 08788, PD Ports, 31st December 2023.

D3958   Derby   1960   ?   ?   P   08790

08790   new (order D2277) 14th May 1960; withdrawn, date not known; acquired by Alstom Transport, Longsight Depot, Manchester, 2007; to Alstom Transport, Oxley Depot, Wolverhampton, 26th April 2007; noted in blue livery with number 08790, Oxley Depot, September 2007; to Arlington Fleet Services Ltd, Eastleigh Works, for overhaul, 18th June 2014; to Alstom Transport, Oxley Depot, Wolverhampton, 11th February 2015; to Arlington Fleet Services Ltd, Eastleigh Works, for repairs, 14th December 2016; noted having traction motor replaced, Eastleigh Works, 11th May 2017; noted in blue livery with black roof and number 08790, Eastleigh Works, 16th September 2017; noted in transit on a low-loader, 18th September 2017; to Alstom Transport, Longsight Depot, about 18th September 2017; noted at Longsight Depot, 13th May 2018; to Alstom Transport, Edge Hill Depot, Liverpool, mid-May 2018; noted at Edge Hill Depot, 22nd May 2018; to Alstom Transport, Longsight Depot, 22nd June 2021.

D3963   **Derby**   **1960**   **87E**   **?**   **P**   **08795**
08795   new (order D2277) 21st May 1960; repainted in black livery at Landore, 2013; withdrawn, date not known; despatched on Alleys low-loader from Landore Depot, Friday 22nd March 2019; last locomotive to leave the site when the depot closed; to Llanelli & Mynydd Mawr Railway, Cynheidre; to Chrysalis Rail, Landore Depot, on hire as depot pilot, 29th March 2021.

D3966   **Derby**   **1960**   **16A**   **11/09**   **P**   **08798**
08798   new (order D2277) 28th May 1960; withdrawn, November 2009; sold to European Metal Recycling, Kingsbury, and moved 15th August 2011; noted in EWS red and yellow livery with number 08798, European Metal Recycling, Kingsbury, February 2012; re-sold to Harry Needle Railroad Company, about June 2012; to European Metal Recycling, Attercliffe, Sheffield, for storage, early July 2012; seen in EWS red and yellow livery with number 08798, European Metal Recycling, Attercliffe, 31st July 2012; to Barrow Hill Engine Shed, Staveley, 14th November 2019; seen in EWS red and yellow livery with number 08798, HNRC, Barrow Hill Engine Shed, 12th March 2022; noted at BHES, 16th December 2023.

D3967   **Derby**   **1960**   **82D**   **?**   **P**   **08799**
08799   new (order D2277) 4th June 1960; noted in EWS red and yellow livery with number 08799 and name FRED, Westbury, 21st July 2015; withdrawn, date not known; put on sale by tender by DB Cargo, with bids due by 5th October 2016; sold to Harry Needle Railroad Company; noted in Westbury yard, 4th March 2017; despatched from Westbury; to East Kent Railway, Shepherdswell, for storage, 8th June 2017; noted in EWS red and yellow livery with number 08799, East Kent Railway, 11th June 2017 and April 2019; to Battlefield Line, Shackerstone (for overhaul by HTRS Ltd), 28th May 2021; to Harry Needle Railroad Company, Barrow Hill Engine Shed, 4th March 2022; outshopped in HNRC orange livery, with nameplates IAN GODDARD 1938-2016 (these previously carried on a Class 20), late January 2023; to Harry Needle Railroad Company, Worksop Depot, 14th February 2023; noted in orange livery with number 08799, Worksop, 24th July 2024.

D3969   **Derby**   **1960**   **86A**   **6/00**   **F**   **801**
08801   new (order D2277) 11th June 1960; withdrawn, 3rd June 2000; noted in BR blue livery with number 801, Canton Depot, Cardiff, 25th August 2003; sold to RT Rail, Crewe; despatched from Canton Depot; to Wabtec, Doncaster, 8th September 2003; used for spares; remains to C.F. Booth Ltd, Rotherham, for scrap, February 2004; noted in BR blue livery with number 801, C.F. Booth Ltd, 14th February 2004; scrapped, 24th March 2004.

D3970   **Derby**   **1960**   **16A**   **?**   **F**   **08802**
08802   new (order D2277) 11th June 1960; withdrawn, date not known; sold to Harry Needle Railroad Company; despatched from DB Schenker, Toton Depot, 18th February 2016; to Railway Support Services, Wishaw, for storage; noted in EWS red and yellow livery with number 08802, Railway Support Services, 23rd February 2016 and 8th August 2020; to Harry Needle Railroad Company, Worksop Depot, 26th January 2022; noted in faded EWS livery, no engine, Worksop, 3rd May 2023; to European Metal Recycling, Attercliffe, Sheffield, for scrap, 4th May 2023; noted at EMR, 16th October 2023; scrapped (wheelsets returned to HNRC), 6th November 2023.

D3972   **Derby**   **1960**   **5A**   **?**   **P**   **08804**
08804   new (order D2277) 18th June 1960; withdrawn, date not known; noted in EWS red and yellow livery with number 08804, Crewe Electric Depot, 3rd December 2015; put

on sale by tender by DB Cargo, with bids due by 5th October 2016; sold to Harry Needle Railroad Company; noted at Crewe, 25th February 2017; despatched from Crewe Electric Depot; to Moveright International, Wishaw, Warwickshire, 1st September 2017; to East Kent Railway, Shepherdswell, for storage, 4th September 2017; to Battlefield Line, Shackerstone, by road, 30th April 2023.

D3975   Derby   1960   66B   2/07   F   08807

08807   new (order D2277) 25th June 1960; noted in blue livery with number 08807, Motherwell, 30th August 2004; withdrawn, 28th February 2007; sold to T.J. Thomson & Son Ltd; despatched from Motherwell Depot; to T.J. Thomson & Son Ltd, Stockton-on-Tees, 26th April 2007; re-sold to A.V. Dawson Ltd, Middlesbrough; to EWS Thornaby Depot, for tyre turning, 30th October 2007; to A.V. Dawson Ltd, Middlesbrough, week-ending 23rd November 2007; noted in blue livery with number 08807, A.V. Dawson Ltd, March 2009; noted with engine removed, May 2016; remains to C.J. Prosser Ltd, Cargo Fleet, for scrap, 7th January 2017; scrapped January 2017.

D3977   Derby   1960   8J   12/93   P   08809 / 24

08809   new (order D2277) 9th July 1960; withdrawn, 1st December 1993; sold to Otis Euro Transrail Ltd, Salford, Manchester, and moved December 1993; to Flixborough Wharf Ltd, Flixborough, Scunthorpe, on hire, week-ending 12th January 1996; returned to Otis Euro Transrail Ltd; re-sold to Harry Needle Railroad Company, December 1999; to Fragonset, Derby, for certification, about March 2000; to Barrow Hill Engine Shed, Staveley, 13th June 2000; to Freightliner, Coatbridge, on hire, 29th June 2000; to Motherwell Depot, for repairs, September 2001; returned to Freightliner, Coatbridge, on hire; to Barrow Hill Engine Shed, Staveley, 24th May 2002; re-sold to Cotswold Rail, Moreton in Marsh, and moved 29th June 2002; to Anglia Railways, Crown Point Depot, Norwich, on hire, 24th October 2002; to Ilford Depot, for tyre turning, 10th December 2003; returned to Crown Point Depot, Norwich, 17th December 2003; to Brush, Loughborough, for overhaul, 17th March 2005; to Allelys Ltd, Studley, Warwickshire, for storage, 7th February 2006; re-sold to RMS Locotec Ltd, Wakefield, October 2007; to Corus, Shotton, for storage, 26th October 2009; to Boden Rail Engineering, Washwood Heath, 29th November 2010; to RMS Locotec Ltd, Wolsingham Depot, 12th November 2013; to PD Ports, Tees Dock, on hire, 27th August 2014; to Tata Steel, Shotton, on hire, June 2017; noted in RMS blue livery with grey cab and numbers 08809 and 24, Shotton, November 2017; to Hanson Cement, Ketton Cement Works, on hire, week-ending 11th May 2019; to Toton Depot, for tyre turning, 19th March 2021; to Hanson Cement, Ketton Cement Works, on hire, 1st April 2021.

D3978   Derby   1960   NC   7/01   P   08810

08810   new (order D2277) 9th July 1960; last worked, November 2000; noted in blue 'Anglia' livery, in store at Crown Point Depot, Norwich, 26th April 2001; withdrawn, July 2001; sold to Cotswold Rail, Moreton in Marsh, and moved about July 2001; to Brush, Loughborough, for overhaul, by road, 15th August 2001; to Anglia Railways, Crown Point Depot, Norwich, on hire, 18th September 2001; to Railway Age, Crewe, 10th December 2003; to London & North Western Railway Company, Carriage Works, Crewe, about March 2004; noted in Anglia blue livery with number 08810, Crewe, 14th November 2004; to Arlington Fleet Services Ltd, Eastleigh Works, December 2011; to Heaton Depot, Newcastle upon Tyne, 25th May 2012; re-sold to London & North Western Railway Company, Traction and Rolling Stock Depot, Eastleigh, and moved 9th December 2014.

D3981   Derby   1960   51L   6/05   F   08813
08813   new (order D2277) 25th July 1960; withdrawn, 30th June 2005; noted at EWS Thornaby Depot, 28th April 2006; sold to Harry Needle Railroad Company; despatched from Thornaby Depot, 26th September 2006; to Harry Needle Railroad Company, Long Marston; noted in grey livery with number 08813, no centre wheels, Long Marston, 12th September 2010; used for spares, January 2011; remains to T.J. Thomson & Son Ltd, Stockton-on-Tees, for scrap, 4th February 2011; scrapped, 9th February 2011.

D3984   Derby   1960   51L   2/86   F   HO25
08816   new (order D2277) 5th August 1960; withdrawn, 1st February 1986; sold to Cobra Railfreight Ltd, Middlesbrough, and moved on 15th February 1986; seen in lime green livery with no number, with BRB NC23 registration plate, Cobra, 12th June 1988 and 4th May 1989; to Thornaby Depot, for repairs, 27th May 1989; returned to Cobra; seen repainted in grey livery with no number, Cobra, 24th November 1990; seen repainted in orange livery with no number, Cobra, 27th September 1994; re-sold to Harry Needle Railroad Company, 1995; to South Yorkshire Railway Preservation Society, Meadowhall, Sheffield, 1995; to Johnson Ltd, Widdrington Disposal Point, Northumberland, on hire, 19th August 1995; noted at Widdrington, 4th September 1996; to RFS (Engineering) Ltd, Doncaster, 10th July 1998; scrapped, August 1999.

D3986   Derby   1960   5A   4/97   P   08818 / 4 / MOLLY
08818   new (order D2277) 13th August 1960; withdrawn, 2nd April 1997; sold to Harry Needle Railroad Company, April 1997; to Railway Age, Crewe, September 1997; to London & North Western Railway Company, Carriage Works, Crewe, on hire, by June 1998; to Port of Felixstowe, on hire, 13th August 1999; to Barrow Hill Engine Shed, Staveley, 3rd November 1999; to Freightliner, Basford Hall, Crewe, on hire, 10th June 2000; to Barrow Hill Engine Shed, Staveley, 12th April 2001; to Freightliner, Stourton, Leeds, on hire, 24th January 2002; to Battlefield Line, Shackerstone, 21st August 2002; to Freightliner, Coatbridge, on hire, 4th November 2002; to Battlefield Line, Shackerstone, 4th December 2002; to Freightliner, Calvert Landfill Site, on hire, 12th May 2003; to MoD Bicester, June 2004; to Severn Valley Railway, Bridgnorth, on hire, 21st September 2004; to Network Rail, Whitemoor Yard, March, on hire, about October 2004; to Harry Needle Railroad Company, Long Marston, 12th October 2005; to Barrow Hill Engine Shed, Staveley, 21st March 2006; noted having a wheel set change, BHES, 2nd February 2007; noted in yellow and grey livery with number 08818 and name MOLLY, BHES, March 2007; to Flixborough Wharf Ltd, Flixborough, Scunthorpe, on hire, 20th May 2007; seen in orange livery with number 08818 and name MOLLY, Flixborough, 15th March 2011; to GBRf, Trafford Park, Manchester, on hire, 30th May 2014; to Celsa, Cardiff, on hire, 16th July 2015; noted in HNRC orange livery with number 08818, Celsa, 14th February 2016; to Barrow Hill Engine Shed, Staveley, 10th January 2017; seen with number 08818, Barrow Hill Engine Shed, 14th January 2017; to Midland Road Depot, Leeds, for tyre turning, 27th January 2017; seen with number 08818 and name MOLLY, Midland Road Depot, 1st February 2017; to Barrow Hill Engine Shed, Staveley, 3rd February 2017; noted repainted in GBRf blue livery with number 4, Barrow Hill Engine Shed, 24th June 2017; to Celsa, Cardiff, on hire, 26th June 2017; to Barrow Hill Engine Shed, Staveley, July 2017; seen in GBRf blue livery, with numbers 08818 and 4, and name MOLLY, Barrow Hill Engine Shed, 18th July 2017; to GBRf, c/o Ford Motor Co Ltd, Speke, Liverpool, on hire, 19th July 2017; noted at Speke, 16th December 2019; to Harry Needle Railroad Company, Worksop Depot, 26th February 2020; seen at Barrow Hill Engine Shed, 14th October 2023; returned to Worksop Depot by 28th October 2023.

D3987   **Derby**   **1960**   **86A**   **9/99**   **F**   **08819**

08819   new (order D2277) 20th August 1960; withdrawn, 17th September 1999; sold to RT Rail, Crewe, September 2003; despatched from Canton Depot, Cardiff; to Wabtec, Doncaster, for storage, about November 2003; to West Yard, Doncaster, for storage, by July 2005; noted in grey livery with number 08819, West Yard, Doncaster, 27th March 2006 and 19th November 2006; to C.F. Booth Ltd, Rotherham, for scrap, 21st May 2008; noted at C.F. Booth Ltd, 30th May 2008; scrapped, August 2008.

D3991   **Derby**   **1960**   **ZF**   **6/95**   **P**   **08823 / KEVLA**

08823   new (order D2277) 9th September 1960; noted in BR blue livery with numbers 08823 and RSO78, Doncaster Works, 10th July 1994; withdrawn, June 1995; to ABB (Customer Support) Ltd, Doncaster Works, June 1995; locomotive included in sale when BRML works privatised; sold to Churnet Valley Railway, Cheddleton, Staffordshire, November 2000; moved to Cheddleton, December 2000; re-sold to LH Group, Barton under Needwood, and moved 3rd August 2007; noted in green livery with number D3991, LH Group, August 2007; to Manchester Ship Canal Company, Trafford Park, Manchester, on hire, 24th March 2008; to LH Group, Barton under Needwood, for repairs, 22nd May 2009; to Sheerness Steel Co Ltd, Sheerness, Kent, on hire, 8th June 2010; to LH Group, Barton under Needwood, for repairs, 13th August 2010; to Sheerness Steel Co Ltd, Sheerness, Kent, on hire, 20th August 2010; to LH Group, Barton under Needwood, 15th February 2012; to Daventry International Rail Freight Terminal, on hire, 27th February 2012; noted in blue, green and yellow livery with number 08823 and name LIBBIE, Daventry International Rail Freight Terminal, March 2012; to LH Group, Barton under Needwood, for repairs, 20th February 2015; returned to Daventry International Rail Freight Terminal, on hire, 6th March 2015; to LH Group, Barton under Needwood, 19th October 2017; noted in blue and orange livery with number 08823, LH Group, October 2018; to Midland Road Depot, Leeds, for tyre turning, 13th December 2018; to LH Group, Barton under Needwood, 17th December 2018; to Tata Steel, Shotton, on hire, 18th January 2019; re-sold to Hunslet Ltd (Ed Murray Group), 1st May 2021; noted in Hunslet royal blue and orange livery with number 08823 and nameplate KEVLA, Shotton Steelworks, 1st November 2022.

D3992   **Derby**   **1960**   **5A**   **?**   **P**   **IEMD 01**

08824   new (order D2277) 22nd September 1960; withdrawn, date not known; put on sale by tender by DB Schenker, with bids due by 22nd September 2015; sold to Harry Needle Railroad Company; despatched from DB Schenker, Crewe Depot, 4th December 2015; to Barrow Hill Engine Shed, Staveley; seen in black livery with number IEMD 01, HNRC, Barrow Hill Engine Shed, Staveley, 19th December 2015 and 12th March 2022.

D3993   **Derby**   **1960**   **81A**   **6/05**   **P**   **08825 / 97808**

08825   new (order D2277) 22nd September 1960; withdrawn, 30th June 2005; sold to Battlefield Line; despatched from EWS Springs Branch Depot, Wigan, 7th October 2005; to Battlefield Line, Shackerstone; noted in blue livery with number 08825, Battlefield Line, April 2007; received various parts salvaged from 08576; re-sold to Chinnor & Princes Risborough Railway, Oxfordshire, and moved 14th May 2013; noted in blue livery with number 08825, Chinnor & Princes Risborough Railway, April 2014; to Railway Support Services, Wishaw, for repairs, 20th May 2022; seen in Network South East livery with number 97808, Railway Support Services, Wishaw, 9th July 2022; to GBRf, Bescot Yard, on hire, 25th July 2022; noted in NSE livery with number 08825, Bescot Yard, 27th November 2022; to Chinnor & Princes Risborough Railway, 19th April 2023.

D3994   Derby   1960   66B   3/96   F   08826

08826   new (order D2277) 1st October 1960; noted in BR blue livery with number 08826, in store, Motherwell Depot, 29th October 1995; withdrawn, 7th March 1996; sold to T.J. Thomson & Son Ltd; despatched from EWS Motherwell Depot, July 2002; noted in blue livery with number 08826, T.J. Thomson & Son Ltd, Stockton-on-Tees, 1st September 2003; re-sold (as a source of spares) to Foster Yeoman Quarries Ltd; to East Somerset Railway, Cranmore, for storage, 8th October 2003; to Merehead Stone Terminal, 9th June 2004; to Whatley Quarry, by 20th February 2005; to Merehead Stone Terminal, by 5th June 2005; noted in blue livery with number 08826, Merehead Stone Terminal, 22nd June 2008; to Mid Hants Railway, Ropley, for storage, October 2008; noted at Ropley, 26th May 2009; to Knights Rail Services, Eastleigh Works, 21st October 2010; used for spares in the overhauls of 08032 and 08933; remains scrapped, April 2011.

D3995   Derby   1960   66B   3/00   F   08827

08827   new (order D2277) 8th October 1960; withdrawn, 17th March 2000; sold to Harry Needle Railroad Company; despatched from EWS Motherwell Depot, 2nd September 2005; to Barrow Hill Engine Shed, Staveley; noted in BR blue livery with number 08827, Barrow Hill Engine Shed, 3rd September 2005 and 17th January 2006; to Harry Needle Railroad Company, Long Marston, week-ending 28th July 2006; noted in faded BR blue livery with number 08827, Long Marston, June 2007 and 12th September 2010; used for spares, January 2011; remains to European Metal Recycling, Kingsbury, for scrap, 7th July 2011; scrapped September 2011.

D3997   Derby   1960   16A   6/93   F   08829

08829   new (order D2277) 8th October 1960; withdrawn, 16th June 1993; sold to European Metal Recycling; despatched from Toton Depot, by road low-loader, 15th August 2000; to European Metal Recycling, Kingsbury, August 2000; re-sold to Harry Needle Railroad Company; to Barrow Hill Engine Shed, Staveley, 19th March 2001; to West Coast Railway Company, Carnforth, 7th October 2002; noted in blue livery with green cab and number painted over, Carnforth, 20th May 2003; used for spares; remains scrapped on site by Harry Needle staff, February 2005.

D3998   Derby   1960   ?   ?   P   08830

08830   new (order D2277) 15th October 1960; privatisation saw it purchased by Cardiff Railways Ltd, Cathays); to East Somerset Railway, Cranmore, on hire, 2nd October 1996; to Foster Yeoman Quarries Ltd, Merehead Stone Terminal, on hire, July 1997; to East Somerset Railway, on hire, about August 1997; may have worked at Merehead for periods in this era before returning to Cardiff Railways Ltd, off hire, 18th September 1999; to L&NWR Ltd, Carriage Works, Crewe, on hire, 30th December 1999; to Crewe Works, for open day, 19th May 2000; returned to L&NWR Ltd, Crewe, late May 2000; noted in black livery with number 08830, Crewe, May 2003; displayed at open day, Crewe Works, 31st May 2003; returned to L&NWR Ltd, Crewe, June 2003; to Freightliner, Port of Felixstowe, on hire, about 12th April 2005; to Heritage Centre, Crewe, 13th July 2005; to Midland Road Depot, Leeds, on hire, by 16th August 2005; to Wabtec, Doncaster, 2007; to Heritage Centre, Crewe, 2007; to Peak Rail, Rowsley, 5th November 2015; seen in BR blue livery with number 08830, Rowsley, 18th June 2022; undergoing repairs, from June 2023.

D4002   Derby   1960   ?   ?   P   08834

08834   new (order D2277) 12th November 1960; withdrawn, date not known; to Transmanche-Link, Dolland Moor, Kent, on hire, 23rd September 1992; to BR Stratford Depot, London, 23rd November 1992; sold to RFS (Engineering) Ltd, Doncaster, and moved 27th April 1993; to GNER, Bounds Green Depot, London, on hire, 1997; returned to RFS (Engineering) Ltd, Doncaster, by 14th February 1998; to GNER, Bounds Green Depot, London, on hire, 2nd August 1999; to Wabtec, Doncaster, 28th May 2003; to Foster Yeoman Quarries Ltd, Merehead Stone Terminal, on hire, 19th January 2004; to Daventry International Rail Freight Terminal, on hire, November 2004; to Wabtec, Doncaster, 26th May 2005; to Foster Yeoman Quarries Ltd, Merehead Stone Terminal, on hire, early 2006; returned to Wabtec, Doncaster, about April 2006; re-sold to Direct Rail Services, May 2006; to DRS, Kingmoor Depot, Carlisle, 2nd October 2006; to DRS, Gresty Bridge Depot, Crewe, 23rd January 2007; noted in DRS blue livery with number 08834, Gresty Bridge, 1st November 2008; re-sold to Harry Needle Railroad Company, January 2009; to Basford Hall, Crewe, 22nd May 2009; to Serco, Old Dalby Test Centre, Leicestershire, 27th May 2009; to Harry Needle Railroad Company, Barrow Hill Engine Shed, Staveley, for overhaul, 10th March 2015; noted being repainted in HNRC orange, 19th November 2015; to GBRf, Trafford Park, Manchester, on hire, 5th January 2016; to Barrow Hill Engine Shed, Staveley, March 2016; to GBRf, Dagenham, on hire, 6th May 2016; to Northern Trains, Allerton Depot, Liverpool, on hire, 11th August 2016; noted in HNRail orange livery with number 08834, Allerton Depot, 2nd March 2017.

D4014   Horwich   1961   8J   10/89   P   003
08846   new (LOT 277) 16th May 1961; withdrawn, 13th October 1989; sold to ABB; despatched from Allerton Depot; to ABB Transportation, Litchurch Lane Works, Derby, October 1989; overhauled; noted in freshly repainted green livery, Litchurch Lane Works, 17th January 1990; during repaint it received erroneous number D4414; put into use on 5th March 1990; noted with number D4414 on 29th June 1990 and 13th October 1990; to ABB Transportation, Crewe, October 1993; to ABB Transportation, York, September 1995; noted carrying its second erroneous number, D4144, ABB York, 27th January 1996; to ABB Transportation, Crewe, for overhaul, 30th August 1996; to Adtranz, Litchurch Lane Works, Derby, 15th May 1998; to Fragonset, Derby, for overhaul, about March 2004; to Bombardier Transportation, Litchurch Lane Works, Derby, about June 2004; to Railway Support Services, Wishaw, for repairs, 9th April 2015; repairs delayed; noted in blue livery with number 003, Wishaw, 25th January 2016 and 23rd June 2017; re-sold to Railway Support Services, Wishaw, 2018; to St Philip's Marsh Depot, Bristol, for tyre turning, 21st June 2019; to Railway Support Services, Wishaw, 24th June 2019; to East Midlands Trains, Neville Hill Depot, Leeds, on hire, 15th July 2019; seen in light blue livery with number 08846, Neville Hill Depot, 4th November 2020; to GBRf, Whitemoor Yard, March, on hire, 28th May 2021; to Tyseley Locomotive Works, Birmingham, for repairs, mid-August 2022; noted at Tyseley, 7th June 2024; to GBRf, Whitemoor Yard, March, on hire, 17th June 2024.

D4015   Horwich   1961   ZG   6/95   P   08847 / LOCO 1
08847   new (LOT 277) 17th May 1961; withdrawn, June 1995; to Wessex Traincare Ltd, Eastleigh Works, June 1995; locomotive included in sale when BREL works privatised; sold to Cotswold Rail, Moreton in Marsh, and moved 18th May 2001; to Brush, Loughborough, for overhaul, 24th July 2001; to Cotswold Rail, about July 2002; to Anglia Railways, Crown Point Depot, Norwich, on hire, September 2002; to GBRf, Railport, Doncaster, on hire, November 2002; to Wabtec, Doncaster, 28th March 2003; to British Gypsum, Mountfield, East Sussex, on hire, 7th May 2003; to Anglia Railways, Crown Point Depot, Norwich, on hire, 11th September 2003; to Horton Road Depot, Gloucester, for storage, 2006; re-sold to RMS Locotec Ltd, Wakefield, October 2007; to Wabtec, Doncaster, early 2008; to Anglia Railways, Crown Point Depot, Norwich, on hire, by 26th March 2008; noted in grey livery with number 08847, Norwich, July 2011; to Allelys, Studley, Warwickshire, 20th December

2012; to Boden Rail Engineering, Washwood Heath, week-ending 28th February 2014; to Tyseley Diesel Depot, Birmingham, for tyre turning, 14th April 2014; to Boden Rail Engineering, Washwood Heath, 18th April 2014; to East Coast Trains, Bounds Green Depot, London, on hire, 3rd December 2014; to Mid-Norfolk Railway, Dereham, 2nd July 2015; to Crown Point Depot, Norwich, on hire, 4th July 2015; noted in grey livery with number 08847, Crown Point Depot, 20th April 2016 and 13th July 2017; to Mid-Norfolk Railway, Dereham, 31st August 2017; to PD Ports, Tees Dock, on hire, 14th May 2019; noted in light grey livery with numbers LOCO 1 and 08847, PD Ports, 31st December 2023.

D4018   Horwich   1961   81D   12/92 P 08850

08850   new (LOT 277) 1st June 1961; withdrawn (in blue livery), 11th December 1992; sold to West Somerset Railway, Minehead, and moved 13th September 1993; renumbered D4018; re-sold to North Yorkshire Moors Railway, Grosmont, and moved 11th March 1998; seen in blue livery with number 08850, Grosmont, 4th May 1998; seen freshly repainted blue with number 4018, North Yorkshire Moors Railway, 30th November 2002; seen in blue livery with number 08850, North Yorkshire Moors Railway, 30th September 2009; regularly used as pilot at Pickering Carriage & Wagon Depot; seen in blue livery with number 08850, PC&WD, 11th September 2020.

D4021   Horwich   1961   EC   1/96   P 08853

08853   new (LOT 277) 19th June 1961; withdrawn, January 1996; sold to RFS (Engineering) Ltd, Doncaster, and moved 30th January 1997; overhauled; to Great North Eastern Railway, Bounds Green Depot, London, on hire, about March 1997; returned to RFS (Engineering) Ltd, Doncaster, 1998; seen at RFS Doncaster, 19th October 2000; to Great North Eastern Railway, Bounds Green Depot, London, on hire, June 2003; to Wabtec, Doncaster, 8th February 2008; seen in Wabtec black livery with number 08853, Wabtec, 29th July 2012; to Midland Road Depot, Leeds, for tyre turning, 7th December 2018; to Wabtec, Doncaster, 12th December 2018; re-sold to Hunslet Ltd (Ed Murray Group), 1st May 2021; seen in Wabtec black livery with number 08853, Wabtec, 31st August 2023.

D4033   Darlington   1960   5A   ?   P 08865 / GILLY

08865   new 9th September 1960; withdrawn, date not known; put on sale by tender by DB Schenker, with bids due by 22nd September 2015; sold to Harry Needle Railroad Company; despatched from Crewe Electric Depot, 30th November 2015; to Moveright International, Wishaw, Warwickshire, for storage; noted in EWS red and yellow livery with number 08865, Wishaw, December 2015; to HNRC, Barrow Hill Engine Shed, Staveley, 20th October 2016; seen with number 08865, Barrow Hill Engine Shed, 19th November 2016; to Hope Cement Works, Derbyshire, on hire, 16th February 2017; noted in red and yellow livery with number 08865 and name GILLY, on hire, Hope Cement Works, November 2017; to Bombardier Transportation, Central Rivers Depot, Barton under Needwood, 22nd November 2019; noted in faded EWS red and yellow livery, Central Rivers Depot, 18th December 2019; to Harry Needle Railroad Company, Worksop Depot, 30th January 2023.

D4035   Darlington   1960   8F   ?   F HL1007

08867   new 17th October 1960; noted in black livery with number 08867 and named 'Ralph Easby', Thornaby Depot, 1st August 1992; withdrawn, date not known; sold to RMS Locotec Ltd, Dewsbury, about 1997; to Brunner Mond, Northwich, on hire, September 1997; suffered collision damage, August 1999; stored at Brunner Mond; to RMS Locotec Ltd, Dewsbury, for repairs, 29th November 1999; to Marcroft Wagon Works, Horbury,

Wakefield, on hire, early 2000; to RMS Locotec Ltd, Dewsbury, by February 2001; to Cobra Railfreight, Wakefield, on hire, 2nd July 2002; to EWS, Ferrybridge Depot, for storage, September 2002; re-sold to T.J. Thomson & Son Ltd, Stockton-on-Tees, for scrap, and moved 4th October 2005; noted at T.J. Thomson & Son Ltd, 8th October 2005; scrapped, 20th June 2007.

D4036   **Darlington**   **1960**   **31B**   **12/92**   **P**   **08868**
08868   new 26th October 1960; withdrawn, 11th December 1992; sold to Harry Needle Railroad Company; to South Yorkshire Railway Preservation Society, Meadowhall, Sheffield, 22nd February 1994; to East Lancashire Railway, Bury, 16th April 1994; to RFS (Engineering) Ltd, Doncaster, on hire, 3rd September 1997; sub-hired to Fastline Track Renewals, Peterborough, 12th September 1997; used at May-Gurney, Connington Tip, near Peterborough; to Railway Age, Crewe, 2nd June 1998; to Freightliner, Basford Hall, Crewe, on hire, January 1999; to Yorkshire Engine Company, Long Marston, on hire, 2nd February 1999; to Railway Age, Crewe, 6th April 1999; to Port of Felixstowe, on hire, 6th August 1999; to London & North Western Railway Company Ltd, Carriage Works, Crewe, on hire, 13th August 1999; to Freightliner, Basford Hall, Crewe, on hire, 18th September 2000; to Freightliner, Trafford Park, Manchester, on hire, 30th April 2001; to Barrow Hill Engine Shed, Staveley, 17th April 2002; to Blue Circle, Hope Cement Works, Derbyshire, on hire, 15th November 2002; to Port of Felixstowe, on hire, 25th July 2003; to Midland Road Depot, Leeds, on hire, 7th December 2003; to Port of Felixstowe, on hire, 28th July 2004; to London & North Western Railway Company Ltd, Carriage Works, Crewe, 24th November 2004; noted in blue livery with number 08868, Crewe, March 2007; noted in grey livery with number 08868, Crewe, May 2011; re-sold to Arriva Traincare, Crewe, date not known; noted in Arriva light blue/dark blue/grey livery with number 08868, Arriva, Crewe Depot, 14th July 2016.

D4037   **Darlington**   **1960**   **NC**   **10/97**   **F**   **08869**
08869   new 29th October 1960; suffered fire damage, Norwich, October 1997; withdrawn, October 1997; noted in green livery, Crown Point Depot, Norwich, 26th April 2001; sold to Cotswold Rail; despatched from Crown Point Depot, Norwich, 14th August 2001; to Cotswold Rail, Moreton in Marsh, but moved direct to Brush, Loughborough, for repairs, 15th August 2001; to Harry Needle Railroad Company, Barrow Hill Engine Shed, Staveley, 27th August 2003; noted in green livery with small-size number 08869, Barrow Hill, 17th October 2005; to HNRC, Long Marston, for storage, September 2006; noted in blue livery with number 08869, Long Marston, June 2008; to European Metal Recycling, Kingsbury, for scrap, 29th September 2010; scrapped, January 2011.

D4038   **Darlington**   **1960**   **55G**   **5/93**   **P**   **08870**
08870   new 5th November 1960; latterly ran on BR with painted name MILLHOUSES; withdrawn, 7th May 1993; sold to Harry Needle Railroad Company; to South Yorkshire Railway Preservation Society, Meadowhall, Sheffield, 1st March 1994; to Cobra Railfreight, Wakefield, on hire, 9th December 1994; seen with number 08870 and name MILLHOUSES, Cobra Railfreight, 18th March 1996; to South Yorkshire Railway Preservation Society, Sheffield, 15th October 1997; re-sold to RMS Locotec Ltd, Dewsbury, September 1998; to Anglian Railways, Crown Point Depot, Norwich, on hire, 26th September 1998; to RMS Locotec Ltd, for repairs, 30th March 1999; to Anglian Railways, Crown Point Depot, Norwich, on hire, February 2000; to RMS Locotec Ltd, for repairs, 1st March 2001; to Anglian Railways, Crown Point Depot, Norwich, on hire, about March 2001; to RMS Locotec Ltd, 14th June 2001; to Bombardier Transportation, Horbury, Wakefield, on hire, 19th June 2001; to Redland, Barrow upon Soar, on hire, by 12th January 2002; to

Ford, Bridgend, on hire, March 2002; to Redland, Barrow upon Soar, on hire, 6th November 2002; to RMS Locotec Ltd, 13th October 2004; to Castle Cement Works, Ketton, on hire, 27th January 2005; to RMS Locotec Ltd, Wolsingham, (for use on hire at Colas coal loading siding), 6th July 2010; to Tyseley Depot, Birmingham, for tyre turning, early February 2015; to Wabtec, Kilmarnock, on hire, 16th February 2015; to RMS Locotec Ltd, Wolsingham Depot, 9th July 2015; to Tata Steel, Trostre Works, Llanelli, on hire, 7th December 2015; noted in green livery with number 08870, Trostre Works, April 2016; to Calkeld Heavy Haulage, Stourton, Leeds, for storage, 26th April 2016; to Castle Cement Works, Ketton, on hire, 7th November 2016; noted in green livery with number 08870, Castle Cement Works, October 2017; to RMS Locotec Ltd, Wolsingham Depot, for repairs and repaint, 5th July 2018; noted with number HO24, Wolsingham, February 2020; to Eastern Railway Services, carriage sidings, Runham, Great Yarmouth, 13th May 2021; noted in black, red and white livery with number 08870, Great Yarmouth, 10th September 2021.

D4039   Darlington   1960   41A   10/90 P HO74
08871   new 17th November 1960; withdrawn, 27th October 1990; sold to Humberside Sea & Land Services Ltd, Royal Dock, Grimsby; moved by rail from Tinsley to Immingham; tripped from BR Immingham Depot to Humberside Sea & Land Services Ltd, Royal Dock, Grimsby, 16th December 1990; lost its number by September 1993; seen in blue livery with no number, HS&LS Ltd, 9th March 1995; re-sold to Cotswold Rail, Moreton in Marsh, and moved 7th April 2001; to Brush, Loughborough, for repairs, 18th April 2001; to Anglia Railways, Crown Point Depot, Norwich, on hire, August 2001; to Wabtec, Doncaster, for repairs, 5th November 2002; to Bombardier Transportation, Ilford, for tyre turning, January 2004; to Network Rail, Whitemoor Yard, March, on hire, 1st April 2004; to Daventry International Rail Freight Terminal, on hire, 8th October 2004; to Brush, Loughborough, for repairs, 9th November 2004; returned to Daventry International Rail Freight Terminal, 14th December 2004; to Brush, Loughborough, for repairs, 11th April 2005; to Anglia Railways, Crown Point Depot, Norwich, on hire, 28th November 2005; to Horton Road Depot, Gloucester, for storage, about August 2006; re-sold to RMS Locotec Ltd, Wakefield, October 2007; to Anglia Railways, Crown Point Depot, Norwich, on hire, by 26th May 2007; to Wabtec, Doncaster, for repairs, by August 2008; noted in grey livery with number 08871, Wabtec, November 2008; to East Coast Trains, Craigentinny Depot, Edinburgh, on hire, 27th November 2010; to Wabtec, Doncaster, 13th September 2011; to Boden Rail Engineering, Washwood Heath, for repairs, December 2011; to Cemex Rail Products, Washwood Heath, on hire, about February 2012; to PD Ports, Tees Dock, on hire, 16th April 2012; to RMS Locotec Ltd, Wolsingham Depot, for repairs, 22nd August 2012; to PD Ports, Tees Dock, on hire, 4th December 2012; to RMS Locotec Ltd, Wolsingham Depot, for repairs, 1st August 2014; noted in blue livery with numbers 08871 and HO74, Wolsingham, April 2016; to Tata Steel, Trostre Works, Llanelli, on hire, 25th April 2016; to Bombardier Transportation, Ilford Depot, on hire, 26th February 2018; noted in RMS Locotec black livery with grey cab and number 08871, Ilford Depot, 5th March 2018; to RMS Locotec Ltd, Wolsingham Depot, 12th June 2021; noted in RMS blue/grey livery with number HO74, on a low-loader at Worksop, 26th July 2022; to Loram UK Ltd, Derby RTC, on hire, 26th July 2022; noted at Loram, 13th August 2023.

D4040   Darlington   1960   40B   2/04   P 08872
08872   new 19th November 1960; withdrawn, February 2004; sold to European Metal Recycling; despatched from DB Schenker, Immingham Depot, 20th August 2010; to European Metal Recycling, Attercliffe, Sheffield; seen on a low-loader on M18 Motorway, passing Junction One (Rotherham) at 11:35, 20th August 2010; seen in EWS red and

yellow livery with number 08872, European Metal Recycling, Attercliffe, 1st November 2010; seen long disused, EMR, 25th February 2024.

D4041   Darlington   1960   5A   12/99   P   08873
08873   new 25th November 1960; to ABB Transportation, Derby, 15th May 1998; withdrawn, December 1999; sold to RT Rail, Crewe, about April 2000; to RFS (Engineering) Ltd, Doncaster, for repairs, May 2000; to L&NWR Ltd, Carriage Works, Crewe, on hire, by 7th September 2000; to Manchester Ship Canal Company, Barton Dock, on hire, May 2005; sold to LH Group, Barton under Needwood, November 2005 (remained on hire to MSC Barton Dock); to LH Group, Barton under Needwood, for repairs, 20th June 2006; to Manchester Ship Canal Company, Barton Dock, on hire, 14th November 2006; to LH Group, Barton under Needwood, for repairs, 24th April 2007; noted in red and black livery with number 08873, LH Group, August 2007; to Port of Felixstowe, on hire, 22nd October 2007; to Manchester Ship Canal Company, Barton Dock, Manchester, on hire, 7th January 2009; to Freightliner, Trafford Park, Manchester, on hire, May 2009; to LH Group, Barton under Needwood, for repairs, 20th August 2009; to Innovative Logistics, Brierley Hill, on hire, November 2009; to Freightliner, Southampton Docks, on hire, 20th May 2010; to LH Group, Barton under Needwood, for repairs, 15th February 2011; to Freightliner, Southampton Docks, on hire, week commencing 4th July 2011; to LH Group, Barton under Needwood, for repairs, 23rd November 2012; to Freightliner, Southampton Docks, on hire, 22nd January 2013; to Hams Hall Rail Freight Terminal, Coleshill, on hire, 20th May 2013; noted in red and black livery with number 08873, Hams Hall, June 2013; to LH Group, Barton under Needwood, 23rd August 2016; to Freightliner, Southampton Docks, on hire, 19th October 2016; noted at Southampton, 15th January 2017; to LH Group, Barton under Needwood, 8th May 2017; noted in red and black livery with number 08873, LH Group, May 2017; re-sold to Hunslet Ltd (Ed Murray Group), 1st May 2021; noted as a stripped shell, Hunslet Ltd, Barton under Needwood, 9th May 2021; noted on low-loader, A5 eastbound, 22nd June 2021; to Beaver Metals, Water Orton, 22nd June 2021; scrapped, late June 2021.

D4042   Darlington   1960   55H   2/92   P   08874
08874   new 26th November 1960; withdrawn, 14th February 1992; sold to RFS (Engineering) Ltd, Kilnhurst, and moved 2nd July 1992; to Trans-Manche Link, Channel Tunnel contract (number 97), on hire, 13th February 1993; returned to RFS (Engineering) Ltd, Doncaster, 11th February 1994; to Teesbulk Handling, Middlesbrough, on hire, 2nd June 1994; returned to RFS (Engineering) Ltd, Doncaster, 13th September 1994; to Sheerness Steel Co Ltd, Sheerness, Kent, on hire, 16th September 1995; returned to RFS (Engineering) Ltd, Doncaster, May 1997; re-sold to RT Rail, Crewe, October 1998; to Crewe Depot, for certification, early October 1998; to Hays Chemicals, Sandbach, on hire, 12th October 1998; to RFS (Engineering) Ltd, Doncaster, for repairs, 8th January 1999; to Silverlink, Bletchley, on hire, 26th February 1999; to LH Group, Barton under Needwood, for fitting with Train Protection Warning System equipment, early April 2004; to Silverlink, Bletchley, on hire, August 2004; noted in blue, green and white livery with number 08874 and name CATHERINE, Bletchley, April 2007; RT Rail was acquired by RMS Locotec Ltd, 8th November 2007; to Dartmoor Rail, Meldon Quarry, on hire, 19th December 2007; to Mid-Norfolk Railway, Dereham, 14th April 2008; to Anglia Railways, Crown Point Depot, Norwich, on hire, late April 2008; to North Norfolk Railway, Sheringham, 20th December 2012; to Allelys, Studley, Warwickshire, 21st December 2012; to Anglia Railways, Crown Point Depot, Norwich, on hire, about October 2013; to Tata Steel, Shotton Works, on hire, 9th July 2015; to RMS Locotec Ltd, Wolsingham Depot, 7th August 2018; noted in Silverlink white/purple/green livery with number 08874, Wolsingham, 21st February 2023; noted at

Wolsingham, 5th January 2024; to Independent Railway Engineering, Chesterfield, 18th January 2024.

D4043　　Darlington　　　　　1960　　51L　　　5/91　　F　08875
08875　　new 3rd December 1960; withdrawn, 3rd May 1991; sold to RFS (Engineering) Ltd, Kilnhurst, and moved 20th November 1991; seen in BR blue livery with number 08875 on cabside and front buffer beam, RFS Kilnhurst, 25th December 1991 and 21st May 1993; used for spares; works closed, May 1993; seen at RFS Kilnhurst, 7th August 1993; remains scrapped on site by C.F. Booth Ltd of Rotherham, August 1993.

D4044　　Darlington　　　　　1960　　36A　　　9/91　　F　08876
08876　　new 9th December 1960; withdrawn, 27th September 1991; sold to RFS (Engineering) Ltd, Kilnhurst; despatched from BR Tinsley Depot, 2nd July 1992; to RFS Kilnhurst; seen at RFS Kilnhurst, 18th October 1992; works closed, May 1993; seen at RFS Kilnhurst, 28th June 1993; to RFS (Engineering) Ltd, Doncaster, autumn 1993; used for spares; remains scrapped on site by Hudson Ltd of Madeley, Telford, April 1994.

D4045　　Darlington　　　　　1960　　8F　　　4/11　　P　08877
08877　　new 15th December 1960; withdrawn, April 2011; put on sale by DB Schenker, 5th May 2015; sold to Harry Needle Railroad Company; despatched from Springs Branch Depot, Wigan, 29th October 2015; to Barrow Hill Engine Shed, Staveley; seen in grey livery with name WIGAN 1, Barrow Hill Engine Shed, Staveley, 19th December 2015; to Celsa, Cardiff, on hire, 6th June 2019; noted in grey livery with number 08877, Celsa, 25th November 2022; to Harry Needle Railroad Company, Worksop Depot, by road, 12th July 2023; noted on A1 northbound, 20th May 2024; to Midland Road Depot, Leeds, for tyre turning, 20th May 2024; to Harry Needle Railroad Company, Worksop Depot, May 2024; noted in HNRC orange livery with number 08877, Worksop Depot, 24th July 2024.

D4047　　Darlington　　　　　1960　　87B　　　?　　P　08879
08879　　new 22nd December 1960; latterly ran on BR with painted name EARLES; withdrawn, date not known; put on sale by tender by DB Cargo, with bids due by 5th October 2016; sold to Raxstar Ltd, Eastleigh Works; despatched from DB Cargo, Margam Depot, 12th December 2016; noted in EWS red and yellow livery with number 08879, Eastleigh Works, 12th December 2016 and 30th June 2017; re-sold to Harry Needle Railroad Company, March 2018; to Railway Support Services, Wishaw, for repairs, 4th May 2018; to Barrow Hill Engine Shed, Staveley, 29th June 2018; seen in EWS red and yellow livery with number 08879, Barrow Hill Engine Shed, 24th January 2019; to Hope Cement Works, for storage, 22nd November 2019; noted near shed, Hope, 21st May 2024.

D4056　　Darlington　　　　　1961　　40B　　　6/72　　F　D4056 / No.55
new 11th February 1961; noted in BR green livery with number D4056, Immingham Depot, 14th May 1972; withdrawn, 11th June 1972; sold to NCB; to Ashington Central Workshops, January 1973; given number No.55; to Shilbottle Colliery, 18th June 1974; noted in BR green livery with number D4056, Shilbottle Colliery, 26th August 1974; seen with number D4056, Shilbottle Colliery, 21st May 1977; noted at Shilbottle Colliery, 10th April 1982; scrapped on site by T.J. Thomson & Son Ltd of Stockton-on-Tees, March 1983.

D4067　　Darlington　　　　　1961　　41J　　　12/70　　P　D4067 / 1802-B4
new 2nd May 1961; withdrawn, 13th December 1970; sold to NCB; to Betteshanger Colliery, Kent, early April 1971; noted at Betteshanger Colliery, 17th April 1971; noted with number 1802/B4, Betteshanger Colliery, 15th October 1972; to Snowdown Colliery, Kent, 27th May 1976; to Nailstone Colliery, Leicestershire, 14th June 1976; noted in blue livery

at Nailstone Colliery, 17th June 1976; to BREL Doncaster Works, for repairs, 28th October 1976; seen in blue livery with number D4067 and lettered 'National Coal Board Kent Colliery No.1802/B4', BREL Doncaster Works, 14th November 1976; to Nailstone Colliery, 23rd December 1976; seen in blue livery, numbered D4067 and 1802/B4, Nailstone Colliery, 10th February 1979; seen, for disposal with burned-out traction motor, Nailstone Colliery, 25th April 1979; privately purchased for £1,350; moved by low-loader to Great Central Railway, Loughborough, 6th February 1980; (one of four preserved Class 10s); noted in blue livery at GCR, Loughborough, 10th February 1980; noted in green livery, GCR, Loughborough, 26th August 1983; noted in blue livery, with number 10119 and name MARGARET ETHEL – THOMAS ALFRED NAYLOR, Loughborough, 29th March 2014 and March 2016; noted repainted in blue livery with number D4067, GCR, 1st September 2022.

D4068   Darlington   1961   40B   6/72   F   No.56 / 9300/116
new 10th May 1961; noted in BR green livery with number D4068, Immingham Depot, 14th May 1972; withdrawn, 11th June 1972; sold to NCB; to Ashington Central Workshops, for repairs, January 1973; to Shilbottle Colliery, 16th February 1973; noted with number No.56, Shilbottle Colliery, 12th June 1973; seen at Shilbottle Colliery, 21st May 1977; to Lambton Engine Works, Philadelphia, 3rd April 1979; noted at Lambton Engine Works, 12th April 1979; seen in blue livery with number No.56, and with RE 1703 of 1953 registration plate, Lambton Engine Works, 26th June 1979; allocated overhaul number 80-601-002 of 1980; to Whittle Colliery, Newton-on-the-Moor, 25th April 1980; noted in dark blue livery with number No.56, Whittle Colliery, 6th September 1980 and 30th September 1983; noted with no engine and no wheels, Whittle Colliery, 3rd November 1985; remains scrapped on site by C.F. Booth Ltd of Rotherham, December 1985.

D4069   Darlington   1961   41J   4/72   F   No.51 / 9300/111
new 25th May 1961; withdrawn, 23rd April 1972; sold to NCB; to Ashington Central Workshops, for repairs, September 1972; given number No.51; to Whittle Colliery, Newton-on-the-Moor, 20th October 1972; noted with number No.51, Whittle Colliery, 22nd October 1972; to Lambton Engine Works, Philadelphia, 28th February 1978; noted at Lambton Engine Works, 19th March 1978 and 25th October 1978; allocated overhaul number 9-601-014 of 1979; to Whittle Colliery, Newton-on-the-Moor, 30th March 1979; noted in dark blue livery with number No.51, Whittle Colliery, 6th September 1980; noted with number 9300/111, Whittle Colliery, 31st May 1981; moved by low-loader to C.F. Booth Ltd, Rotherham, November 1985; scrapped, 9th and 10th December 1985.

D4070   Darlington   1961   41J   4/72   F   No.52 / 9300/112
new 5th June 1961; withdrawn, 23rd April 1972; sold to NCB; to Ashington Central Workshops, for repairs, September 1972; given number No.52; to Ashington Colliery, 22nd October 1972; to Shilbottle Colliery, 9th February 1973; noted in green livery with number No.52, Shilbottle Colliery, 12th June 1973; to Ashington Central Workshops, 23rd September 1974; noted at Ashington, 28th October 1974 and 20th July 1975; to Lambton Engine Works, Philadelphia, for rebuild, 11th October 1975; allocated overhaul number 6-601-009 of 1976; allocated NCB plant number 9300/112; to Bates Colliery, Blyth, 12th November 1976; to Whittle Colliery, Newton-on-the-Moor, 16th April 1977; to Lambton Engine Works, 29th April 1980; noted in dark blue livery, Lambton Engine Works, 6th September 1980; to Whittle Colliery, Newton-on-the-Moor, 5th June 1981; noted with number No.52, Whittle Colliery, 16th May 1982; noted with no engine and no wheels, Whittle Colliery, 3rd November 1985; remains scrapped on site by C.F. Booth Ltd of Rotherham, December 1985.

D4072   Darlington   1961   31B   4/72   F   No.53 / 93100/114

new 26th July 1961; withdrawn, 23rd April 1972; sold to NCB; to Ashington Central Workshops, for repairs, 7th October 1972; given number No.53; to Ashington Colliery, 22nd October 1972; to Whittle Colliery, Newton-on-the-Moor, 1st November 1972; noted with number No.53, Whittle Colliery, 12th June 1973; to Lambton Engine Works, Philadelphia, 14th June 1977; noted at Lambton Engine Works, 20th September 1977; to Whittle Colliery, 28th February 1978; to Lambton Engine Works, Philadelphia, 22nd June 1978; allocated overhaul number 8-601-005 of 1978; to Whittle Colliery, 16th August 1978; to Lambton Engine Works, Philadelphia, 29th September 1978; noted at Lambton Engine Works, Philadelphia, 25th October 1978; to Whittle Colliery, 26th November 1978; to Lambton Engine Works, Philadelphia, 19th December 1979; to Whittle Colliery, 5th June 1980; noted in dark blue livery with number No.53, Whittle Colliery, 6th September 1980; to Lambton Engine Works, Philadelphia, 14th April 1981; noted at Lambton Engine Works, 18th June 1981; to South Hetton Colliery, 10th May 1982; noted at South Hetton Colliery, 14th June 1982; seen in blue livery with number No.53, and cast plate 'Overhauled Wk. No. 6-601-005, 1978, Philadelphia Workshops', South Hetton Colliery, 28th July 1982; to Lambton Engine Works, Philadelphia, 27th September 1982; to Ashington Colliery, July 1983; to Lambton Coking Plant, about September 1983; noted at Lambton Coking Plant, 26th October 1983; to Lambton Engine Works, Philadelphia, by 2nd December 1983; noted with number 93100/114, Philadelphia Locomotive Shed, 20th October 1985; noted being scrapped on site by C.Herring & Son Ltd of Hartlepool, 3rd November 1985.

D4074   Darlington   1961   31B   4/72   F   No.54

new 24th August 1961; withdrawn, 23rd April 1972; sold to NCB; to Ashington Central Workshops, for repairs, October 1972; given number No.54; to Ashington Colliery, 22nd October 1972; to Whittle Colliery, Newton-on-the-Moor, 15th December 1972; noted with number No.54, Whittle Colliery, 12th June 1973; to Lambton Engine Works, Philadelphia, 8th February 1977; scrapped on site, August 1978.

D4092   Darlington   1962   34E   9/68   P   D4092

new 2nd May 1962; withdrawn, 1st September 1968; sold to NCBOE; despatched from 34E New England Depot, October 1968; to Powell Duffryn Fuels Ltd, NCBOE Gwaun-cae-Gurwen Disposal Point, Glamorgan; noted at Gwaun-cae-Gurwen, 26th October 1968; to BR Canton Depot, Cardiff, for repairs, August 1977; displayed at open day, Canton Depot, 1st October 1977; returned to Gwaun-cae-Gurwen Disposal Point, October 1977; noted at Gwaun-cae-Gurwen, 3rd April 1979; seen in blue and white livery with number D4092 and name CHRISTINE, Gwaun-cae-Gurwen, 12th June 1980; noted in blue and white livery, Gwaun-cae-Gurwen, 6th July 1988; re-sold to Harry Needle Railroad Company, 1988; to South Yorkshire Railway Preservation Society, Meadowhall, Sheffield, 26th October 1988; noted in blue and white livery, with number D4092 and name CHRISTINE, Meadowhall, 27th August 1989; to Barrow Hill Engine Shed, Staveley, 26th July 2001; noted in blue and white livery, with number painted over, Barrow Hill Engine Shed, 4th October 2003 and 11th June 2005; noted repainted in green livery with number D4092, Barrow Hill Engine Shed, 14th April 2007; seen in green livery with number D4092 (one of four preserved Class 10s), Barrow Hill Engine Shed, 12th March 2022.

D4095   Horwich   1961   66B   2/04   P   D4095

08881   new (LOT 294) 23rd August 1961; withdrawn, February 2004; noted in grey livery, Motherwell Depot, February 2007; sold to Railway Support Services; despatched from Motherwell Depot; to Railway Support Services, Wishaw, and moved 20th April 2007; to Alstom Transport, Stonebridge Park Heavy Repair Depot, Wembley, for tyre turning,

15th August 2007; returned to Wishaw, 19th September 2007; to Gloucestershire Warwickshire Railway, Toddington, for storage, early 2008; to Lafarge Cement, Barrow-on-Soar, on hire, 5th October 2008; to Gloucestershire Warwickshire Railway, Toddington, for storage, December 2008; re-sold to Somerset & Dorset Railway Museum Trust, Midsomer Norton, Somerset, and moved 31st January 2012; to Gloucestershire Warwickshire Railway, Toddington, 5th December 2013; to Somerset & Dorset Railway Museum Trust, Midsomer Norton, 17th March 2014; noted in green livery with number D4095, Midsomer Norton, 4th September 2016 and 31st October 2020.

D4115   Horwich   1962   36A   5/93   P   08885 / 18 / HO42
08885   new (LOT 294) 18th January 1962; withdrawn, 7th May 1993; sold to Great Central Railway (Nottingham), Ruddington, Nottingham, and moved 15th June 1994; to Midland Railway, Butterley, 7th November 2004; re-sold to RT Rail, Crewe, August 2005; to RMS Locotec Ltd, Dewsbury, for overhaul, 25th August 2005; to Network Rail, Whitemoor Yard, March, on hire, 31st October 2005; RT Rail was acquired by RMS Locotec Ltd, 8th November 2007; to PD Ports, Tees Dock, on hire, 3rd April 2009; to RMS Locotec Ltd, Wolsingham Depot, about 15th April 2013; to Celtic Energy, Onllwyn Disposal Point, on hire, 14th May 2013; to RMS Locotec Ltd, Wolsingham Depot, about 1st June 2013; noted in blue livery with numbers 08885, HO42 and 18, Wolsingham Depot, 21st February 2023; noted at Wolsingham, 5th January 2024; to Independent Railway Engineering, Chesterfield, 17th January 2024.

D4116   Horwich   1962   5A   ?   F   08886
08886   new (LOT 296) 26th January 1962; withdrawn, date not known; noted in EWS red and yellow livery, Crewe Depot, 30th January 2016; sold to Railway Support Services; despatched from Crewe Electric Depot, 12th May 2016; to Railway Support Services, Wishaw; used for spares; remains to European Metal Recycling, Kingsbury, for scrap, 20th May 2016; scrapped.

D4117   Horwich   1962   ?   ?   P   08887
08887   new (LOT 296) 2nd February 1962; withdrawn, date not known; acquired by Alstom Transport, Longsight Depot, Manchester, 2007; noted in black livery with number 08887, Longsight, April 2007; to Alstom Transport, Wembley Depot, London, by 25th January 2012; noted in blue livery with number 08887, Wembley Depot, November 2013 and June 2015; to Arlington Fleet Services Ltd, Eastleigh Depot, for repairs, 20th July 2016; to Alstom Transport, Wembley Depot, London, 19th December 2017; to Alstom Transport, Polmadie Depot, Glasgow, by 3rd May 2018; to Allely's, Studley, 28th October 2022; to Alstom Transport, Wembley Depot, London, 28th October 2022; to Alstom Transport, Longsight Depot, 11th May 2023.

D4118   Horwich   1962   ?   ?   P   D4118
08888   new (LOT 296) 16th February 1962; withdrawn, date not known; put on sale by tender by DB Cargo, with bids due by 5th October 2016; sold to K&ESR; despatched from DB Cargo, Hoo Junction, 15th December 2016; to Kent & East Sussex Railway, Tenterden; noted in EWS red and yellow livery with number 08888, Kent & East Sussex Railway, 15th December 2016; noted in green livery with number D4118, Kent & East Sussex Railway, April 2019; re-sold and moved to Avon Valley Railway, Bitton, 23rd September 2021; to Gwili Railway, on hire, 31st October 2022; to Avon Valley Railway, Bitton, 6th February 2024; to Railway Support Services, Wishaw, 6th August 2024; to Wolverton Works, on hire, 13th August 2024.

D4121   Horwich   1962   ?   ?   P   08891
08891   new (LOT 296) 7th March 1962; withdrawn, date not known; sold to LH Group, Barton under Needwood, and moved 3rd May 2008; noted in green livery with yellow cab and number 08891, LH Group, 28th July 2012; to Nemesis Rail, Burton upon Trent, 16th August 2017; noted in green livery with yellow cab and number 08891, Nemesis Rail, 16th August 2017; to LH Group, Barton under Needwood, late 2021; re-sold to Freightliner; noted in orange and yellow Freightliner livery with number 08891, LH Group, 21st January 2022 and 5th March 2022; initially moved to Southampton Maritime, 4th April 2022.

D4122   Horwich   1962   70D   12/96 P   08892
08892   new (LOT 296) 16th March 1962; withdrawn, December 1996; sold to RFS (Engineering) Ltd, Doncaster, and moved 11th December 1996; to GNER, Bounds Green Depot, London, on hire, 18th April 1997; returned to RFS (Engineering) Ltd, Doncaster, 2nd August 1999; to GNER, Bounds Green Depot, London, on hire, by 21st April 2001; to Wabtec, Doncaster, 10th April 2003; noted in blue livery with number 08892, Wabtec, July 2003; to Soho EMU Depot, Birmingham, on hire, October 2004; to Central Trains, Tyseley, on hire, 12th November 2004; returned to Wabtec, Doncaster, June 2005; re-sold to Direct Rail Services, May 2006; to DRS, Kingmoor Depot, Carlisle, 25th July 2006; to DRS, Gresty Bridge Depot, Crewe, about January 2007; to North Pole Depot, near Kensal Green, London, on hire, October 2007; to DRS, Gresty Bridge Depot, Crewe, 14th November 2007; to Fastline, Doncaster, on hire, early July 2008; re-sold to Harry Needle Railroad Company, about August 2008; to Lafarge, Hope Cement Works, Derbyshire, on hire, by 6th September 2008; seen in dark blue livery with number 08892, Hope Cement Works, September 2008; to Bombardier Transportation, Litchurch Lane Works, Derby, on hire, 14th October 2010; to Nemesis Rail, Burton upon Trent, for storage, 4th November 2011; noted in dark blue livery with number 08892, Nemesis Rail, 7th April 2012; to Bombardier Transportation, Litchurch Lane Works, Derby, on hire, 20th June 2013; to HNRC, Barrow Hill Engine Shed, Staveley, 24th July 2013; to First Capital Connect, Hornsey Depot, London, on hire, 20th December 2013; to Serco, Old Dalby Test Centre, on hire, 19th February 2015; noted in blue livery with number 08892, Old Dalby, March 2016; to Tyseley Diesel Depot, Birmingham, for tyre turning, 29th July 2016; returned to Serco, Old Dalby Test Centre, on hire, 12th August 2016; noted at Serco, Old Dalby Test Centre, 6th September 2020; to HNRC, Worksop Depot, 31st March 2021; noted in blue livery, Worksop, 20th May 2023; to Very Light Rail Innovation Centre, Dudley, for trials, 24th February 2024, to HNRC, Worksop Depot, 15th May 2024.

D4126   Horwich   1962   86A   2/04   P   08896 / STEVEN DENT
08896   new (LOT 296) 12th April 1962; withdrawn, 1st February 2004; noted at Toton Depot, 11th March 2006; sold to Railway Support Services, Wishaw; despatched from EWS Toton Depot, 8th March 2007; noted in EWS red and yellow livery with number 08896, Wishaw, July 2007; re-sold (without its centre wheels) to Severn Valley Railway, and moved November 2009; placed in store at Kidderminster and used for spares; noted in EWS red and yellow livery with number 08896, Severn Valley Railway, May 2017 and October 2018; placed inside Kidderminster carriage shed, for further storage, spring 2024.

D4129   Horwich   1962   ?   ?   P   08899
08899   new (LOT 296), (ex East Midlands Railway, Derby); acquired by Railway Support Services, Wishaw, April 2022; noted at Etches Park Depot, 17th April 2022; moved to Chaddesden hauled by 37418, 3rd May 2022; then moved to Loram UK Ltd, Derby RTC, on hire, early May 2022; noted in red livery with white roof and number 08899, Loram UK Ltd, 7th May 2022; to Electro-Motive Diesel Ltd, Longport, on hire, 1st July 2022; to GBRf,

Whitemoor Yard, March, on hire, 18th August 2022; noted at Whitemoor Yard, 19th August 2022; to Railway Support Services, Wishaw, 23rd December 2022; noted on A14 on a low-loader, 25th January 2023; to GBRf, Whitemoor Yard, March, on hire, 25th January 2023; to Railway Support Services, Wishaw, 29th June 2023; to Nemesis Rail, Burton upon Trent, 2nd July 2023; to Railway Support Services, Wishaw, 28th November 2023; to Hams Hall Rail Freight Terminal, Coleshill, on hire, 16th February 2024.

D4133   Horwich   1962   36A   9/95   P   08903
08903   new (LOT 295) 1st June 1962; withdrawn, 19th September 1995; sold to ICI Billingham Works, Stockton-on-Tees, and moved by road on 9th May 1996; to RFS (Engineering) Ltd, Doncaster, for repairs, March 1997; to ICI Wilton Works, Middlesbrough, 4th August 1997; to ICI Billingham Works, September 1999; to Thornaby Depot, for tyre turning, 30th January 2000; returned to ICI Billingham, about 2nd February 2000; to ICI Wilton Works, 2005; to ICI Billingham Works, 17th April 2007; to ICI Wilton Works, Middlesbrough, 2008; to Wensleydale Railway, Leeming Bar, for gala, early September 2016; noted in blue livery with number 08903 and name JOHN W. ANTILL, Wensleydale Railway, 13th September 2016; to Sembcorp Utilities, Wilton Works, 27th September 2016.

D4134   Horwich   1962   55G   ?   P   08904
08904   new (LOT 295) 7th June 1962; withdrawn, date not known; put on sale by tender by DB Cargo, with bids due by 5th October 2016; sold to Harry Needle Railroad Company; despatched from Knottingley Depot, 3rd May 2016; to Eastleigh Works, on hire; noted in EWS red and yellow livery with number 08904, Eastleigh Works, 31st October 2016; noted at Eastleigh Works, 29th January 2017; to Celsa, Cardiff, on hire, 23rd February 2017; noted in EWS red and yellow livery with number 08904, Celsa, 26th February 2017 and 7th June 2019; to Harry Needle Railroad Company, Worksop Depot, 13th June 2019.

D4135   Horwich   1962   2F   1/05   P   08905
08905   new (LOT 295) 18th June 1962; withdrawn, January 2005; sold to Harry Needle Railroad Company, 2011; to Hope Cement Works, Derbyshire, on hire, October 2011; seen in EWS red and yellow livery with number 08905, Hope Cement Works, 15th February 2013; noted in faded EWS livery, Hope Cement Works, June 2019; noted on a low-loader, A42, 11th November 2022; to HTRS Ltd, Battlefield Line, Shackerstone, for overhaul, 11th November 2022.

D4137   Horwich   1962   2F   ?   P   D4137
08907   new (LOT 295) 29th June 1962; was latterly Crewe celebrity shunter, operating in LNWR black with a cast numberplate; noted at Bescot Depot, January 2015; withdrawn, date not known; noted in DB Cargo red livery, Bescot Depot, 11th August 2016; put on sale by tender by DB Cargo, with bids due by 5th October 2016; sold to GCR; despatched from Bescot Depot; to Great Central Railway, Loughborough, 28th November 2016; noted in DB Cargo red livery with number 08907, Great Central Railway, 27th December 2016; noted in green livery with number D4137, Great Central Railway, 24th July 2019 and 3rd October 2021.

D4141   Horwich   1962   51L   5/04   P   08911 / MATEY
08911   new (LOT 295) 8th August 1962; noted in grey livery with number 08911, Thornaby Depot, 6th March 2004; withdrawn, 1st May 2004; sold to NRM; despatched from Thornaby Depot, 15th May 2004; to National Railway Museum, York; noted at NRM York, 24th March 2005; noted in blue livery with number 08911, NRM York, January 2007; seen at NRM York, 24th July 2008; to Southall Railway Centre, London (for use in 'Railway Children' production), 23rd May 2010; to National Railway Museum, York, 8th January

2011; to Southall Railway Centre, London (for use in 'Railway Children' production), 26th May 2011; to National Railway Museum, York, 24th January 2012; to Freightliner, York, on hire, 28th January 2013; noted at National Railway Museum gala weekend, 10th March 2013; returned to Freightliner, York; to National Railway Museum, York, March 2014; noted at NRM York, 11th May 2016; to National Railway Museum, Shildon, 9th June 2016; used as yard shunter; seen in blue livery with number 08911 and name MATEY, National Railway Museum, Shildon, 11th November 2023.

**D4142	Horwich	1962	8J	2/07	P	b**

08912	new (LOT 295) 17th August 1962; withdrawn, 28th February 2007; sold to T.J. Thomson & Son Ltd; despatched from EWS Toton Depot, 6th March 2007; to T.J. Thomson & Son Ltd, Stockton-on-Tees; re-sold to A.V. Dawson Ltd, Middlesbrough; to EWS Thornaby Depot, for repairs, 1st February 2008; to A.V. Dawson Ltd, Middlesbrough, 20th March 2008; noted in BR blue livery, dismantled, no wheels, off track, A.V. Dawson Ltd, Ayrton Store, 12th August 2023.

**D4143	Horwich	1962	66B	11/05	F	08913**

08913	new (LOT 295) 23rd August 1962; withdrawn, 1st November 2005; noted in EWS red and yellow livery with number 08913 and name HYWEL, Motherwell Depot, February 2007; sold to LH Group, Barton under Needwood; despatched from EWS Motherwell Depot, 20th March 2007; to Barton under Needwood; to Manchester Ship Canal Company, Barton Dock, on hire, 24th April 2007; to Cleveland Potash Ltd, Boulby Mine, on hire, May 2009; to Barton under Needwood, 30th October 2009; to Daventry International Rail Freight Terminal, on hire, 10th December 2009; to Barton under Needwood, for repairs, week-ending 24th June 2011; to Daventry International Rail Freight Terminal, on hire, July 2011; noted in yellow, blue and green livery with number 08913 (but no longer named), Daventry International Rail Freight Terminal, May 2012; to Barton under Needwood, 4th March 2013; used for spares; noted in Malcolm Rail yellow, blue and green livery with number 08913, with no engine, Barton under Needwood, 12th March 2017; remains to European Metal Recycling, Kingsbury, for scrap, 18th January 2018; scrapped at unknown date in 2018.

**D4145	Horwich	1962	8J	2/04	P	08915 / HERCULES**

08915	new (LOT 295) 11th September 1962; withdrawn, 1st February 2004; noted in grey livery with number 08915, Toton Depot, 4th March 2007; sold to Railway Support Services, Wishaw; despatched from EWS Toton Depot and moved direct to Colne Valley Railway, Castle Hedingham, 13th March 2007; noted in black livery with number 08915, Colne Valley Railway, April 2009; re-sold to Stephenson Railway Museum, Chirton, near Newcastle upon Tyne, and moved 5th November 2009; seen in black livery with number 08915, Chirton, 12th April 2011 and 1st February 2017; noted in blue livery with number 08915, Chirton, 17th September 2022.

**D4148	Horwich	1962	TM	7/05	P	08918**

08918	new (LOT 295) 25th September 1962; withdrawn, July 2005; sold to Harry Needle Railroad Company; to Nemesis Rail, Burton upon Trent, for storage, October 2011; noted in grey livery with number 08918, Nemesis Rail, May 2016 and 2nd July 2023.

**D4151	Horwich	1962	86A	4/11	P	08921 / PONGO**

08921	new (LOT 295) 26th October 1962; latterly in EWS red and yellow livery and named PONGO after Andy Lynch, retired Canton Depot fitter; withdrawn, April 2011; sold to European Metal Recycling, Kingsbury, and moved 5th August 2011; re-sold to Railway

Support Services; moved to Railway Support Services, Wishaw, 1st February 2018; noted in EWS red and yellow livery with number 08921 and name PONGO, with no engine, Railway Support Services, Wishaw, 2nd January 2019 and 29th April 2024.

D4152   Horwich   1962   16A   ?   P   08922
08922   new (LOT 295) 2nd November 1962; withdrawn, date not known; put on sale by tender by DB Cargo, with bids due by 5th October 2016; sold to Great Central Railway (Nottingham); despatched from DB Cargo, Toton Depot, 14th December 2016; to Great Central Railway (Nottingham), Ruddington, Nottingham; noted in grey livery with number 08922, Ruddington, 27th December 2016; to DB Cargo Maintenance Ltd, Wheildon Road Wagon Works, Stoke on Trent, on hire, 28th March 2019; to Electro-Motive Diesel, Longport, Staffordshire, on hire, 22nd September 2019; to Railway Support Services, Wishaw, 13th March 2020; to Hitachi, Newton Aycliffe, on hire, 16th March 2020; noted in grey livery with number 08922, Newton Aycliffe, 18th March 2020; to Electro-Motive Diesel, Longport, on hire, 25th October 2020; to Serco, Old Dalby Test Centre, on hire, 31st March 2021; noted at Old Dalby, 7th February 2022; to Railway Support Services, Wishaw, early March 2022; noted northbound on M42, 7th April 2022; to Great Central Railway (Nottingham), Ruddington, Nottingham, 7th April 2022; noted in grey livery with number 08922, Ruddington, 28th May 2023; to Spa Valley Railway, by road, 7th September 2023; noted at Spa Valley Railway, in service, 6th July 2024.

D4154   Horwich   1962   66B   8/06   P   08924 / CELSA 2
08924   new (LOT 295) 13th December 1962; withdrawn, August 2006; sold to C.F. Booth Ltd, Rotherham; despatched from DB Schenker, Tyne Yard, 27th January 2011; re-sold to Harry Needle Railroad Company; to Barrow Hill Engine Shed, Staveley, 15th February 2011; seen in EWS red and yellow livery with number 08924, Barrow Hill Engine Shed, Staveley, 21st June 2011; to Lafarge, Hope Cement Works, Derbyshire, on hire, 2nd April 2013; to Barrow Hill Engine Shed, Staveley, 12th February 2014; to GBRf, Garston Railport, Liverpool, on hire, 21st February 2014; to Barrow Hill Engine Shed, Staveley, 21st December 2015; repainted in GBRf blue and yellow livery with number 2, February 2016; to Celsa, Cardiff, on hire, 15th March 2016; noted at Celsa, 19th March 2016; to Barrow Hill Engine Shed, Staveley, 3rd October 2016; to Tyseley Diesel Depot, Birmingham, for tyre turning, 12th October 2016; to Barrow Hill Engine Shed, Staveley, 17th October 2016; to Celsa, Cardiff, on hire, 25th November 2016; to Barrow Hill Engine Shed, Staveley, 13th February 2017; to Celsa, Cardiff, on hire, 3rd March 2017; noted in GBRf blue livery with number 08924, Celsa, 5th March 2017; to Barrow Hill Engine Shed, Staveley, 31st March 2017; seen at Barrow Hill Engine Shed, 5th April 2017; to Celsa, Cardiff, on hire, 7th April 2017; to Barrow Hill Engine Shed, Staveley, for repairs, 10th June 2019; to Celsa, Cardiff, on hire, June 2019; noted at Celsa, Cardiff, June 2019; to Barrow Hill Engine Shed, Staveley, 11th October 2019; to Celsa, Cardiff, on hire, 8th December 2020; noted on a low-loader, M1 northbound, 12th May 2021; to Barrow Hill Engine Shed, Staveley, for repairs, 12th May 2021; to Celsa, Cardiff, on hire, 3rd August 2021; to Barrow Hill Engine Shed, Staveley, for repairs, 20th January 2022; noted in GBRf blue and yellow livery with number 2, under repair, 3rd February 2022; to Celsa, Cardiff, on hire, 2nd March 2022; noted at Celsa, 25th November 2022; to Midland Road Depot, Leeds, for tyre turning, 2nd February 2024; to Celsa, Cardiff, on hire, 5th February 2024.

D4155   Horwich   1962   ?   ?   P   08925
08925   new (LOT 295) 14th December 1962; withdrawn, date not known; sold to GBRf, but to be maintained by Harry Needle Railroad Company; noted in green livery with number 08925, Celsa, Cardiff, 14th February 2016; to Barrow Hill Engine Shed, Staveley, 23rd

August 2016; seen with number 08925, Barrow Hill Engine Shed, 1st September 2016; to Railway Support Services, Wishaw, 18th November 2016; to Midland Road Depot, Leeds, for tyre turning, 20th November 2016; noted at Midland Road, 28th November 2016; to GBRf, Immingham, about 29th November 2016; to Barrow Hill Engine Shed, Staveley, December 2016; to GBRf, Whitemoor Yard, March, 3rd February 2017; to Barrow Hill Engine Shed, Staveley, for repairs, May 2018; to GBRf, Whitemoor Yard, March, 13th June 2018; noted in green livery with number 08925, March, 23rd August 2018; to Harry Needle Railroad Company, Worksop Depot, 10th November 2020.

D4156   Horwich   1962   8J   ?   F   08926
08926   new (LOT 295) 21st December 1962; withdrawn, date not known; noted in blue livery with number 08926, Allerton Depot, 30th October 2005; sold to Railway Support Services, Wishaw, and moved 28th March 2007; noted at Wishaw, 17th April 2007; used for spares; remains to European Metal Recycling, Kingsbury, for scrap, 17th June 2007; scrapped, July 2007.

D4157   Horwich   1962   66B   6/05   P   08927
08927   new (LOT 295) 28th December 1962; was the last locomotive to be built at Horwich; withdrawn, 1st June 2005; sold to Railway Support Services, Wishaw, and moved direct to Gloucestershire Warwickshire Railway, Toddington, for storage, 28th March 2007; to Alstom Transport, Stonebridge Park Heavy Repair Depot, Wembley, for tyre turning, 19th September 2007; returned to Railway Support Services (GWR), 24th November 2007; noted in green livery with blue cab and with number 08927, Toddington, 6th April 2008; noted in green livery with number D4157, Toddington, December 2009; to Pontypool & Blaenavon Railway, on hire, 28th April 2010; to Gloucestershire Warwickshire Railway, Toddington, for storage, 31st January 2011; to Southall Railway Centre, London (for use in Railway Children production), May 2011; to Gloucestershire Warwickshire Railway, Toddington, for storage, 31st January 2012; to National Railway Museum, Shildon, on hire, 16th July 2012; to Electro-Motive Diesel Ltd, Roberts Road Depot, Doncaster, on hire, 5th November 2013; seen in green livery with dual-numbers D4157 and 08927, Roberts Road, 12th March 2014 and 30th September 2018; to Railway Support Services, Wishaw, 8th October 2018; noted at Railway Support Services, 16th October 2018; to DB Cargo Maintenance Ltd, Wheidon Road Wagon Works, Stoke on Trent, on hire, 15th December 2018; to Bescot Yard, Walsall, on hire, 28th March 2019; started work at Bescot, 1st April 2019; seen in green livery, Bescot Yard, 1st April 2019; to Railway Support Services, Wishaw, 30th October 2019; to Bounds Green Depot, London, on hire, 31st January 2020; to Railway Support Services, Wishaw, 30th June 2020; to GBRf, Bescot Yard, on hire, 1st July 2020; noted in faded green livery with number 08927, Bescot, 24th July 2022; to Railway Support Services, Wishaw, 25th July 2022; to Avon Valley Railway, for repairs, 24th February 2023.

D4158   Darlington   1962   NC   7/01   F   08928
08928   new 16th April 1962; withdrawn, July 2001; noted in grey livery, Crown Point Depot, Norwich, 26th April 2001; sold to Cotswold Rail, Moreton in Marsh, about July 2001; despatched by rail from Crown Point Depot, Norwich, 14th August 2001; to Brush, Loughborough, for repairs, 15th August 2001; re-sold to Harry Needle Railroad Company; to Barrow Hill Engine Shed, Staveley, 17th July 2003; noted in grey livery with yellow cab, with number 08928, Barrow Hill Engine Shed, 4th October 2003; noted on a low-loader at Tibshelf Services on the M1, 21st March 2006; to Harry Needle Railroad Company, Long Marston, for storage, 21st March 2006; noted in grey livery with yellow cab and number

08928, Long Marston, June 2007; used for spares; to European Metal Recycling, Kingsbury, for scrap, 3rd December 2010; scrapped, December 2010.

D4163   Darlington   1962   81A   9/08   P   08933
08933   new 5th May 1962; withdrawn, 30th September 2008; sold to T.J. Thomson & Son Ltd, Stockton-on-Tees (but did not move), early 2009; re-sold to Foster Yeoman Quarries Ltd, Somerset, early 2009; despatched from Hoo Junction, 13th March 2009; noted on low-loader on A361, 13th March 2009; arrived at Merehead Stone Terminal, 13th March 2009; to Knights Rail Services, Eastleigh Works, for repairs, 19th October 2010; received spares from 08826; to Merehead Stone Terminal, Somerset, 3rd July 2012; noted in MRL blue livery with number 08933, Merehead Stone Terminal, March 2017; to Whatley Quarry, February 2020; to LH Group, Barton under Needwood, 3rd June 2023.

D4166   Darlington   1962   31B   12/92   P   08936 / HO75
08936   new 17th May 1962; withdrawn, 11th December 1992; sold to Harry Needle Railroad Company; to South Yorkshire Railway Preservation Society, Meadowhall, Sheffield, 31st January 1994; to Railway Age, Crewe, on hire, 22nd January 1999; to Fragonset, Derby, for overhaul, 9th June 2000; to Barrow Hill Engine Shed, Staveley, 4th December 2001; to Fragonset, Derby, May 2002; to Barrow Hill Engine Shed, Staveley, 4th February 2003; noted in yellow and grey livery with number 08936, Barrow Hill Engine Shed, October 2003; to Network Rail, Whitemoor Yard, March, on hire, about May 2004; re-sold to Cotswold Rail, Moreton in Marsh, about July 2004; to Allelys Ltd, Studley, Warwickshire, for storage, by 31st January 2006; to Alstom Transport, Stonebridge Park Heavy Repair Depot, Wembley, on hire, by 23rd February 2006; to Willesden Depot, 28th May 2006; to Horton Road Depot, Gloucester, for storage, early November 2006; noted in orange and grey livery with number 08936, Horton Road Depot, 21st March 2007 and 1st September 2007; re-sold to RMS Locotec Ltd, 2007; to RMS Locotec Ltd, Wakefield, for repairs, 4th October 2007; to Corus, Shotton, on hire, November 2007; noted in blue livery with number 08936, Shotton, May 2016; to RMS Locotec Ltd, Wolsingham Depot, 17th August 2016; noted in black livery with numbers 08936 and HO75, Wolsingham Depot, 21st February 2023; noted at Wolsingham, 5th January 2024; to Independent Railway Engineering, Chesterfield, 19th January 2024.

D4167   Darlington   1962   84A   12/93   P   D4167
08937   new 24th May 1962; withdrawn, 23rd December 1993; sold to English China Clays; to English China Clays, Meldon Quarry, Devon, 4th March 1994; used at the quarry and by the Dartmoor Railway; to RMS Locotec Ltd, Dewsbury, for repairs, 25th November 2005; to Wabtec, Doncaster, for further repairs, May 2006; to Meldon Quarry, Devon, 24th April 2007; noted in green livery with name BLUEBELL MEL, Meldon Quarry, October 2007; quarry mothballed, July 2011; locomotive continued to be used by Dartmoor Railway; noted in green livery with number D4167, Dartmoor Railway, 12th March 2016 and 9th September 2019.

D4169   Darlington   1962   16A   ?   P   08939
08939   new 29th May 1962; withdrawn, date not known; put on sale by DB Schenker, 5th May 2015; sold to Railway Support Services; despatched from DB Schenker, Toton Depot, August 2015; to Railway Support Services, Wishaw; to Colne Valley Railway, Castle Hedingham, for storage, 22nd September 2015; noted in Euro Cargo Rail grey livery with number 08939, Colne Valley Railway, 14th February 2016; to Railway Support Services, Wishaw, 21st March 2017; noted in grey livery with number 08939, Railway Support Services, Wishaw, 23rd March 2017; to East Midlands Trains, Neville Hill Depot, Leeds, on

hire, 29th May 2019; to GBRf, Port of Felixstowe, on hire, 12th July 2019; to Railway Support Services, Wishaw, for repairs, 23rd October 2019; to GBRf, Felixstowe, on hire, 30th October 2019; to Railway Support Services, Wishaw, 6th February 2020; noted at Wishaw, 1st June 2020; to Chasewater Railway, 20th July 2020; to Springs Branch, Wigan, on hire, 20th November 2020; to Railway Support Services, Wishaw, about 3rd April 2023; seen in Euro Cargo Rail grey livery with number 08939, Wishaw, 23rd September 2023; to Direct Rail Services, Garston, Liverpool, on hire, 16th November 2023; to Railway Support Services, Wishaw, 12th July 2024.

D4173   Darlington   1962   1A   7/88   P   08943
08943   new 19th June 1962; latterly the regular engine for the St Pancras Station pilot duty; withdrawn, 20th July 1988; arrived at Crewe Works, 25th October 1988; acquired by ABB Transportation, Crewe Works, April 1989; locomotive included in sale when BREL works privatised in April 1989; to ABB Transportation, Derby, September 1989; noted with number 002, December 1992; to ABB Transportation, York, February 1993; noted with running number 002, ABB York, 27th January 1996; to ABB British Wheelset Ltd, Trafford Park, Manchester, about August 1996; to ABB Transportation, Crewe, May 1998; carried running number PET II; noted in green livery with number D4173, ABB Crewe, May 2000; re-sold to Harry Needle Railroad Company, 31st July 2009; to Barrow Hill Engine Shed, Staveley, 12th February 2010; to Southall Railway Centre, London (for use in connection with Railway Children play at Waterloo Station), on hire, 25th May 2010; to Bombardier Transportation, Central Rivers Depot, Barton under Needwood, on hire, 7th January 2011; noted in HNRC orange livery with silver roof and number 08943, Central Rivers Depot, 12th April 2017; to Barrow Hill Engine Shed, Staveley, 25th November 2019; seen in HNRC orange and silver livery with number 08943, Barrow Hill Engine Shed, 12th March 2022; to Midland Road Depot, Leeds, for tyre turning, 8th November 2022; noted at Midland Road, 24th November 2022; to Harry Needle Railroad Company, Worksop Depot, about 29th November 2022; to Alstom Transport, Central Rivers Depot, Barton under Needwood, about 30th January 2023.

D4174   Darlington   1962   81A   5/98   P   08944
08944   new 20th June 1962; withdrawn, 1st May 1998; sold to Mike Darnall, Newton Heath, Manchester, November 2000; to Wabtec, Doncaster, for overhaul, about December 2000; re-sold to East Lancashire Railway, Bury, and moved by April 2001; noted in black livery with number 08944, East Lancashire Railway, Bury, March 2005; to Crewe Electric Depot, for tyre turning, 7th March 2007; to East Lancashire Railway, Bury, 15th March 2007; noted in black livery with number 08944, East Lancashire Railway, Bury, July 2014 and April 2016; stored unserviceable and being used for spares, Bury, 10th June 2023.

D4176   Darlington   1962   8J   2/07   F   08946
08946   new 22nd August 1962; noted in two-tone grey livery with number 08946, Allerton Depot, 22nd January 2006; withdrawn, 28th February 2007; sold to Railway Support Services, Wishaw, and moved 23rd March 2007; noted in two-tone grey livery with number 08946, Wishaw, 17th April 2007; used for spares; scrapped, Wishaw, June 2008.

D4177   Darlington   1962   81A   2/07   P   08947
08947   new 28th August 1962; withdrawn, 28th February 2007; sold to T.J. Thomson & Son Ltd, Stockton-on-Tees (but did not move), about March 2007; re-sold to Foster Yeoman Quarries Ltd, about March 2007; despatched from EWS Westbury, 13th April 2007; arrived at Merehead Stone Terminal, 13th April 2007; to Whatley Quarry, by 9th September 2007; noted at Whatley, 22nd June 2008; to Merehead Stone Terminal, by 9th

June 2010; to Isle of Grain Stone Terminal, 6th June 2012; to Arlington Fleet Services Ltd, Eastleigh Works, for overhaul, 3rd May 2013; to Whatley Quarry, 2013; to Arlington Fleet Services Ltd, Eastleigh Works, for repairs, by 3rd October 2015; noted in Mendip Rail blue livery with number 08947, on a low-loader leaving Eastleigh Works, 23rd December 2015; noted on low-loader on A36, 23rd December 2015; to Whatley Quarry, 23rd December 2015; noted at Whatley Quarry, 28th December 2015; to LH Group, Barton under Needwood, by road, for repairs, 26th January 2022; noted in Mendip Rail blue livery with number 08947, Barton under Needwood, 2nd June 2023; to Whatley Quarry, by road, 3rd June 2023.

D4178   Darlington   1962   ?   ?   P   08948
08948   new 31st August 1962; withdrawn, date not known; sold to Eurostar UK Ltd, 1994; fitted with a Scharfenberg coupler; based at Eurostar, North Pole Depot, London; to Barrow Hill Engine Shed, Staveley, for repairs, 17th September 2004; noted in white, grey and blue livery with number 08948, Barrow Hill Engine Shed, 19th February 2006; to Eurostar, North Pole Depot, London, 21st February 2006; to Eurostar, Temple Mills Depot, Leyton, London, October 2007; noted at Temple Mills Depot, April 2008; noted in grey livery with number 08948, Temple Mills Depot, 18th June 2020.

D4183   Darlington   1962   16A   1/06   F   08953
08953   new 17th October 1962; withdrawn, January 2006; noted in grey livery at Doncaster Depot, 18th June 2006; sold to EMR; despatched from DB Schenker, Doncaster Depot, 16th May 2010; to European Metal Recycling, Attercliffe, Sheffield; seen in grey livery with number 08953, European Metal Recycling, Attercliffe, 1st November 2010 and 28th November 2011; scrapped, early May 2012.

D4184   Darlington   1962   8H   5/04   P   08954
08954   new 23rd October 1962; withdrawn from Allerton, May 2004; moved to Toton Depot, 8th September 2009; sold to Harry Needle Railroad Company; despatched from DB Schenker, Toton Depot, 15th February 2011; to Boden Rail Engineering, Washwood Heath, for repairs; to Nemesis Rail, Burton upon Trent, for storage, 18th October 2011; re-sold to Alstom Transport about April 2013; to Alstom Transport, Longsight Depot, Manchester, 29th May 2013, to Alstom Transport, Polmadie Depot, Glasgow, 1st May 2014; noted in blue livery with number 08954, Polmadie Depot, 16th February 2016; noted in green livery with number 08954, Polmadie Depot, 6th June 2016; to Arlington Fleet Services Ltd, Eastleigh Works, for repairs, 10th August 2016; to Alstom Transport, Edge Hill Depot, Liverpool, 21st September 2016; to Alstom Transport, Wembley Depot, London, 15th December 2016; noted on a low-loader on the M6, 25th January 2017; to Alstom Transport, Polmadie Depot, Glasgow, 25th January 2017; noted on southbound M6, 12th August 2020; to Alstom Transport, Longsight Depot, 12th August 2020; noted on a low loader, M40 near Warwick, 22nd June 2021; to Arlington Fleet Services Ltd, Eastleigh Works, for overhaul, 23rd June 2021; noted on a low loader, M6 near Stoke, 16th February 2022; to Alstom Transport, Polmadie Depot, 16th February 2022.

D4186   Darlington   1962   ?   ?   P   08956
08956   new 27th June 1962; withdrawn, date not known; sold to Serco Railtest, Derby, and moved about August 2001; to Fragonset Rail, Derby, by May 2005; noted in blue livery with number 08956, Fragonset Rail, June 2005; to Serco Railtest, Derby, about July 2007; to Metronet Rail, Old Dalby Test Centre, 9th December 2008; to Railway Support Services, Wishaw, for repairs, 19th February 2016; noted in Serco Railtest blue livery with number 08956, Wishaw, 23rd February 2016; to Serco Railtest, Old Dalby Test Centre, June 2016;

noted in green livery, with Serco branding and number 08956, Old Dalby Test Centre, 18th July 2017; whilst working at Old Dalby it carried a 92D shed-plate; to Harry Needle Railroad Company, Barrow Hill Engine Shed, 31st March 2021; seen in green livery with SERCO on side and number 08956, Barrow Hill Engine Shed, 31st August 2022.

SECTION 17:

British Railways built 0-6-0 diesel electric locomotives, numbered D3665-D3671, D3719-D3721, and D4099-D4114. Basically the same as a 08 shunter, but capable of a higher speed (27.5mph instead of 20mph). Later classified TOPS Class 09.

D3665   **Darlington**   **1959**   **36A**   **9/10**   **P**   **09001**
09001   new 2nd February 1959; withdrawn, 30th September 2010; sold to Heritage Shunters Trust; despatched from DB Schenker, Doncaster Depot, 28th January 2011; to Heritage Shunters Trust, Rowsley; repaired and began use in March 2012; noted in EWS red and yellow livery, Rowsley, 2nd September 2012; seen at Rowsley, 18th June 2022; repainted in red livery with number 09001, Rowsley, autumn 2022; major rebuilding of its engine began in 2023.

D3666   **Darlington**   **1959**   **75C**   **9/92**   **P**   **09002**
09002   new 3rd February 1959; withdrawn, 25th September 1992; sold to South Devon Railway, Buckfastleigh, and moved 11th June 1993; noted in green livery with number D3666, South Devon Railway, 16th July 1995; noted in green livery with number 09002, South Devon Railway, June 2005; re-sold to GBRf, early 2011; to LH Group, Barton under Needwood, for repairs, 2nd March 2011; to Great Central Railway, Ruddington, Nottingham, for running-in, 29th September 2011; to Freightliner, Trafford Park, Manchester, on hire, 1st October 2011; to GBRf, Whitemoor Yard, March, January 2013; to Moveright International, Wishaw, Warwickshire, 3rd February 2017; to Barrow Hill Engine Shed, Staveley, 6th February 2017; seen in green livery with number 09002, Barrow Hill Engine Shed, 5th April 2017; to GBRf, Whitemoor Yard, March, 7th April 2017; noted at Whitemoor Yard, 1st October 2017; to Heanor Haulage, 30th October 2020; to Midland Road Depot, Leeds, for tyre turning, 2nd November 2020; seen in green livery with number 09002, Midland Road Depot, 4th November 2020; to Harry Needle Railroad Company, Worksop Depot, for overhaul, 12th November 2020; to Barrow Hill Engine Shed, Staveley, 7th December 2020; seen in green livery with number 09002, HNRC, Barrow Hill Engine Shed, 12th March 2022.

D3667   **Darlington**   **1959**   **87B**   **5/08**   **F**   **09003**
09003   new 19th February 1959; withdrawn from Margam Depot, May 2008; sold to Harry Needle Railroad Company, July 2010; despatched from Margam Depot; noted on low-loader, 24th August 2010; arrived at Barrow Hill Engine Shed, Staveley, 26th August 2010; used for spares; remains to European Metal Recycling, Kingsbury, for scrap, 19th September 2011; scrapped September 2011.

D3668   **Darlington**   **1959**   **75C**   **4/99**   **P**   **D3668**
09004   new 17th February 1959; suffered accident damage, 1998; withdrawn (in blue livery), April 1999; sold to Lavender Line, Isfield, and moved 14th December 2000; to Spa Valley Railway, Tunbridge Wells, 12th March 2003; noted in blue livery with number 09004, Tunbridge Wells, 10th August 2003; re-sold to Swindon & Cricklade Railway, and moved 27th June 2009; noted in blue livery with number D3668, Swindon & Cricklade Railway, October 2013; to Avon Valley Railway, Bitton, 10th April 2014; to St Philip's Marsh Depot,

Bristol, for tyre turning, 29th May 2014; to Swindon & Cricklade Railway, 5th June 2014; noted in blue livery with number D3668, Swindon & Cricklade Railway, 1st April 2016; to Avon Valley Railway, Bitton, 26th January 2023; to Railway Support Services, Wishaw, 22nd February 2023; to Hams Hall Rail Freight Terminal, Coleshill, on hire, about 25th February 2023; to Avon Valley Railway, Bitton, 16th February 2024.

D3670   Darlington   1959   16A   4/11   P   09006
09006   new 25th March 1959; withdrawn, April 2011; put on sale by DB Schenker, 5th May 2015; sold to Harry Needle Railroad Company; despatched from DB Schenker, Toton Depot, 15th September 2015; to Nemesis Rail, Burton upon Trent, for storage; noted in EWS red and yellow livery with number 09006, Nemesis Rail, 13th January 2016 and 10th September 2022.

D3671   Darlington   1959   16A   12/09   P   D3671
09007   new 26th March 1959; moved to Toton Depot, 4th December 2009; withdrawn, December 2009; sold to London Overground, Willesden Depot, London, and moved in stages between 17th and 23rd September 2010; noted in green livery with number D3671, Willesden Depot, December 2012 and 3rd January 2019; to Alstom Transport, Stonebridge Park, on hire to cover for Unilok away for repairs, April 2023 to March 2024; to Willesden Depot, April 2024; noted back at Stonebridge Park, 31st July 2024.

D3719   Darlington   1959   87B   5/04   F   09008
09008   new 13th April 1959; noted in EWS red and yellow livery with number 09008 and unofficial name SMUDGER (after John Smith, retired Canton Depot plant fitter), working at Tremorfa, 7th May 2004; withdrawn, May 2004; noted at Canton Depot, 16th July 2005; sold to Harry Needle Railroad Company; tripped from Bescot Depot to Boden Rail Engineering, Washwood Heath, 7th January 2011; used for spares; noted in EWS red and yellow livery with number 09008, Washwood Heath, June 2011; remains to European Metal Recycling, Kingsbury, for scrap, 19th September 2011; scrapped September 2011.

D3720   Darlington   1959   81A   5/04   P   09009
09009   new 15th April 1959; withdrawn, May 2004; moved to Toton Depot, by road, 10th March 2009; sold to C.F. Booth Ltd, Rotherham; despatched from DB Schenker, Toton Depot, 25th January 2011; re-sold to GBRf; to LH Group, Barton under Needwood, for repairs, 14th February 2011; noted in EWS red and yellow livery with number 09009, LH Group, 1st March 2011; noted in grey livery with number 09009, Barrow Hill Engine Shed, May 2017; to Biffa Waste Services Ltd, Collyhurst Street, Manchester, by 6th October 2017; seen at Biffa, 20th March 2022; to Railway Support Services, Wishaw, 23rd August 2023; seen at RSS Wishaw, 23rd September 2023; to Gemini Rail, Wolverton Works, on hire, 22nd November 2023; to Railway Support Services, Wishaw, 13th August 2024.

D3721   Darlington   1959   81A   2/04   P   NPT
09010   new 13th April 1959; withdrawn, 5th February 2004; noted in grey livery with number 09010, Hither Green Depot, 21st January 2006; put up for sale by tender by DB Schenker, 19th April 2010; sold to South Devon Railway; despatched on road low-loader from DB Schenker, Hither Green Depot, 28th September 2010; arrived at South Devon Railway, Buckfastleigh, 30th September 2010; noted in green livery with number D3721, 9th August 2011; to St Philip's Marsh Depot, Bristol, for tyre turning, 30th August 2012; to South Devon Railway, 5th September 2012; noted in green livery with number D3721, South Devon Railway, 30th June 2013 and 24th April 2017; noted in black livery with no number, South Devon Railway, 25th September 2022.

D4100 Horwich 1961 81A 7/10 P D4100 / DICK HARDY
09012   new (LOT 294) 26th September 1961; noted in departmental grey livery with number 09012, Hither Green Depot, 21st January 2006; withdrawn, 31st July 2010; sold to Harry Needle Railroad Company; despatched from DB Schenker, Hither Green Depot, 22nd September 2010; to Barrow Hill Engine Shed, Staveley; noted at Barrow Hill Engine Shed, 9th December 2012; re-sold (price £40,000) to Severn Valley Railway, Bridgnorth, and arrived by road on 22nd February 2013; overhauled; repainted in green livery with number D4100 and DICK HARDY (lifelong railwayman and enthusiast) nameplates, Severn Valley Railway, spring 2015; re-entered service, June 2015; used as Kidderminster pilot, 2024.

D4102 Horwich 1961 16A 3/09 P 09014
09014   new (LOT 294) 12th October 1961; withdrawn, March 2009; moved to Doncaster Depot, 11th May 2009; sold to Harry Needle Railroad Company; despatched from DB Schenker, Doncaster Depot, 21st January 2011; to Barrow Hill Engine Shed, Staveley; to Boden Rail Engineering, Washwood Heath, for repairs; to Nemesis Rail, Burton upon Trent, for storage, 13th September 2012; noted in grey livery with number 09014, Nemesis Rail, February 2019 and 10th September 2022.

D4103 Horwich 1961 87B 3/07 P 09015 / ROB
09015   new (LOT 294) 26th October 1961; withdrawn, March 2007; sold to T.J. Thomson & Son Ltd; seen on a low-loader, northbound on M1 near Rotherham, at 09:10 on 21st February 2011; to T.J. Thomson & Son Ltd, Stockton-on-Tees, 21st February 2011; re-sold to Railway Support Services, Wishaw, and moved 29th March 2011; noted in EWS red and yellow livery with number 09015, Wishaw, 17th March 2012 and 1st February 2017; used for spares; remains to Avon Valley Railway, Bitton, 10th July 2019; seen in red and yellow livery with number 09015 and name ROB, Avon Valley Railway, 16th July 2021.

D4105 Horwich 1961 16A 3/10 P 09017 / LEO
09017   new (LOT 294) 7th November 1961; withdrawn, March 2010; sold to National Railway Museum, York; despatched from DB Schenker, Toton Depot; stored initially at Network Rail, Klondyke Yard, York, 10th August 2011; noted being moved by rail (with steam locomotive 42085), York, 22nd September 2011; noted in workshops at NRM York, for repairs and repaint in red livery, 10th May 2012; to Freightliner Maintenance Ltd, York, on hire, March 2014; to National Railway Museum, York, by early June 2016; seen in red livery with number 09017, National Railway Museum, York, 8th June 2016.

D4106 Horwich 1961 81A 7/04 P D4106
09018   new (LOT 294) 13th November 1961; withdrawn, July 2004; noted in EWS red and yellow livery with number 09018, Hither Green Depot, 21st January 2006; sold to Harry Needle Railroad Company; despatched from DB Schenker, Hither Green Depot, 29th September 2010; to Boden Rail Engineering, Washwood Heath, for repairs; to Metronet Rail, Old Dalby Test Centre, Leicestershire, on hire, 19th October 2011; to Lafarge, Hope Cement Works, Derbyshire, on hire, 7th June 2012; seen in orange & grey livery with number 09018, Hope Cement Works, 15th February 2013; re-sold to Bluebell Railway, Horsted Keynes, East Sussex, and moved 23rd April 2013; noted in orange and grey livery with number 09018, Bluebell Railway, 30th June 2015; to Bombardier Transportation, Ilford Depot, for tyre turning, 9th December 2015; to Bluebell Railway, Horsted Keynes, 10th December 2015; noted in green livery with number 09018, Horsted Keynes, 15th August 2022.

D4107   **Horwich**   **1961**   **16A**   **6/10**   **P**   **D4107**
09019   new (LOT 294) 16th November 1961; withdrawn, June 2010; sold to Harry
Needle Railroad Company; despatched from DB Schenker, Toton Depot, 16th February
2011; to Barrow Hill Engine Shed, Staveley; seen in 'Mainline' blue livery with number
09019, Barrow Hill Engine Shed, Staveley, 4th May 2011; to Nemesis Rail, Burton upon
Trent, for storage, 10th February 2012; noted in 'Mainline' blue livery with number 09019,
Nemesis Rail, 28th July 2012; re-sold to West Somerset Railway, Minehead, and moved
5th March 2013; noted in green livery with number D4107, West Somerset Railway, 11th
June 2016 and September 2019.

D4110   **Horwich**   **1961**   **16A**   **2/11**   **P**   **09022 / PB144**
09022   new (LOT 294) 8th December 1961; withdrawn, February 2011; sold to Harry
Needle Railroad Company; to Boden Rail Engineering, Washwood Heath, for repairs, 8th
August 2011; re-sold to Port of Boston, 28th July 2011; to Port of Boston, 16th April 2012;
to Tyseley Locomotive Works, Birmingham, for repairs, 8th February 2016; to Port of
Boston, 16th February 2016; noted in blue livery with number 09022, Port of Boston, 17th
December 2018; later given extra number PB144.

D4111   **Horwich**   **1961**   **40B**   **4/11**   **P**   **09023**
09023   new (LOT 294) 15th December 1961; withdrawn, April 2011; despatched from
DB Schenker, Immingham Depot, 24th August 2011; to European Metal Recycling,
Kingsbury; noted at Kingsbury, 29th March 2012; to European Metal Recycling Ltd,
Attercliffe, Sheffield, 18th September 2014; seen in EWS red and yellow livery with number
09023, European Metal Recycling Ltd, Attercliffe, 7th October 2014; seen long disused,
EMR, 25th February 2024.

D4112   **Horwich**   **1961**   **81A**   **5/08**   **P**   **09024**
09024   new (LOT 294) 22nd December 1961; withdrawn, May 2008; sold to C.F. Booth
Ltd; despatched from Eastleigh, 26th October 2011; noted on a low-loader on M40, 26th
October 2011; to C.F. Booth Ltd, Rotherham; seen in Mainline blue livery with number
09024, C.F. Booth Ltd, 11th November 2011; to Railway Support Services, Wishaw, for
storage, 10th December 2011; noted in Mainline blue livery with number 09024, Wishaw,
26th March 2012; re-sold to East Lancashire Railway, Bury, and moved 23rd January 2014;
noted in Mainline blue livery with number 09024, Bury, April 2016; regularly used as
Buckley Wells pilot; repainted Railfreight grey,with number 09024, 2017; noted at Bury,
10th June 2023.

D4113   **Horwich**   **1962**   **75C**   **8/05**   **P**   **09025 / D4113**
09025   new (LOT 294) 4th January 1962; withdrawn, 31st August 2005; sold to EKR;
despatched from Selhurst Depot, 25th September 2005; to East Kent Railway,
Shepherdswell; noted in Connex yellow and grey livery with number 09025, Shepherdswell,
23rd September 2008; re-sold to Lavender Line, Isfield, and moved 10th October 2014;
noted in green livery with number D4113, Lavender Line, July 2016.

D4114   **Horwich**   **1962**   **?**   **?**   **P**   **09026/CEDRIC WARES**
09026   new (LOT 294) 21st January 1962; resident shunter for several years at Lovers
Walk Depot, Brighton; believed withdrawn in 2015; despatched by low-loader from
Brighton, 22nd May 2016; to Spa Valley Railway, Tunbridge Wells; noted in green livery
with number 09026 and nameplate CEDRIC WARES, Spa Valley Railway, May 2016;
Cedric Wares was for many years a fitter at Brighton Works; noted at Spa Valley Railway,
under overhaul, 6th July 2024.

D3927   Horwich   1962   55G   10/16   P   09106 / 6

09106   new 16th February 1962; former number 08759; overhauled and re-geared by RFS, Kilnhurst, 1993, and re-numbered 09106; withdrawn, October 2016; put on sale by tender by DB Cargo, with bids due by 5th October 2016; sold to Harry Needle Railroad Company; noted in red livery with grey roof, Knottingley Depot, 26th October 2016; despatched from DB Cargo, Knottingley Depot; to DB Cargo, Ferrybridge, for storage, March 2017; seen in red livery with number 09106, Ferrybridge, 23rd March 2017 and 10th July 2017; to Barrow Hill Engine Shed, Staveley, 10th July 2017; to GBRf, Dagenham, on hire, 20th October 2017; seen on low-loader, M1 Motorway near Junction 26, 7th December 2017; to Barrow Hill Engine Shed, Staveley, 7th December 2017; noted repainted in HNRC orange livery with black roof, 17th April 2018; seen at Barrow Hill Engine Shed, 6th April 2019; to Railway Support Services, Wishaw, 7th June 2019; to Celsa, Cardiff, on hire, 10th June 2019; noted in orange livery with large number 6, Celsa, 15th June 2019; to Barrow Hill Engine Shed, 8th February 2021; noted at Barrow Hill Engine Shed, 27th February 2021; to Celsa, Cardiff, on hire, 11th May 2021.

D4013   Horwich   1961   36A   5/11   P   09107

09107   new (LOT 277) 2nd May 1961; former number 08845; seen at Doncaster Depot, in red and yellow livery with no number, 6th February 2011; withdrawn, 31st May 2011; sold to European Metal Recycling, Kingsbury, and moved 14th September 2011; noted in EWS red and yellow livery with no number, European Metal Recycling, Kingsbury, 29th March 2012; re-sold to Severn Valley Railway, Bridgnorth, and moved 15th June 2017; underwent lengthy overhaul; noted in BR blue livery with number 09107, Severn Valley Railway, 31st October 2018; entered service, 18th January 2019; overhauled in 2023; to Telford Steam Railway, on hire to work 'Polar Express' Christmas trains, 16th November 2023; to Severn Valley Railway, 16th January 2024.

D3536   Derby   1958   16A   ?   P   09201

09201   new (order D658) 8th August 1958; former number 08421; regeared, repainted grey and renumbered 09201, RFS, Kilnhurst, October 1992; withdrawn, date not known; put on sale by DB Schenker, 5th May 2015; sold to Harry Needle Railroad Company; despatched from DB Schenker, Toton Depot; moved direct to Hope Cement Works, Derbyshire, for storage, 14th September 2015; noted in grey livery with number 09201, Hope Cement Works, 25th November 2016; to Harry Needle Railroad Company, Worksop Depot, 6th June 2023.

D3884   Crewe   1960   ?   4/11   P   09204

09204   new (order E503) 19th May 1960; former number 08717; withdrawn, April 2011; to London & North Western Railway Company Ltd, Tyne Yard Depot, Lamesley, Gateshead (ex DB Schenker, with site), 21st April 2011; to London & North Western Railway Company Ltd, Carriage Works, Crewe, 3rd August 2012; to Crewe Electric Depot, for tyre turning, by 13th August 2012; to Carriage Works, Crewe, 29th August 2012; re-sold to Arriva TrainCare, Crewe, by 2019; noted in Arriva two-tone blue livery with number 09204, Crewe, 3rd June 2017 and 11th August 2020.

SECTION 18:

Clayton Equipment Co Ltd built Type 1 Bo-Bo diesel electric locomotives, numbered D8500-D8616, and introduced 1962. Built in two batches, the first for the Scottish Region, and the second for the Eastern and North Eastern Regions. Fitted with two Paxman 6ZHXL six-cylinder engines (each developing 450bhp at 1500rpm) and driving wheels of 3ft 3½in diameter. Later classified TOPS Class 17, with three sub-divisions of which D8568 was 17/1.

D8568   CE   4365/U69   1964   66A   10/71 P D8568

new 6th January 1964; withdrawn, 6th October 1971; sold to Hemel Hempstead Lightweight Concrete Co Ltd, 1972; it became the sole survivor of its class; received some repairs at Polmadie Depot, Glasgow, early September 1972; despatched from Polmadie Depot, under its own power (special notice No.2596), 08:20, Monday 11th September 1972; noted passing through Rotherham, under its own power, 11th September 1972; to Hemel Hempstead Lightweight Concrete Co Ltd, Cupid Green, Hertfordshire (due Harpenden Junction 20:28, 11th September 1972); thereafter worked mineral trains on the six-mile long Hemel Hempstead branch to Harpenden Junction; noted working at Hemel Hempstead Lightweight Concrete Co Ltd, 12th October 1972; fitted with replacement engine (received from Polmadie Depot), 1975; put up for sale, early 1977; re-sold to Ribblesdale Cement Co Ltd, Horrocksford Works, near Clitheroe, Lancashire; left Hemel Hempstead Lightweight Concrete Co Ltd, 15th June 1977; moved initially to Cricklewood Depot, for attention on 16th June 1977; left in a freight train leaving Cricklewood, 20th June 1977; arrived at Horrocksford Works, 24th June 1977; thereafter worked cement trains on the one-mile branch from the works to the main line at Pimlico; was repainted in 1977 into the Ribblesdale livery of light grey body and dark green underframe with red castle logo on both bonnet ends and both cabsides; noted with number No.8, Ribblesdale Cement Co Ltd, 18th August 1981; retired from service and put up for sale, 1982; re-sold to North Yorkshire Moors Railway, December 1982; left Clitheroe on a low-loader, 9th February 1983; arrived at North Yorkshire Moors Railway, Pickering yard, 11th February 1983; noted at North Yorkshire Moors Railway, 20th March 1983; its first run in preservation was on 1st April 1983; noted in original Ribble grey and green livery, Grosmont, 28th April 1984 and 23rd April 1988; seen temporarily numbered with Deutsche Bahn style number 290-068-6, Grosmont, 2nd August 1987; repainted in green livery with number D8568, spring 1989; noted at North Yorkshire Moors Railway, Grosmont, 24th July 1989 and 23rd June 1991; to BR Gloucester, 2nd August 1991; exhibited at BR Gloucester Depot open day, 18th August 1991; to BR Old Oak Common Depot, London, open day, August 1991; to BR Stonebridge Park Sidings, London, November 1991; to BR Willesden Depot, for tyre turning, December 1991; re-sold to Chinnor and Princes Risborough Railway, Oxfordshire, where it arrived on 25th April 1992; to Severn Valley Railway, Bridgnorth, for gala, October 1998; noted in green livery with number D8568, Severn Valley Railway, 11th October 1998; returned to Chinnor and Princes Risborough Railway; to Old Oak Common Depot, London, for open day on 5th August 2000; returned to Chinnor and Princes Risborough Railway; to Barrow Hill Engine Shed, Staveley, for open day, October 2001; returned to Chinnor and Princes Risborough Railway, 9th October 2001; noted in green livery with number D8568, Chinnor and Princes Risborough Railway, April 2014; to Severn Valley Railway, Bridgnorth, for gala, 8th September 2015; returned to Chinnor and Princes Risborough Railway, 13th October 2015; to Severn Valley Railway, Bridgnorth, for gala, 16th May 2017; returned to Chinnor and Princes Risborough Railway; noted on low-loader, M5 southbound, 9th September 2017; to Dean Forest Railway, for gala held on 15th to 17th September 2017; returned to Chinnor and Princes Risborough Railway; to South Devon Railway,

Buckfastleigh, for gala, early October 2017; returned to Chinnor and Princes Risborough Railway, 7th November 2017; to Spa Valley Railway, for gala held on 4th and 5th August 2018; to Epping Ongar Railway, Essex, for gala held on 15th and 16th September 2018; returned to Chinnor and Princes Risborough Railway; to Severn Valley Railway, for gala, September 2019; to Nene Valley Railway, for gala, 8th October 2019; to Severn Valley Railway, about December 2019; noted in blue livery, Severn Valley Railway, 16th May 2021; noted leaving SVR Kidderminster, on a low-loader, 8th June 2022; to West Somerset Railway, for gala, 8th June 2022; returned to Chinnor and Princes Risborough Railway; to Swanage Railway, for gala, 9th May 2023; returned to Chinnor and Princes Risborough Railway; to Severn Valley Railway, 17th May 2023; noted at Severn Valley Railway, 28th September 2023; to Kent & East Sussex Railway, for gala, early April 2024; to Severn Valley Railway, about 17th April 2024.

SECTION 19:

British Railways built Type 1 0-6-0 diesel hydraulic locomotives, numbered D9500-D9555, built at Swindon Works and introduced July 1964. Fitted with a Paxman 'Ventura' 6YJX engine developing 650bhp at 1500rpm, and driving wheels of 4ft 0in diameter. Later classified TOPS Class 14.

D9500   Swindon   1964   86A   4/69   P   D9500
new 24th July 1964; moved to Worcester Depot, for storage, 5th December 1967; withdrawn 26th April 1969; moved to Canton Depot; sold to NCB; despatched from BR Canton Depot, Cardiff, working 8X50, departing at 09:40 on 17th November 1969; noted at York Depot, 22nd November 1969; noted at Gateshead Depot, 24th November 1969; to NCB Ashington Colliery, 25th November 1969; to Ashington Central Workshops, for repainting in blue, early June 1972; to Ashington Colliery, 7th June 1972; noted in blue livery with numbers No.1 and plant number 9312/92, Ashington Colliery, 25th June 1972; to BR South Gosforth Depot, for tyre turning, 7th December 1972; to Ashington Central Workshops, for replacement engine to be fitted, December 1972; to Ashington Colliery, October 1973; to BR South Gosforth Depot, for tyre turning, early April 1977; to Ashington Colliery, 12th April 1977; to Lambton Engine Works, Philadelphia, 12th July 1978; to Ashington Colliery, September 1978; to Lambton Engine Works, Philadelphia, for major overhaul, 10th November 1978; seen in blue livery with numbers No.1 and 9312/92, with RE 1433 of 1953 registration plate, Lambton Engine Works, 26th June 1979; noted at Lambton, 5th July 1979; to Ashington Colliery, 14th August 1979; to BR Thornaby Depot, for tyre turning, 14th May 1982; to Ashington Colliery, 24th May 1982; by road low-loader to Lambton Engine Works, 12th May 1983; to Ashington Colliery, 25th July 1983; to Lambton Engine Works, for repairs to propshaft, September 1986; to Ashington Colliery, autumn 1986; seen at Ashington Colliery, 20th April 1987; put up for sale, August 1987; re-sold to Llangollen Railway, and moved 25th September 1987; to Heritage Centre, Swindon, April 1988; noted in blue livery with numbers No.1 and 9312/92, Swindon, May 1989; to West Somerset Railway, Minehead, 4th November 1989; re-sold to South Yorkshire Railway Preservation Society, Meadowhall, Sheffield, and moved August 1992; to Barrow Hill Engine Shed, Staveley, for static display, 2nd August 2001; to Heritage Shunters Trust, Rowsley, 13th May 2008; seen with number D9500, Rowsley, 12th December 2008; acquired by Harry Needle Railway Company, in part exchange for D9525, summer 2010; later re-sold to Andrew Briddon and moved to Darley Dale, 29th May 2015; re-sold and

departed on a Heanor Fleet Services Ltd low-loader, 5th November 2021; to a private site at Port Elphinstone, Inverurie, for overhaul, 9th November 2021.

D9502   Swindon   1964   86A   4/69   P   D9502
new 27th July 1964; withdrawn 26th April 1969; sold to NCB; despatched in four-locomotive convoy from BR Canton Depot, Cardiff, hauled by D6607, 30th June 1969; noted at Horton Road Depot, Gloucester, 3rd July 1969; continued to Washwood Heath; departed Washwood Heath, hauled by D102, 16th July 1969; delivered to NCB Ashington Colliery, 19th July 1969; allocated plant number 9312/97, but this number never carried externally; to Burradon Colliery, Dudley, 6th September 1969; seen at Burradon Colliery, 21st August 1970; to Ashington Central Workshops, for overhaul, 24th July 1973; to Burradon Colliery, Dudley, 14th March 1974; noted in original BR two-tone green livery, Burradon Colliery, 25th October 1974; Burradon Colliery closed on 22nd November 1975; to Weetslade Coal Preparation Plant, early January 1976; to Backworth Colliery, 26th January 1976; noted at Backworth Colliery, 17th February 1976; to Weetslade Coal Preparation Plant, June 1976; noted with number 9312/97, Weetslade, 9th April 1980; to Ashington Colliery, 24th April 1981; to BR Thornaby Depot, for tyre turning, 28th April 1983; by road low-loader to Ashington Colliery, 12th May 1983; disused with worn-out engine, March 1986; officially became a source of spares, 15th April 1986; seen in original BR two-tone green livery with number D9502, Ashington Colliery, 20th April 1987; re-sold to Llangollen Railway, and moved 25th September 1987; re-sold to South Yorkshire Railway Preservation Society, Meadowhall, Sheffield, and moved 12th March 1992; to Heritage Shunters Trust, Rowsley, March 2002; seen with number D9502, Heritage Shunters Trust, Rowsley, 12th December 2003; re-sold to East Lancashire Railway, Bury, and moved 20th November 2014; noted at Bury, partly dismantled, under overhaul, 10th June 2023.

D9503   Swindon   1964   50B   4/68   F   65 / 8411/25
new 30th July 1964; withdrawn 1st April 1968; sold to Stewarts & Lloyds Minerals Ltd, Harlaxton Quarries, Lincolnshire; despatched from BR Hull Dairycoates Depot, working 8G38, 11th November 1968; allocated numbers 25 and 8411/25; No.25 applied, June 1970; 8411 was the identification number for North Welland locomotives; noted with number 25, Harlaxton, 24th September 1973; production ceased at Harlaxton, February 1974; noted at Harlaxton, 19th March 1974; moved by low-loader to Corby Quarries, 29th July 1974; renumbered 65, October 1974; noted out of use and being used for spares, 1st May 1976; scrapped on site by Shanks & McEwan Ltd of Corby, September 1980.

D9504   Swindon   1964   50B   4/68   P   D9504
new 8th August 1964; withdrawn 1st April 1968; sold to NCB; despatched from BR Dairycoates Depot, Hull, 27th November 1968; to NCB Philadelphia Locomotive Shed, County Durham; arrived at Philadelphia, 29th November 1968; noted at Philadelphia shed, 1st February 1969; repainted emerald green and numbered 506, August 1969; seen at Philadelphia, 21st August 1970; noted with number 506, Philadelphia, 13th June 1973; to Boldon Colliery, 21st August 1973; to Lambton Engine Works, Philadelphia, 7th September 1973; to Boldon Colliery, February 1974; to Burradon Colliery, 17th December 1974; to BR Cambois Depot, for repairs, 3rd January 1975; to Burradon Colliery, Dudley, 29th January 1975; to Weetslade Coal Preparation Plant, 3rd January 1976; noted with numbers 506 and 2233/506, Weetslade, 26th June 1979; to Lambton Engine Works, Philadelphia, 21st April 1981; noted at Lambton Engine Works, 18th June 1981; to Ashington Colliery, 11th September 1981; seen in green livery with number 506, Ashington Colliery, 20th April 1987; re-sold to Kent & East Sussex Railway, Tenterden, and moved 25th September 1987; to Nene Valley Railway, Wansford, for overhaul, 25th February 1998; returned to Kent & East

Sussex Railway, Tenterden, by road, 21st April 1999; given number No.48; to Channel Tunnel Rail Link (CTRL1), Beechbrook Farm, near Ashford, on hire, 1st August 2001; noted in green livery with number No.48, Beechbrook Farm, September 2002; to Medway Ports, Chatham Dockyard, on hire, February 2003; to Nene Valley Railway, Wansford, for repairs, 4th April 2003; to EWS Toton Depot, for tyre turning, 7th January 2004; to Nene Valley Railway, Wansford, 9th January 2004; to Victa Rail, March, on hire, 15th January 2004; to Nene Valley Railway, Wansford, for repairs, 8th April 2004; to Channel Tunnel Rail Link (CTRL2), Swanscombe, Kent, on hire, 15th June 2004; to Channel Tunnel Rail Link, Dagenham, on hire, 2nd November 2004; to Nene Valley Railway, Wansford, for repairs, 24th February 2005; to Channel Tunnel Rail Link, Dagenham, on hire, by 3rd May 2005; to Nene Valley Railway, Wansford; to Channel Tunnel Rail Link, Swanscombe, on hire, 7th November 2005; to Nene Valley Railway, Wansford, 13th January 2006; to Channel Tunnel Rail Link, Dagenham, on hire, 2nd March 2006; to Nene Valley Railway, Wansford, 25th January 2007; to Aggregate Industries Ltd, Bardon Hill, on hire, 16th January 2008; to Nene Valley Railway, Wansford, 10th February 2009; re-sold to Kent & East Sussex Railway, Tenterden, and moved 5th May 2010; noted in green livery with number D9504, Kent & East Sussex Railway, April 2019; has registration 01566; fitted with JON GRIMWOOD nameplates, on 6th April 2024; attended the Class 14's 60th anniversary gala, held at the Ecclesbourne Valley Railway on 25th to 28th July 2024.

D9505   Swindon   1964   50B   4/68   F   MICHLOW
new 2nd August 1964; withdrawn 1st April 1968; sold to APCM, Hope Cement Works, Derbyshire (reputedly at a price of £4,000); despatched from BR Hull Dairycoates Depot, 26th September 1968; arrived at Hope, 26th September 1968; named MICHLOW (cast plate on cabside) in 1970; noted at Hope Cement Works, 17th March 1973; seen at Hope Cement Works, 16th June 1974; noted disused at Hope Cement Works, 13th October 1974; re-sold to Hunslet Engine Company, Leeds (and allocated works number 7496), 22nd April 1975; left Hope, on a low-loader, 5th May 1975; noted on a low-loader, March Depot, 9th May 1975; noted at March Depot, 10th May 1975; left March Depot by rail; noted being towed through Ipswich by 37044, working 9X35 to Harwich Docks, 10th May 1975; exported from Harwich Docks (see Appendix C).

D9507   Swindon   1964   50B   3/68   F   55 / 8311/35
new 28th August 1964; withdrawn 1st April 1968; sold to S&L; despatched from BR Hull Dairycoates Depot, to Stewarts & Lloyds Minerals Ltd, Corby Quarries, 25th November 1968; initially intended as a source of spares, but made into a 'runner', by Stewarts & Lloyds Minerals Ltd and allocated numbers 35 and 8311/35; 8311 was the identification number for Corby and Glendon locomotives; seen at Gretton Brook shed, 23rd June 1973; renumbered 55, October 1974; to BSC Steelworks Disposal Site, Heavy Mills, (in sidings adjacent to the now closed Bessemer Plant), Corby, 29th December 1980; noted at Heavy Mills, 4th June 1981; noted on 25th August 1982; scrapped on site by Shanks & McEwan Ltd of Corby, September 1982.

D9508   Swindon   1964   87E   10/68   F   No.9 / 9312/99
new 7th September 1964; to store at Canton Depot, September 1968; withdrawn 5th October 1968; sold to NCB; despatched from BR Canton Depot, Cardiff, 6th March 1969; noted at Derby, hauled by D161, 7th March 1969; noted at Toton Depot, 9th March 1969; arrived at NCB Ashington Colliery, 14th March 1969; noted in BR two-tone green with plant number 9312/99 painted over BR number, Ashington Colliery, 29th June 1969; to Ashington Central Workshops, for repainting, June 1972; to Ashington Colliery; noted in blue livery with number No.9, Ashington Colliery, 22nd October 1972; to BR South Gosforth

Depot, for tyre turning, 26th June 1973; to Ashington Colliery, 29th June 1973; seen at Ashington Colliery, 13th July 1974; fitted with overhauled engine, May 1975; to BR South Gosforth Depot, for tyre turning, May 1977; to Ashington Colliery, 2nd June 1977; noted with RE 1708 of 1953 registration plate, Ashington Colliery, 7th May 1980; out of use, 7th May 1980; noted derelict, July 1983; officially withdrawn October 1983; scrapped on site by D. Short Ltd of North Shields, 17th January 1984.

D9510　　Swindon　　　　　　1964　　50B　　　4/68　　F　60 / 8411/23

new 18th September 1964; withdrawn 1st April 1968; sold to S&L; despatched from BR Hull Dairycoates Depot; to Stewarts & Lloyds Minerals Ltd, Buckminster Quarries, Lincolnshire, 4th December 1968; allocated numbers 23 and 8411/23, September 1969; Buckminster Quarries ceased production, 27th January 1972; noted at BR Grantham Station, 31st August 1972; to Corby Quarries, 6th September 1972; seen at Gretton Brook shed, 23rd June 1973; renumbered 60, October 1974; noted at Gretton Brook shed, 18th April 1979; to BSC Tube Works, Corby, July 1980; to BSC Steelworks Disposal Site, Heavy Mills, Corby, January 1981; noted at Heavy Mills, 8th June 1982; scrapped on site by Shanks & McEwan Ltd of Corby, August 1982.

D9511　　Swindon　　　　　　1964　　50B　　　4/68　　F　9312/98

new 17th September 1964; withdrawn 1st April 1968; sold to NCB; despatched from BR Hull Dairycoates Depot; noted at Tyne Yard, 5th January 1969; to Ashington Colliery, 7th January 1969; it was the first Class 14 to arrive at Ashington Colliery; noted at Ashington Colliery, early April 1969; given plant number 9312/98; to Bates Colliery, Blyth, for trial, 18th April 1969; trial not successful due to tight curves; to Burradon Colliery, Dudley, 5th May 1969; noted in BR two-tone green with number 9312/98, Burradon Colliery, 10th June 1969; noted with Railway Executive plate 1750 of 1953, Burradon Colliery, 3rd September 1970; suffered severe fire damage, Havannah Drift, July 1972; to Ashington Central Workshops, for assessment, October 1972; declared beyond economic repair; moved to Ashington Colliery; used as a source of spares; seen out of use at Ashington Colliery, 13th July 1974; noted dismantled, Ashington Colliery, 23rd September 1979; remains scrapped, about October 1979.

D9512　　Swindon　　　　　　1964　　50B　　　4/68　　F　63 / 8411/24

new 29th September 1964; moved to Hull Dairycoates, hauled by D6893, 20th January 1967; withdrawn 1st April 1968; sold to S&L for £1,000; purchased as a source of spares; despatched from BR Hull Dairycoates Depot; to Stewarts & Lloyds Minerals Ltd, Buckminster Quarries, Lincolnshire, December 1968; allocated numbers 24 and 8411/24; Buckminster Quarries ceased production, 27th January 1972; noted at BR Grantham Station, 31st August 1972; to BSC Corby Quarries, 6th September 1972; used for spares; seen in semi-dismantled condition, in BR two-tone green livery with number D9512, Gretton Brook shed, 23rd June 1973; renumbered 63, October 1974; noted at Gretton Brook, 18th April 1979; to BSC Steelworks Disposal Site, Heavy Mills, Corby, 29th December 1980; noted at Heavy Mills, 7th February 1982; scrapped on site by Shanks & McEwan Ltd of Corby, February 1982.

D9513　　Swindon　　　　　　1964　　86A　　　3/68　　P　NCB 38

new 6th October 1964; withdrawn 10th March 1968; noted at BR Worcester Depot, 16th April 1968; sold per tender document dated 23rd May 1968; sold per 'Confirmation of sale' document dated 7th June 1968 to W.H. Arnott, Young & Co Ltd; despatched from BR Worcester Depot, 18th July 1968; to W.H. Arnott, Young & Co Ltd, scrap merchants, Parkgate, Rotherham; seen in BR two-tone green livery with number D9513, with odometer reading of 60,965 miles, Parkgate, 5th August 1968; seen at Parkgate, 24th November 1968; re-sold to Hargreaves (West Riding) Ltd, NCBOE British Oak Disposal Point, Crigglestone, late November 1968; noted at British Oak Disposal Point, Crigglestone, 24th December 1968; took over from steam locomotives, 28th January 1969; seen in green livery with number D1/9513, Crigglestone, 9th March 1969; noted re-painted orange and black, Crigglestone, 11th May 1969; to NCBOE Bowers Row Disposal Point, Astley, 5th September 1969; re-sold to NCB North East Area; to Allerton Bywater Central Workshops, West Yorkshire, for overhaul, October 1973; noted in a northbound freight train, York, 12th January 1974; to Ashington Central Workshops, for overhaul, 21st January 1974; to Ashington Colliery, May 1974; to Backworth Colliery, for repainting in dark blue, 10th June 1974; given number 38; to Burradon Colliery, Dudley, July 1974; noted with number 38, Burradon Colliery, 28th October 1974; Burradon Colliery closed on 22nd November 1975; to Backworth Colliery, 5th January 1976; noted at Backworth Colliery, 17th February 1976; to Weetslade Coal Preparation Plant, late February 1976; to Lambton Engine Works, Philadelphia, 22nd November 1976; to Ashington Colliery, 14th February 1977; seen at Ashington Colliery, 21st May 1977; to Lambton Engine Works, 8th September 1977; noted at Lambton Engine Works, 20th September 1977; to Ashington Colliery, 14th November 1977; overhauled engine fitted, June 1978; noted with number 38, Ashington Colliery, 7th May 1980; overhauled engine fitted, 29th June 1982; to Lambton Engine Works, 30th June 1982; to Ashington Colliery, 17th February 1983; was in collision with D9555 at Lynemouth Colliery, 16th October 1985; noted in use, January 1987; engine failure, late January 1987; seen in navy blue livery, with numbers 38 and D1/9513, Ashington Colliery, 20th April 1987; re-sold to C.F. Booth Ltd of Rotherham, September 1987; did not leave Ashington site and was re-sold by Booth's for preservation; noted at Ashington, 8th October 1987; re-sold to Embsay & Bolton Abbey Railway, 12th October 1987; noted in blue livery with number NCB 38, Embsay & Bolton Abbey Railway, June 1988; to East Lancashire Railway, Bury, for gala, 22nd July 2014; noted in blue livery with number NCB 38, East Lancashire Railway, July 2014; to Embsay & Bolton Abbey Railway, 1st August 2014; noted in blue livery with number NCB 38, Embsay & Bolton Abbey Railway, 9th September 2022; re-sold to Wensleydale Railway and moved on 5th October 2023.

D9514   Swindon   1964   86A   4/69   F   No.4 / 9312/96
new 6th October 1964; withdrawn 26th April 1969; sold to NCB; despatched in a four-locomotive convoy from BR Canton Depot, Cardiff, hauled by D6607, 30th June 1969; noted at Horton Road Depot, Gloucester, 4th July 1969; continued to Washwood Heath; departed Washwood Heath, hauled by D102, 16th July 1969; delivered to NCB Ashington Colliery, 19th July 1969; noted with plant number 9312/96, Ashington Colliery, 25th October 1971; fitted with overhauled engine, April 1972; to Ashington Central Workshops, for repainting, June 1972; painted blue with number No.4; to Ashington Colliery, June 1972; to BR Gosforth Depot, for tyre turning, 12th October 1972; to Ashington Colliery, 26th October 1972; noted in blue livery with number No.4, Ashington Colliery, 29th October 1972; seen at Ashington Colliery, 13th July 1974; to BR Gosforth Depot, for tyre turning, 11th October 1975; to Ashington Colliery, November 1975; seen at Ashington Colliery, 21st May 1977; to Lambton Engine Works, Philadelphia, for engine repairs, about August 1977; to Ashington Colliery, 23rd December 1977; noted with RE 1474 of 1953 registration plate, Ashington Colliery, 7th May 1980; to Lambton Engine Works, Philadelphia, September 1980; to Ashington Colliery, 3rd December 1980; fitted with overhauled engine, January 1981; to BR Thornaby Depot, for tyre turning, early October 1982; noted at Thornaby Depot, 10th October 1982; returned to Ashington Colliery; out of use at Ashington Colliery from 24th November 1983; scrapped on site by Robinson & Hannan Ltd of Blaydon,

commencing 5th December 1985; engine and various parts sold to Rutland Railway Museum, 13th December 1985.

D9515   Swindon   1964   50B   4/68   F   62 / 8411/22

new 17th October 1964; moved from Cardiff to Hull, hauled by D1572, 18th January 1967; withdrawn 1st April 1968; sold to Stewarts & Lloyds Minerals Ltd; despatched from BR Hull Dairycoates Depot, hauled by D6737, 2nd November 1968; to Stewarts & Lloyds Minerals Ltd, Buckminster Quarries, Lincolnshire, 2nd November 1968; allocated number 8411/22; painted dark green with number 22, September 1969; worked Wirral Railway Circle rail tour, 1st January 1972; Buckminster Quarries ceased production, 27th January 1972; noted at BR Grantham Station, 31st August 1972; arrived at Corby Quarries, 6th September 1972; seen with number 22, Gretton Brook shed, 23rd June 1973; renumbered 62, October 1974; noted with number 62, Corby Quarries, 21st July 1980; to BSC Steelworks Disposal Site, Heavy Mills, Corby, 29th December 1980; noted at Heavy Mills, 17th November 1981; re-sold to The Hunslet Engine Co Ltd, Leeds, and moved December 1981; chosen due to having a recently rebuilt engine and new tyres fitted; overhauled and converted to 5ft 6in gauge; Aiken Espanola SA of Madrid had placed the order on Hunslet on 4th November 1981 and the conversion was carried out to Hunslet works order number GM1024; locomotive repainted yellow; despatched from The Hunslet Engine Co Ltd, 11th June 1982; noted in yellow livery with no number, on a ship in Goole Docks, 16th June 1982; exported from Goole Docks (see Appendix C).

D9516   Swindon   1964   50B   4/68   P   D9516

new 21st October 1964; withdrawn 1st April 1968; sold to S&L for £1,000; despatched from BR Hull Dairycoates Depot, 25th November 1968; purchased as a source of spares, by Stewarts & Lloyds Minerals Ltd, Corby Quarries; but repaired and returned to traffic; allocated numbers 36 and 8311/36; repainted green with number 36; seen at Gretton Brook shed, 23rd June 1973; renumbered 56, October 1974; to BSC Steelworks Disposal Site, Heavy Mills, Corby, 29th December 1980; noted at Heavy Mills, 4th June 1981; re-sold to Great Central Railway, Loughborough, and moved 17th October 1981; seen at Loughborough, 14th March 1982; to Severn Valley Railway, Bridgnorth, for diesel weekend, 15th October 1988; returned to Great Central Railway; noted in green livery, Great Central Railway, 21st August 1988; re-sold to Nene Valley Railway, Wansford, and moved by road on 8th December 1988; seen in green livery with number D9516, Wansford, 25th February 1993; to Boden Rail Engineering, Washwood Heath, 1st April 2011; to Wensleydale Railway, Leeming Bar, 11th April 2011; to Midland Road Depot, Leeds, for tyre turning, 21st February 2014; to Nene Valley Railway, Wansford, for repairs, 14th March 2014; re-sold to Didcot Railway Centre, Oxfordshire, and moved 14th May 2014; noted in green livery with number D9516, Didcot, May 2016; to Old Oak Common Depot, London, for open day, September 2017; returned to Didcot Railway Centre, September 2017; noted in BR green livery with cast D9516 numberplates, Didcot, August 2018; to Epping Ongar Railway, 13th September 2023; to Didcot Railway Centre, 7th October 2023.

D9517   Swindon   1964   86A   10/68   F   No.8 / 9312/93

new 12th November 1964; withdrawn 5th October 1968; sold to NCB; despatched from BR Canton Depot, Cardiff, working 8X50, departing at 09:40 on 17th November 1969; noted at York Depot, 22nd November 1969; noted at Gateshead Depot, 24th November 1969; to NCB Ashington Colliery, 25th November 1969; noted with plant number 9312/93, Ashington Colliery, 24th May 1970; noted with Railway Executive 1597 of 1953 plate, Ashington Colliery, 1st October 1970; fitted with overhauled engine, April 1971; to Ashington Central Workshops, for repainting, June 1972; painted blue with number No.8;

to Ashington Colliery, June 1972; to BR Gosforth Depot, for tyre turning, 30th January 1974; to Ashington Colliery, 4th February 1974; fitted with overhauled engine, June 1976; seen with numbers No.8 and 9312/93, Ashington Colliery, 21st May 1977; to Lambton Engine Works, Philadelphia, 14th June 1977; to Ashington Colliery, 5th September 1977; to NCBOE Butterwell Opencast Site, on loan, September 1977; to Ashington Colliery, September 1977; to BR Gosforth Depot, for tyre turning, early September 1979; to Ashington Colliery, 12th September 1979; noted with RE 1597 of 1953 registration plate, Ashington Colliery, 7th May 1980; out of use by 24th April 1981; officially withdrawn November 1983; scrapped on site by D. Short Ltd of North Shields, January 1984.

D9518   Swindon   1964   86A   4/69   P D9518
new 30th October 1964; held the distinction of being the last of the class to operate on BR, from Radyr Depot until 19th April 1969; withdrawn 26th April 1969; sold to NCB; despatched in a four-locomotive convoy from BR Canton Depot, Cardiff, hauled by D6607, 30th June 1969; noted at Horton Road Depot, Gloucester, 4th July 1969; continued to Washwood Heath; departed Washwood Heath, hauled by D102, 16th July 1969; delivered to NCB Ashington Colliery, 19th July 1969; to Ashington Central Workshops, 15th September 1969; to Ashington Colliery, October 1969; fitted with overhauled engine, October 1970; received Railway Executive registration plates, 1430 of 1953; noted with plant number 9312/95, Ashington Colliery, 24th May 1970; to Ashington Central Workshops, for repainting, June 1972; painted blue with number No.7; to Ashington Colliery, June 1972; noted at Ashington Colliery, 22nd October 1972; to BR Gosforth Depot, for tyre turning, 22nd May 1973; to Ashington Colliery, late May 1973; to Lambton Engine Works, Philadelphia, 1st May 1975; noted at Lambton Engine Works, Philadelphia, 29th August 1975; fitted with overhauled engine; to Ashington Colliery, September 1975; seen at Ashington Colliery, 21st May 1977; to Lambton Engine Works, Philadelphia, for new tyres, 29th August 1980; noted in dark blue livery, Lambton Engine Works, 6th September 1980 and 4th December 1980; to Ashington Colliery, 5th December 1980; noted with numbers No.7 and 9312/95, and RE 1430 of 1953 registration plate, Ashington Colliery, 7th May 1980; noted in 'Ashington blue' livery, 25th March 1986; withdrawn, 17th April 1986; seen at Ashington Colliery, 25th September 1987; re-sold to Rutland Railway Museum, Cottesmore, and moved by road, 26th September 1987; seen in blue livery with numbers No.7 and 9312/95, Cottesmore, 31st May 1990; re-sold to Nene Valley Railway, Wansford, and moved 8th September 2006; noted in blue livery with number No.7, Nene Valley Railway, October 2011; re-sold to West Somerset Railway, Minehead, and moved 1st December 2011; still undergoing long-term overhaul, 2024.

D9520   Swindon   1964   50B   4/68   P D9520 / 45
new 12th November 1964; moved to Hull Dairycoates, hauled by D6873, 23rd January 1967; withdrawn 1st April 1968; sold to Stewarts & Lloyds Minerals Ltd; despatched from BR Hull Dairycoates Depot, 16th December 1968; arrived at Glendon East Quarries, Northamptonshire, 16th December 1968; allocated numbers 24 and 8311/24; noted at Glendon East, 11th January 1969; to Corby Quarries, 12th January 1970; seen at Gretton Brook shed, 23rd June 1973; given number 45, October 1974; transferred to BSC Tube Works stock, for shunting steel strip trains from Lackenby, October 1980; noted at Tube Works, 21st January 1981; re-sold to North Yorkshire Moors Railway, Grosmont, and moved 16th March 1981; noted at NYMR, 20th March 1983; re-sold to Rutland Railway Museum, Cottesmore, and moved by road, 21st February 1984; to Great Central Railway, Loughborough, on loan, 5th October 1985; returned to Rutland Railway Museum, Cottesmore, 2nd December 1985; seen with number 45, Cottesmore, 31st May 1990; to Great Central Railway, Ruddington, Nottingham, for restoration, 6th March 1998; re-sold to

Nene Valley Railway, Wansford, and moved 21st April 2004; to West Somerset Railway, Minehead, for gala, 15th June 2007; returned to Nene Valley Railway, Wansford; to Appleby Frodingham RPS, Scunthorpe, for gala, 9th to 11th May 2008; to National Railway Museum, York, May 2008; to West Somerset Railway, for gala, 13th to 15th June 2008; to Barrow Hill Engine Shed, Staveley, for gala, August 2008; to Lafarge, 'Hope Cement Works Rail Weekend', Derbyshire, 6th and 7th September 2008; to Nene Valley Railway, Wansford, 8th September 2008; to West Somerset Railway, Minehead, for gala, June 2009; to Nene Valley Railway, Wansford, 14th June 2009; to West Somerset Railway, Minehead, for gala, 7th June 2010; to Nene Valley Railway, Wansford, 14th June 2010; noted in green livery with number D9520, Nene Valley Railway, June 2011; to East Lancashire Railway, Bury, for gala, 23rd July 2014; to Nene Valley Railway, Wansford, 29th July 2014; worked temporarily renumbered as D9505 (which see) at Hope Cement Works open day, 6th September 2014; returned to Nene Valley Railway; to Mid-Norfolk Railway, 9th June 2021; to Keighley & Worth Valley Railway (with number D9520 on one side and 45 on the other), for gala, 30th May 2023; remained at KWVR and received repairs in January 2024; displayed at KWVR Diesel Gala, 20th to 23rd June 2024.

D9521   Swindon   1964   87E   4/69   P   D9521

new 20th November 1964; withdrawn 26th April 1969; moved to Canton Depot, 1969; sold to NCB, February 1970; seen at Canton Depot, Cardiff, 2nd March 1970; despatched from BR Canton Depot, 4th March 1970; noted at Tyne Yard, 5th March 1970; arrived at NCB Ashington Colliery, 6th March 1970; noted with plant number 9312/90, Ashington Colliery, 24th May 1970; fitted with overhauled engine, December 1971; to Ashington Central Workshops, for repainting, early June 1972; painted blue with number No.3; to Ashington Colliery, June 1972; to BR Gosforth Depot, for tyre turning, 10th June 1974; to Ashington Colliery, June 1974; seen at Ashington Colliery, 13th July 1974; to BR Gosforth Depot, Newcastle upon Tyne, for repairs, 11th March 1976; to Ashington Colliery, 17th April 1976; to Lambton Engine Works, Philadelphia, 30th November 1977; to Ashington Colliery, 6th April 1978; noted with number 9312/90, Ashington Colliery, 7th May 1980; fitted with overhauled engine, September 1980; to Lambton Engine Works, Philadelphia, for overhaul and repaint, 7th January 1982; to Ashington Colliery, 30th June 1982; to Lambton Engine Works, Philadelphia, for repairs to transmission, March 1983; noted at Lambton Engine Works, 18th March 1983; to Ashington Colliery, 24th March 1983; seen in dark blue livery with numbers 9312/90 and No.3, Ashington Colliery, 20th April 1987; withdrawn; re-sold to C.F. Booth Ltd of Rotherham, September 1987; did not leave Ashington Colliery site and was re-sold by Booth's for preservation; noted at Ashington Colliery, 8th October 1987; re-sold to Rutland Railway Museum, Cottesmore, and moved 14th October 1987; seen in navy blue livery with numbers 9312/90 and No.3, Cottesmore, 31st May 1990; re-sold to Swanage Railway, Dorset, and moved 29th January 1992; seen in 'Railfreight grey' livery with yellow cab and number 14021, Swanage Railway, 13th September 1992; to Wimbledon Depot, for tyre turning, 8th June 1995; to Swanage Railway, 12th June 1995; re-sold to D9521 Group and moved to Barry Rail Centre, Barry Island, 12th November 2004; to Mid-Norfolk Railway, Dereham, 20th May 2008; to Quainton Railway Society, near Aylesbury, Buckinghamshire, 9th July 2008; to Barry Rail Centre, Barry Island, autumn 2008; to Dean Forest Railway, Lydney, 16th January 2009; to Swindon & Cricklade Railway, for gala, September 2009; returned to Dean Forest Railway, Lydney, October 2009; to Gwili Railway, Bronwydd Arms, on loan, 10th August 2010; noted in green livery with number D9521, Gwili Railway, August 2010; to Llangollen Railway, on loan, 25th August 2010; returned to Dean Forest Railway, Lydney, by 24th April 2011; to Avon Valley Railway, Bitton, 6th April 2013; to Dean Forest Railway, Lydney, May 2013; to East

Lancashire Railway, Bury, for gala, 8th July 2014; noted in green livery with number D9521, East Lancashire Railway, Bury, 8th July 2014; to Dean Forest Railway, Lydney, 5th August 2014; noted in BR blue livery with number D9521, Dean Forest Railway, 25th September 2022.

D9523　　Swindon　　　　1964　50B　　4/68　P　D9523
new 1st December 1964; moved to Hull Dairycoates, hauled by D6893, 20th January 1967; withdrawn 1st April 1968; sold to S&L; despatched from BR Hull Dairycoates Depot; to Stewarts & Lloyds Minerals Ltd, Glendon East Quarries, Northamptonshire, 16th December 1968; noted at Glendon East, 11th January 1969; allocated numbers 25 and 8311/25; noted at Glendon East, 11th August 1976; seen in original BR green livery, with numbers 25 and D9523, Glendon East, 3rd August 1979; to Corby Quarries, 28th May 1980; to BSC Steelworks Disposal Site, Heavy Mills, Corby, 29th December 1980; noted at Heavy Mills, 19th September 1981; re-sold to Great Central Railway, Loughborough, and moved 17th October 1981; noted in two-tone green livery with number D9523, Loughborough, 18th April 1982; to Nene Valley Railway, Wansford, 7th December 1988; seen in green livery with number D9523, Wansford, 25th February 1993; to Boden Rail Engineering, Washwood Heath, 8th April 2011; to Derwent Valley Light Railway, Murton, York, 21st April 2011; seen in maroon livery with cast D9523 plate, on low-loader, southbound on M1 Motorway, near Rotherham, at 11:34 on 25th April 2013; to Cholsey & Wallingford Railway, Oxfordshire, for gala, 25th April 2013; to Derwent Valley Light Railway, Layerthorpe, York, 17th May 2013; to Wensleydale Railway, Leeming Bar, 18th July 2013; to East Lancashire Railway, Bury, for gala, 15th July 2014; noted in maroon livery with number D9523, East Lancashire Railway, Bury, 15th July 2014; to Nene Valley Railway, Wansford, for repairs, 13th August 2014; re-sold to Wensleydale Railway, Leeming Bar, and moved 2nd February 2017; noted in maroon livery with number D9523, Wensleydale, 17th July 2022; attended the Class 14's 60th anniversary gala, held at the Ecclesbourne Valley Railway on 25th to 28th July 2024.

D9524　　Swindon　　　　1964　87E　　4/69　P　14901
new 11th December 1964; withdrawn 26th April 1969; moved to Canton Depot, January 1970; seen at BR Canton Depot, Cardiff, 2nd March 1970; sold to BP Ltd; despatched from Canton Depot, Cardiff, early July 1970; noted stabled at Hereford Station, 4th July 1970; to BP Refinery Ltd, Grangemouth, July 1970; to BR Grangemouth Depot, for repairs, November 1971; noted in green livery with no number, BR Grangemouth Depot, 21st November 1971; returned to BP Refinery Ltd; to Andrew Barclay, Sons & Co Ltd, Kilmarnock, November 1973; fitted with a Dorman type 8QT 500hp engine; returned to BP Refinery Ltd; to BR Grangemouth Depot, for repairs, November 1971; returned to BP Refinery Ltd; noted with numbers No.8 and 144-8, BP Grangemouth, 16th March 1975; noted at Cadder Yard (about three miles north of Eastfield on the Falkirk line), 15th March 1978; to BR Eastfield Depot, for repairs, March 1978; noted at Eastfield Depot, 25th March 1978; returned to BP Refinery Ltd; noted working, June 1978; withdrawn from use, January 1980; noted in light blue livery with number 144-8, BP Grangemouth, 10th September 1980 and 15th August 1981; re-sold to Scottish Railway Preservation Society, Falkirk, and moved 9th September 1981; to Scottish Railway Preservation Society, Bo'ness, 7th February 1988; overhauled, fitted with a Rolls-Royce type DV8 750hp engine, 1992; repainted blue, and numbered 14901 (reclassified after new engine fitted) in 1992; re-sold to Middle Peak Railways, about June 2006; to RMS Locotec Ltd, Wakefield, for overhaul, 1st July 2006; re-sold to RMS Locotec Ltd, early 2007; to Elsecar Steam Railway, near Barnsley, 17th April 2007; re-sold to Andrew Briddon, early 2008; seen in blue livery with yellow cab and number 14901, Elsecar, April 2010; to Midland Railway, Butterley, for gala, 18th May 2010; to Peak Rail, Rowsley, for gala, 1st July 2010; noted at Rowsley, 10th July

2010; to Gwili Railway, Bronwydd Arms, on hire, 2nd April 2011; to Peak Rail, Rowsley, 22nd March 2013; to East Lancashire Railway, Bury, for gala, 22nd July 2014; noted in blue and yellow livery with number 14901, 26th July 2014; to Peak Rail, Rowsley, 28th July 2014; to Great Central Railway, Loughborough, for gala, 27th August 2014; to Peak Rail, Rowsley, 3rd September 2014; to Andrew Briddon, Darley Dale, 1st May 2015; to Churnet Valley Railway, Cheddleton, 26th May 2016; to Andrew Briddon, Darley Dale, 1st March 2017; to Colne Valley Railway, 25th May 2017; to Old Oak Common Depot, London, for open day held on 2nd September 2017; to Colne Valley Railway, Castle Hedingham, 11th September 2017; to Andrew Briddon, Darley Dale, 21st March 2019.

D9525   Swindon   1965   50B   4/68   P   D9525
new 13th January 1965; to Hull Dairycoates, hauled by D6903, 9th January 1967; withdrawn 1st April 1968; sold to NCB; despatched from BR Dairycoates Depot, Hull, 28th November 1968; to NCB Philadelphia Locomotive Shed, County Durham, 2nd December 1968; noted at Philadelphia, 23rd January 1969; repainted green, and renumbered 507, August 1969; seen at Philadelphia, 15th February 1969 and 21st August 1970; noted with number 507, Philadelphia, 13th June 1973; to Burradon Colliery, Dudley, 7th March 1975; to Ashington Colliery, 14th March 1975; noted at Ashington Colliery, 28th August 1975; to Backworth Colliery, 15th December 1975; noted at Backworth Colliery, 20th February 1976; noted with numbers 507 and 2233/507, Backworth Colliery, 9th April 1980; to Weetslade Coal Preparation Plant, 15th August 1980; noted at Weetslade, 1st January 1981; to Ashington Colliery, 24th April 1981; to Lambton Engine Works, Philadelphia, for overhaul and repaint in 'Ashington blue', 21st July 1983; noted at Lambton Engine Works, Philadelphia, 22nd July 1983 and 11th January 1984; to Ashington Colliery, 7th February 1984; fitted with overhauled engine, 16th January 1986; seen with number 507, Ashington Colliery, 20th April 1987; noted at Ashington Colliery, 25th September 1987; re-sold to Kent & East Sussex Railway, Tenterden, and moved 29th September 1987; to Great Central Railway (Nottingham), Ruddington, Nottingham, 26th June 2000; re-sold to Harry Needle Railroad Company; to Barrow Hill Engine Shed, Staveley, 13th August 2001; to Battlefield Line, Shackerstone, 2nd March 2002; to South Devon Railway, Buckfastleigh, 27th May 2004; to Heritage Shunters Trust, Rowsley, 13th April 2005; re-sold to Heritage Shunters Trust (with D9500 in part exchange), summer 2010; seen in green livery with number D9525, Heritage Shunters Trust, Rowsley, 10th July 2010; re-sold to a group of enthusiasts, and moved to Ecclesbourne Valley Railway, Wirksworth, 18th February 2022; seen in BR two-tone green livery with number D9525, Ecclesbourne Valley Railway, 8th September 2022; lengthy overhaul, 2023; had test run, 29th October 2023; attended the Class 14's 60th anniversary gala, held at the Ecclesbourne Valley Railway on 25th to 28th July 2024.

D9526   Swindon   1965   86A   11/68   P   D9526
new 6th January 1965; withdrawn 30th November 1968; sold to APCM, Westbury, Wiltshire; despatched from Canton Depot, by rail, December 1969; spent at least a week at Westbury Depot, late December 1969; delivered to APCM, January 1970; noted at APCM, Westbury, 26th August 1970; to APCM, Dunstable, under its own power, 28th May 1971; repainted dark green, October 1971; noted passing through Reading, under its own power, 26th November 1971; to APCM, Westbury, late November 1971; noted at APCM, Westbury, 11th December 1973; seen in dark green livery with no number, APCM, Westbury, 26th February 1977; noted at Westbury, 23rd March 1979; became the first of the class to be secured for preservation when donated to The Diesel & Electric Preservation Group in 1980; noted on a low-loader on M5, 2nd April 1980; arrived at West Somerset Railway, 3rd April 1980; restored by the D&EPG; noted at West Somerset Railway, 25th

June 1983; entered service in 1984; noted in green livery with number D9526, West Somerset Railway, May 2005; to East Lancashire Railway, Bury, for gala, 14th July 2014; to West Somerset Railway, Minehead, September 2014; noted in green livery with number D9526, West Somerset Railway, May 2016; to South Devon Railway, Buckfastleigh, for gala, 7th October 2018; to West Somerset Railway, Minehead, 6th November 2018; to Kent & East Sussex Railway, Tenterden, 2nd July 2021; to West Somerset Railway, Minehead, 19th January 2023.

D9527   Swindon   1965   86A   4/69   F   No.6 / 9312/94
new 18th January 1965; withdrawn 26th April 1969; sold to NCB; despatched in a four-locomotive convoy from BR Canton Depot, Cardiff, hauled by D6607, 30th June 1969; noted at Horton Road Depot, Gloucester, 4th July 1969; continued to Washwood Heath; departed Washwood Heath, hauled by D102, 16th July 1969; delivered to NCB Ashington Colliery, 19th July 1969; noted with plant number 9312/94, Ashington Colliery, 3rd November 1969; fitted with overhauled engine, August 1971; to Ashington Central Workshops, for repainting, June 1972; to Ashington Colliery, June 1972; noted in blue livery with number No.6, Ashington Colliery, 22nd October 1972; to BR Gosforth Depot, for tyre turning, 3rd May 1973; to Ashington Colliery, 5th May 1973; to BR Gosforth Depot, for tyre turning, 19th March 1975; to Ashington Colliery, March 1975; fitted with overhauled engine, November 1975; to Lambton Engine Works, Philadelphia, for overhaul, 28th May 1977; noted at Lambton Engine Works, 20th September 1977; to Ashington Colliery, 20th September 1977; noted on a low-loader on A1, 5th September 1978; to Lambton Engine Works, Philadelphia, 5th September 1978; to Ashington Colliery, 21st September 1978; fitted with overhauled engine, 19th December 1979; out of use from 12th September 1983; withdrawn October 1983; noted at Ashington Colliery, 20th January 1984; scrapped on site by D. Short Ltd of North Shields, 20th to 23rd January 1984.

D9528   Swindon   1965   86A   3/69   F   No.2 / 9312/100
new 21st January 1965; withdrawn 2nd March 1969; sold to NCB; despatched from BR Canton Depot, Cardiff, 6th March 1969; noted at Derby, hauled by D161, 7th March 1969; noted at Toton Depot, 9th March 1969; arrived at NCB Ashington Colliery, 14th March 1969; noted with plant number 9312/100, Ashington Colliery, 29th June 1969; to Ashington Central Workshops, for repainting, June 1972; to Ashington Colliery, June 1972; noted in blue livery with numbers No.2 and 9312/100, Ashington Colliery, 22nd October 1972; to BR Gosforth Depot, for tyre turning, 31st October 1972; to Ashington Colliery, 23rd November 1972; fitted with overhauled engine, December 1972; fitted with overhauled engine, September 1974; to BR Gosforth Depot, for tyre turning, 11th March 1976; to Ashington Colliery, 10th April 1976; seen at Ashington Colliery, 21st May 1977; to Lambton Engine Works, Philadelphia, for repairs, 14th June 1977; to Ashington Colliery, 12th October 1977; out of use by February 1978; noted with RE 1404 of 1953 registration plate, Ashington Colliery, 7th May 1980; used for spares; noted in blue livery with number No.2, with no engine, Ashington Colliery, 19th June 1981; noted at Ashington Colliery, 12th September 1981; scrapped on site, December 1981.

D9529   Swindon   1965   50B   4/68   P   9529
new 26th January 1965; withdrawn 1st April 1968; despatched from BR Hull Dairycoates Depot, August 1968; demonstrated at Stewarts & Lloyds Minerals Ltd, Buckminster Quarries, Lincolnshire (while still a BR locomotive), from 26th August 1968; after successful trial was sold to Stewarts & Lloyds Minerals Ltd for Buckminster; this was the first Paxman to go into industry; allocated numbers 20 and 8411/20; Buckminster Quarries ceased production, 27th January 1972; noted at BR Grantham Station, 31st August 1972; to Corby

Quarries, 6th September 1972; renumbered 61, October 1974; noted with number 61, 1st July 1975; seen working IRS railtour, in green livery with number 61, Corby, 17th May 1980; to BSC Steelworks Disposal Site, Heavy Mills, Corby, 29th December 1980; noted at Heavy Mills, 21st January 1981; re-sold to North Yorkshire Moors Railway, Grosmont, and moved by rail, 16th March 1981; noted at North Yorkshire Moors Railway, 6th April 1984; re-sold to Great Central Railway, Loughborough, and moved 11th December 1984; to BR Coalville Depot, open day, 5th June 1988; returned to Great Central Railway; overhauled and renumbered 14029 in 1985; re-sold to Nene Valley Railway, Wansford, and moved 8th December 1988; seen in blue livery with number 14029, Wansford, 25th February 1993; to Battlefield Line, Shackerstone, on loan, 10th April 1995; to Nene Valley Railway, Wansford, 9th November 1995; to Kent & East Sussex Railway, Tenterden, Kent, 23rd June 2000; to Channel Tunnel Rail Link(CTRL1), Beechbrook Farm, near Ashford, on hire, 10th July 2001; to Nene Valley Railway, Wansford, for repairs, April 2002; to Channel Tunnel Rail Link, on hire, 28th June 2002; noted in blue livery with number 14029, Beechbrook Farm, September 2002; to Tilbury Docks, on hire, 2nd January 2003; to Medway Ports, Chatham Dockyard, on hire, 1st March 2003; to Nene Valley Railway, Wansford, 14th July 2003; to Channel Tunnel Rail Link (CTRL2), Dagenham, on hire, 19th July 2004; allocated Network Rail number 01599 for use on the main line, 2004; to Nene Valley Railway, Wansford, for repairs, 19th September 2004; to EWS Toton Depot, for tyre turning, 17th November 2004; to Nene Valley Railway, Wansford, 29th November 2004; to Channel Tunnel Rail Link, Dagenham, on hire, 23rd February 2005; noted at Dagenham, 11th March 2005; to Channel Tunnel Rail Link, Swanscombe, on hire, September 2005; to Nene Valley Railway, Wansford, 14th April 2006; to Channel Tunnel Rail Link, Dagenham, on hire, 7th September 2006; to Nene Valley Railway, Wansford, for repairs, 29th September 2006; to Channel Tunnel Rail Link, Swanscombe, on hire, 24th January 2007; noted at Swanscombe, 20th April 2007; to Nene Valley Railway, Wansford, for repairs, 24th April 2007; to Kent & East Sussex Railway, Tenterden, 30th May 2007; to Nene Valley Railway, Wansford, 29th December 2008; to Aggregate Industries Ltd, Bardon Hill, on hire, 7th January 2009; to Nene Valley Railway, Wansford, 9th October 2010; noted in BR blue livery with number 9529, Nene Valley Railway, April 2017; still numbered 9529, June 2021; attended the Class 14's 60th anniversary gala, held at the Ecclesbourne Valley Railway on 25th to 28th July 2024.

D9530   Swindon   1965   86A   10/68 F NFT
new 2nd February 1965; seen at 86A Canton Depot, 28th July 1968; withdrawn 5th October 1968; to Gulf Oil Co Ltd, Cardiff Roath Docks, for trials, for short period in early September 1969; sold to Gulf Oil; to Gulf Oil Co Ltd, Waterston, Pembrokeshire, 26th September 1969; noted at Waterston, still in BR two-tone green livery, 4th July 1970; to BREL Swindon Works, for overhaul, by rail, 5th August 1971; noted in 'A' Shop, 13th August 1971; repainted light blue while at Swindon; to Gulf Oil, Waterston, 7th October 1971; re-sold to Hunslet Engine Company of Leeds and allocated Hunslet works number 8508; moved initially from Gulf Oil to BR Canton Depot, Cardiff, for repairs, 22nd October 1975; noted at Canton Depot, 23rd October 1975; re-sold to NCB; to NCB Mardy Colliery, Glamorgan, 6th January 1976; the colliery was named Mardy although the nearby town was Maerdy; noted in light blue livery with 'NCB Mardy, South Wales Area' on cab-side, Mardy Colliery, 26th March 1976; to BR Canton Depot and BR Ebbw Junction Depot, for tyre turning/repairs, July/August 1976; noted at Ebbw Junction Depot, 25th July 1976 and 16th August 1976; despatched from Ebbw Junction Depot, under own power, to Mardy Colliery, 16th August 1976; noted at Mardy Colliery, 12th October 1976; to BR Canton Depot, Cardiff, for open day held on 1st October 1977; stayed at depot for repairs; noted in blue livery at BR Canton

Depot, Cardiff, 22nd October 1977 and 7th November 1977; to Mardy Colliery, about December 1977; noted out of use, Mardy Colliery, 1st June 1980, 6th July 1980 and 9th September 1981; scrapped on site, March 1982.

D9531   Swindon   1965   86A   12/67   P   D9531 / ERNEST
new 2nd February 1965; moved to Worcester Depot, for storage, 5th December 1967; withdrawn 30th December 1967 when less than three years old!; noted at BR Worcester Depot, 1st July 1968; put out to tender per document dated 23rd May 1968; sold per 'Confirmation of sale' document dated 7th June 1968; despatched from BR Worcester Depot, 18th July 1968; to W.H. Arnott, Young & Co Ltd, scrap merchants, Parkgate, Rotherham; seen in scrapyard, in BR green livery with number D9531, with odometer reading of 56,680 miles, Parkgate, 5th August 1968; re-sold to NCBOE; to Hargreaves (West Riding) Ltd, NCBOE British Oak Disposal Point, Crigglestone, November 1968; seen at Parkgate, 24th November 1968; noted at British Oak Disposal Point, Crigglestone, 24th December 1968; took over from steam locomotives on 28th January 1969; seen in green livery with number D2/9531, Crigglestone, 9th March 1969; noted re-painted orange and black, Crigglestone, 11th May 1969; re-sold to NCB North East Area, 1973; to NCB Burradon Colliery, Dudley, 10th October 1973; to Ashington Colliery, by March 1974; seen with number D2/9531, Ashington Colliery, 15th July 1974; fitted with overhauled engine, 20th January 1977; to South Gosforth Depot, for tyre turning, early July 1977; to Ashington Colliery, 9th July 1977; seen with number D2/9531 at Tyne & Wear Metro, South Gosforth shed, for repairs, 27th June 1979; to Ashington Colliery, 29th June 1979; to Lambton Engine Works, Philadelphia, for repairs and repaint, 26th September 1981; repainted dark blue with number 2100/523; to Ashington Colliery, 27th October 1981; to BR Thornaby Depot, for axle repairs, early May 1982; noted at Thornaby Depot, 8th May 1982; returned to Ashington Colliery; fitted with overhauled engine, February 1984; seen with numbers 2100/523 and No.31, Ashington Colliery, 20th April 1987; re-sold to C.F. Booth Ltd of Rotherham, September 1987; did not leave Ashington Colliery site and was re-sold by Booth's for preservation; noted at Ashington Colliery, 1st October 1987; to East Lancashire Railway, Bury, 2nd October 1987; noted in green livery with number D9531, East Lancashire Railway, Bury, 6th October 1991; to North Norfolk Railway, Sheringham, for gala, 11th July 2014; to East Lancashire Railway, Bury, for gala, July 2014; to Severn Valley Railway, Bridgnorth, for gala, 28th September 2015; to East Lancashire Railway, Bury, 8th October 2015; noted in green livery with number D9531 and name ERNEST, East Lancashire Railway, Bury, August 2018.

D9532   Swindon   1965   50B   4/68   F   57 / 8311/37
new 18th February 1965; moved to Hull Dairycoates, hauled by D6919, 20th May 1967; withdrawn 1st April 1968; purchased from BR as a 'runner', by Stewarts & Lloyds Minerals Ltd, for £3,250; despatched from BR Hull Dairycoates Depot, November 1968; to Stewarts & Lloyds Minerals Ltd, Corby Quarries, 25th November 1968; allocated numbers 37 and 8311/37; seen at Gretton Brook shed, 23rd June 1973; renumbered 57, October 1974; noted out of use, 18th April 1979; to BSC Steelworks Disposal Site, Heavy Mills, Corby, 29th December 1980; noted at Heavy Mills, 4th June 1981; noted with no engine, 23rd January 1982; scrapped on site by Shanks & McEwan Ltd of Corby, February 1982.

D9533   Swindon   1965   50B   4/68   F   47 / 8311/26
new 19th February 1965; withdrawn 1st April 1968; purchased from BR as a 'runner', by Stewarts & Lloyds Minerals Ltd, for £3,250; despatched from BR Hull Dairycoates Depot; to Stewarts & Lloyds Minerals Ltd, Corby Quarries, by rail, December 1968; allocated numbers 26 and 8311/26; seen at Gretton Brook shed, 23rd June 1973; renumbered 47,

October 1974; noted in green livery with number 47, 1st July 1975; noted out of use, 18th April 1979; used for spares; to BSC Steelworks Disposal Site, Heavy Mills, Corby, 29th December 1980; noted at Heavy Mills, 4th June 1981; noted in green livery with number 47, 21st May 1982; noted at Heavy Mills, 25th August 1982; scrapped on site by Shanks & McEwan Ltd of Corby, September 1982.

D9534   Swindon   1965   50B   4/68   F ECCLES
new 2nd March 1965; withdrawn 1st April 1968; sold to APCM, Hope Cement Works, Derbyshire (reputedly at a price of £4,000); despatched from BR Hull Dairycoates Depot, October 1968; noted in BR two-tone green livery with number D9534, 12th January 1969; seen at Hope Cement Works, 7th November 1970; seen in dark green livery with ECCLES cast plates, Hope Cement Works, 16th June 1974; noted disused at Hope Cement Works, 13th October 1974; re-sold to Hunslet Engine Company, Leeds (and allocated works number 7497), 22nd April 1975; left Hope, on a low-loader, 5th May 1975; noted on a low-loader, March Depot, 9th May 1975; noted at March Depot, 10th May 1975; left March Depot by rail; noted being towed through Ipswich by 37044, working 9X35 to Harwich Docks, 10th May 1975; exported by sea from Harwich Docks (see Appendix C).

D9535   Swindon   1965   86A   12/68   F 37 / 9312/59
new 8th March 1965; withdrawn 21st December 1968; seen at BR Canton Depot, Cardiff, 2nd March 1970; despite its status was depot pilot at Canton, August 1970; sold to NCB; despatched from Canton Depot, working 8E38, 9th November 1970; delivered to NCB Ashington Colliery, November 1970; to Burradon Colliery, Dudley, January 1971; to Ashington Central Workshops, for repairs, 25th May 1973; to BR Gosforth Depot, for tyre turning, 11th October 1973; noted at Gosforth Depot, 13th October 1973; to Backworth Colliery, for repaint, by 24th October 1973; painted dark blue with red buffer beams and number 37, by 23rd November 1973; to Weetslade, for test run, late November 1973; to Backworth Colliery; to Burradon Colliery, Dudley, by 26th March 1974; given number 9312/59 by August 1974; to Weetslade Coal Preparation Plant, 5th January 1976; to Backworth Colliery, by 15th May 1976; noted with numbers 37 and 9312/59, Backworth Colliery, 9th April 1980; to Ashington Colliery, 13th September 1980; to Lambton Engine Works, Philadelphia, 5th December 1980; to Ashington Colliery, 23rd April 1981; withdrawn November 1983; noted at Ashington Colliery, 12th January 1984; scrapped on site by D. Short Ltd of North Shields, 18th to 20th January 1984.

D9536   Swindon   1965   87E   4/69   F No.5 / 9312/91
new 16th March 1965; withdrawn 26th April 1969; sold to NCB; moved to Canton Depot, January 1970; seen at BR Canton Depot, Cardiff, 2nd March 1970; despatched from BR Canton Depot, 4th March 1970; noted at Tyne Yard, 5th March 1970; arrived at NCB Ashington Colliery, 6th March 1970; noted with plant number 9312/91 painted over BR number, Ashington Colliery, 15th May 1970; to Ashington Central Workshops, for repainting, June 1972; to Ashington Colliery, June 1972; noted in blue livery with number No.5, Ashington Colliery, 22nd October 1972; to BR Gosforth Depot, for tyre turning, 16th August 1973; noted at Gosforth Depot, 18th August 1973; returned to Ashington Colliery, 18th August 1973; fitted with overhauled engine, June 1974; seen at Ashington Colliery, 13th July 1974; cab damaged in an accident, January 1977; to Lambton Engine Works, Philadelphia (where received replacement cab from D9545), 28th January 1977; to Ashington Colliery, 15th May 1977; to Lambton Engine Works, Philadelphia, January 1978; to Ashington Colliery, 31st January 1978; seen with RE 1544 of 1953 registration plate, Ashington Colliery, 7th May 1980; to Lambton Engine Works, Philadelphia, for new tyres, 16th September 1981; to Ashington Colliery, 5th January 1982; fitted with overhauled

engine, 1983; out of use by March 1984; used for spares, 1985; noted at Ashington Colliery, 1st December 1985; scrapped on site by Robinson & Hannan Ltd of Blaydon, commencing 5th December 1985; engine and various parts sold to Rutland Railway Museum, 13th December 1985.

D9537   Swindon   1965   50B   4/68   P   D9537 / ERIC
new 16th March 1965; withdrawn 1st April 1968; purchased from BR as a 'runner', by Stewarts & Lloyds Minerals Ltd, for £3,250; despatched from BR Hull Dairycoates Depot, November 1968; noted at BR Toton Depot, 18th November 1968; to Stewarts & Lloyds Minerals Ltd, Corby Quarries, November 1968; allocated numbers 32 and 8311/32; seen at Gretton Brook shed, 23rd June 1973; renumbered 52, October 1974; seen in green livery with number 52, Corby Quarries, 8th November 1975; seen working IRS 'Farewell to Corby' rail tour, 17th May 1980; officially withdrawn, December 1980; noted at Gretton Brook, 4th June 1981; to BSC Pen Green Crane Depot, for use helping to clear the site, September 1981; noted at Pen Green, 27th October 1982; re-sold to Gloucestershire Warwickshire Railway Society, Toddington, and moved 23rd November 1982; noted at Toddington, 29th June 1984; seen in green livery with number D9537, Toddington, 1st October 1988; to John Scholes, The Old Station, Rippingale, Lincolnshire, for storage and restoration work, 8th May 2003; re-sold to East Lancashire Railway, Bury, and moved 4th March 2013; repainted in desert sand livery and fitted with cast numberplates, 2014; resumed working, July 2014; to Dean Forest Railway, Lydney, 24th August 2015; returned to East Lancashire Railway, Bury, September 2015; to Ribble Steam Railway, Preston, for gala, 30th September 2015; to East Lancashire Railway, Bury, 6th October 2015; to Spa Valley Railway, Tunbridge Wells, 19th July 2016; returned to East Lancashire Railway, Bury; to Ecclesbourne Valley Railway, Wirksworth, for gala, 15th March 2017; returned to East Lancashire Railway, Bury, 22nd March 2017; to North Norfolk Railway, Sheringham, June 2018; to East Lancashire Railway, Bury, 10th September 2018; re-sold to Ecclesbourne Valley Railway, Wirksworth, and moved, 26th October 2018; noted in black livery with number D9537, Ecclesbourne Valley Railway, December 2018; to Great Central Railway, Loughborough, for gala, 10th April 2019; returned to Ecclesbourne Valley Railway, Wirksworth, by 26th April 2019; seen in black livery with number D9537 and name ERIC, Ecclesbourne Valley Railway, 1st June 2023; to North Norfolk Railway, on hire, 2nd June 2023; to Ecclesbourne Valley Railway, 3rd October 2023; to North Yorkshire Moors Railway, for gala held on 14th to 16th June 2024; returned to Ecclesbourne Valley Railway; attended the Class 14's 60th anniversary gala, held at the Ecclesbourne Valley Railway on 25th to 28th July 2024.

D9538   Swindon   1965   87E   4/69   F   160
new 26th March 1965; withdrawn 26th April 1969; moved to Canton Depot, December 1969; seen at BR Canton Depot, Cardiff, 2nd March 1970; sold to Shell-Mex & BP Ltd, (intended for use at Shell Haven, Essex), April 1970; despatched from Canton Depot; to BREL Swindon Works, for repairs plus costing for flameproofing, April 1970; noted at Swindon Works, 29th July 1970; noted in 'A' Shop, Swindon Works, 5th August 1970; the quote for flameproofing was prohibitive, so re-sold to British Steel Corporation, January 1971; noted in Reading yard, 29th January 1971; arrived at BSC Ebbw Vale Steelworks, Monmouthshire, 22nd February 1971; given number 160; noted in dark green livery with number 160, Ebbw Vale, February 1973; re-sold to BSC Corby Quarries; to BSC Corby Quarries, by 25th April 1976; was in poor condition and used as a source of spares only; not allocated a plant number; noted with number 160, Corby Quarries, 14th July 1979 and 21st July 1980; to BSC Steelworks Disposal Site, Heavy Mills, Corby, 29th December 1980; noted at Heavy Mills, 25th August 1982; scrapped on site by Shanks & McEwan Ltd of Corby, September 1982.

D9539   Swindon   1965   50B   4/68   P   D9539

new 9th April 1965; withdrawn 1st April 1968; demonstrated at Stewarts & Lloyds Minerals Ltd, Gretton Brook shed, August 1968; returned to BR Dairycoates Depot, Hull; purchased from BR as a 'runner', by Stewarts & Lloyds Minerals Ltd, for £3,250; despatched from BR Dairycoates Depot, 7th October 1968; noted southbound at Toton, 18th November 1968; to Stewarts & Lloyds Minerals Ltd, Corby Quarries, November 1968; allocated numbers 30 and 8311/30; seen at Gretton Brook shed, 23rd June 1973; renumbered 51, October 1974; to BSC Steelworks Disposal Site, Heavy Mills, Corby, 29th December 1980; noted at Heavy Mills, 19th February 1983; re-sold to Gloucestershire Warwickshire Railway, Toddington, and moved 23rd February 1983; noted at Toddington, 10th March 1983; seen in green livery with number D9539, Toddington, 5th May 1991; re-sold to Ribble Steam Railway, Preston, and moved 26th July 2005; to EWS Crewe Depot, for tyre turning, 16th February 2009; to Ribble Steam Railway, Preston, 27th April 2009; to East Lancashire Railway, Bury, on hire, 24th April 2014; noted in green livery with number D9539, East Lancashire Railway, Bury, July 2014; to Ribble Steam Railway, Preston, 7th August 2014; to Spa Valley Railway, Tunbridge Wells, 11th June 2015; to Epping Ongar Railway, Essex, 16th September 2015; to Ribble Steam Railway, Preston, 21st September 2015; to Peak Rail, Rowsley, on hire, 5th May 2016; to Ribble Steam Railway, Preston, 12th August 2016; noted in two-tone green livery with cast D9539 numberplate, Ribble Steam Railway, 23rd April 2022.

D9540   Swindon   1965   50B   4/68   F   36 / 508 / 2233/508

new 22nd April 1965; withdrawn 1st April 1968; sold to NCB; despatched from BR Dairycoates Depot, Hull, 29th November 1968; to NCB Philadelphia Locomotive Shed, County Durham, 2nd December 1968; seen at Philadelphia, 15th February 1969; repainted green and renumbered 508, August 1969; to Burradon Colliery, Dudley, 25th November 1971; to Ashington Colliery, June 1972; noted with number 508, Ashington Colliery, 19th June 1972; to Ashington Central Workshops, for repaint, late June 1972; repainted blue with number No.36; to Burradon Colliery, Dudley, 9th September 1972; noted at Burradon Colliery, 18th August 1973; to Ashington Central Workshops, about January 1974; to Burradon Colliery; noted at Burradon Colliery, 26th March 1974; to Weetslade Coal Preparation Plant, 3rd January 1976; to Gosforth Depot, for tyre turning, July 1977; noted at Gosforth Depot, 24th July 1977; to Weetslade CPP; noted with number 36 and plant number 2233/508 at Weetslade CPP, 26th June 1979; to Ashington Colliery, 24th April 1981; withdrawn November 1983; noted at Ashington Colliery, 6th January 1984; scrapped on site by D. Short Ltd of North Shields, 10th to 11th January 1984.

D9541   Swindon   1965   50B   4/68   F   66 / 8411/26

new 27th April 1965; moved from Cardiff to Hull, hauled by D1572, 18th January 1967; withdrawn 1st April 1968; sold to Stewarts & Lloyds Minerals Ltd; despatched from BR Dairycoates Depot, Hull, working 8G38, 11th November 1968; delivered to Harlaxton Quarries, Lincolnshire; allocated numbers 26 and 8411/26; repainted green with number No.26, June 1970; worked Wirral Railway Circle rail tour, 19th June 1971; production ceased at Harlaxton, February 1974; noted at Harlaxton, 24th April 1974; moved by low-loader to Corby Quarries, 4th August 1974; renumbered 66, October 1974; noted with number 66, Corby Quarries, 21st July 1980; to BSC Steelworks Disposal Site, Heavy Mills, Corby, 29th December 1980; noted in green livery with number 66, 8th June 1982; scrapped on site by Shanks & McEwan Ltd of Corby, August 1982.

D9542   Swindon   1965   50B   4/68   F   48 / 8311-27

new 6th May 1965; moved to Hull Dairycoates, hauled by D6873, 23rd January 1967; withdrawn 1st April 1968; purchased from BR as a 'runner', by Stewarts & Lloyds Minerals Ltd, for £3,250; despatched from BR Dairycoates Depot, Hull; to Stewarts & Lloyds Minerals Ltd, Corby Quarries, December 1968; allocated numbers 27 and 8311/27; seen at Gretton Brook shed, 23rd June 1973; renumbered 48, October 1974; to BSC Steelworks Disposal Site, Heavy Mills, Corby, 29th December 1980; noted in green livery with number 48, 8th June 1982; scrapped on site by Shanks & McEwan Ltd of Corby, August 1982.

D9544   Swindon   1965   50B   4/68   F   D9544

new 29th May 1965; moved to Hull Dairycoates, hauled by D6919, 20th May 1967; withdrawn 1st April 1968; purchased from BR for £1,000 as a source of spares, by Stewarts & Lloyds Minerals Ltd; despatched from BR Dairycoates Depot, Hull, hauled by D6737, 2nd November 1968; delivered to Corby Quarries, 2nd November 1968; allocated numbers 31 and 8311/31, but these were never carried; dismantled and used for spares from 1970; seen at Gretton Brook shed, 23rd June 1973; given number 53 in October 1974, but this never carried; seen dumped near Gretton Brook shed, still in BR two-tone green livery, with no wheels, 17th May 1980; remains scrapped on site by Shanks & McEwan Ltd of Corby, September 1980.

D9545   Swindon   1965   50B   4/68   F   D9545 / 9312/101

new 8th June 1965; to Hull Dairycoates, hauled by D6903, 9th January 1967; withdrawn 1st April 1968; sold to NCB; despatched from BR Dairycoates Depot, Hull; noted in Tyne Yard, 13th April 1969; delivered to NCB Ashington Colliery, 15th April 1969; noted at Ashington Colliery, 27th June 1969; given number 9312/101; seen at Ashington Colliery, 13th July 1974; used for spares, including wheels to D9527 and cab to D9536; noted in dismantled state, Ashington Colliery, 21st May 1977; remains scrapped, early July 1979.

D9547   Swindon   1965   50B   4/68   F   49 / 8311/28

new 20th July 1965; withdrawn 1st April 1968; purchased from BR as a 'runner', by Stewarts & Lloyds Minerals Ltd, for £3,250; despatched from BR Dairycoates Depot, Hull; to Stewarts & Lloyds Minerals Ltd, Corby Quarries, December 1968; allocated numbers 28 and 8311/28; seen working IRS rail tour, in green livery with number 28, Corby Quarries, 23rd June 1973; renumbered 49, October 1974; withdrawn in 1980; to BSC Steelworks Disposal Site, Heavy Mills, Corby, 29th December 1980; noted in green livery with number 49, Heavy Mills, 8th June 1982; scrapped on site by Shanks & McEwan Ltd of Corby, August 1982.

D9548   Swindon   1965   50B   4/68   F   67 / 8411/27

new 26th July 1965; moved from Cardiff to Hull, hauled by D1572, 18th January 1967; withdrawn 1st April 1968; sold to Stewarts & Lloyds Minerals Ltd; despatched from BR Dairycoates Depot, Hull, working 8G38, 11th November 1968; delivered to Harlaxton Quarries, Lincolnshire; allocated numbers 27 and 8411/27; its cab height reduced in order to negotiate the tunnel under Grantham Road; repainted dark green with number No.27, Harlaxton, June 1970; production ceased at Harlaxton, February 1974; noted at Harlaxton, 24th April 1974; moved by low-loader to Corby Quarries, 4th August 1974; renumbered 67, October 1974; seen in green livery with number 67, Corby Quarries, 17th May 1980; to BSC Steelworks Disposal Site, Heavy Mills, Corby, 29th December 1980; noted at Heavy Mills, 17th November 1981; re-sold to Hunslet Engine Co Ltd, Leeds, and moved 19th November 1981; chosen due to having a recently rebuilt engine and new tyres fitted; overhauled and rebuilt to 5ft 6in gauge; Aiken Espanola SA of Madrid had placed the order on Hunslet on 4th November 1981 and the conversion was carried out to Hunslet works order number GM1025; locomotive repainted yellow; exported from Goole Docks, June 1982 (see Appendix C).

D9549   Swindon   1965   50B   4/68   F   64 / 8311/33

new 19th August 1965; moved from Cardiff to Hull, hauled by D1572, 18th January 1967; withdrawn 1st April 1968; purchased from BR as a 'runner', by Stewarts & Lloyds Minerals Ltd, for £3,250; despatched from BR Dairycoates Depot, Hull, November 1968; noted at BR Toton Depot, 18th November 1968; to Stewarts & Lloyds Minerals Ltd, Corby Quarries, November 1968; allocated numbers 33 and 8311/33; seen at Gretton Brook shed, 23rd June 1973; to Glendon East Quarries, 8th October 1973; to Corby Quarries, 26th June 1974; renumbered 64, October 1974; repainted dark green; transferred to BSC Tube Works stock, for shunting steel strip trains from Lackenby, September 1980; to BSC Steelworks Disposal Site, Heavy Mills, Corby, by 9th May 1981; noted at Heavy Mills, 4th June 1981; re-sold to Hunslet Engine Co Ltd, Leeds, and moved 14th November 1981; chosen due to having a recently rebuilt engine and new tyres fitted; overhauled and rebuilt to 5ft 6in gauge; Aiken Espanola SA of Madrid had placed the order on Hunslet on 4th November 1981 and the conversion was carried out to Hunslet works order number GM1026; locomotive repainted yellow; exported from Goole Docks, June 1982 (see Appendix C).

D9551   Swindon   1965   50B   4/68   P   D9551

new 2nd September 1965; moved to Hull Dairycoates, hauled by D6893, 20th January 1967; withdrawn 1st April 1968; purchased from BR as a 'runner', by Stewarts & Lloyds Minerals Ltd, for £3,250; despatched from BR Dairycoates Depot, Hull, December 1968; to Stewarts & Lloyds Minerals Ltd, Corby Quarries, December 1968; allocated numbers 29 and 8311/29; seen at Gretton Brook shed, Corby, 23rd June 1973; renumbered 50, October 1974; noted in green livery with number 50, Corby, October 1976; transferred to BSC Tube Works, for shunting steel strip trains from Lackenby, July 1980; to BSC Steelworks Disposal Site, Heavy Mills, Corby, by 9th May 1981; noted at Heavy Mills, 4th June 1981; re-sold to a consortium of West Somerset Railway, Minehead, members and moved (by Wynns low-loader) on 5th June 1981; promptly entered service on 7th June 1981; noted in green livery, West Somerset Railway, 8th July 1984; re-sold to a private owner at the Royal Deeside Railway, Banchory, and moved 9th November 2000; re-sold to Severn Valley Railway, Bridgnorth, and moved 25th November 2013; was initially given the nickname 'Angus' after its previous home; underwent lengthy overhaul; entered service, May 2017; noted in desert sand (or golden ochre) livery with number D9551, Severn Valley Railway, May 2017; to Didcot Railway Centre, Oxfordshire, 9th March 2019; to Severn Valley Railway, Bridgnorth, 14th May 2019; to Mid Hants Railway, Ropley, for gala held on 25th to 27th June 2021; returned to Severn Valley Railway; to Epping Ongar Railway, Essex, for gala held on 25th and 26th September 2021; returned to Severn Valley Railway, late September 2021; repainted in BR two-tone green livery, May 2023; to West Somerset Railway, for gala held 8th to 10th June 2023; to Severn Valley Railway, 14th June 2023; to Mid Hants Railway, Ropley, for gala held on 14th and 15th July 2023; returned to Severn Valley Railway; attended the Class 14's 60th anniversary gala, held at the Ecclesbourne Valley Railway on 25th to 28th July 2024.

D9552   Swindon   1965   50B   4/68   F   59 / 8411/21

new 6th September 1965; moved to Dairycoates Depot, Hull, hauled by D6893, 20th January 1967; withdrawn 1st April 1968; despatched from Dairycoates Depot, September 1968; to Stewarts & Lloyds Minerals Ltd, Buckminster Quarries, Lincolnshire, for initial trial and then purchased, September 1968; allocated numbers 21 and 8411/21; repainted dark green with number 21, September 1969; worked an IRS rail tour, 28th September 1969; Buckminster Quarries ceased production, 27th January 1972; noted at BR Grantham Station, 31st August 1972; to Corby Quarries, 6th September 1972; renumbered 59, October 1974; out of use by April 1976; used for spares; noted with number 59, Corby Quarries, 22nd August 1980; scrapped on site by Shanks & McEwan Ltd of Corby, September 1980.

D9553   Swindon   1965   50B   4/68   P   D9553
new 18th September 1965; moved to Dairycoates Depot, Hull, hauled by D6873, 23rd January 1967; withdrawn 1st April 1968; purchased from BR as a 'runner', by Stewarts & Lloyds Minerals Ltd, for £3,250; despatched from Dairycoates Depot, November 1968; noted southbound at Toton, 18th November 1968; to Stewarts & Lloyds Minerals Ltd, Corby Quarries, November 1968; allocated numbers 34 and 8311/34; seen at Gretton Brook shed, 23rd June 1973; renumbered 54, October 1974; to BSC Steelworks Disposal Site, Heavy Mills, Corby, 29th December 1980; noted in green livery with number 54, 20th November 1982; re-sold to Gloucestershire Warwickshire Railway, Toddington, and moved 23rd February 1983; noted at Toddington, 10th March 1983; seen in green livery, Toddington, 1st October 1988; noted in green livery with number D9553, Toddington, July 2006; re-sold to Vale of Berkeley Railway, 2015; to Moveright International, Wishaw, Warwickshire, 14th December 2015; to Vale of Berkeley Railway, Sharpness, 6th January 2016; to Berkeley Power Station, Gloucestershire, September 2017; to Railway Support Services, Wishaw, 6th August 2019; seen in green livery with numbers D9553 and 54, RSS Wishaw, 9th July 2022; to Caledonian Railway, Brechin, 14th November 2022; noted on low-loader, M1 southbound, 25th July 2024; attended the Class 14's 60th anniversary gala, held at the Ecclesbourne Valley Railway on 25th to 28th July 2024; noted at gala in new livery of chocolate and cream bodywork with pale green cab and a cast BR lion and wheel emblem on the cabside; noted M18 Hellaby, 12th August 2024.

D9554   Swindon   1965   50B   4/68   F   58 / 8311/38
new 6th October 1965; moved to Dairycoates Depot, Hull, hauled by D6873, 23rd January 1967; withdrawn 1st April 1968; purchased from BR as a 'runner', by Stewarts & Lloyds Minerals Ltd, for £3,250; despatched from Dairycoates Depot, November 1968; to Stewarts & Lloyds Minerals Ltd, Corby Quarries, 25th November 1968; allocated numbers 38 and 8311/38; noted numbered 38, 11th January 1969; seen at Gretton Brook shed, 23rd June 1973; renumbered 58, October 1974; to Pen Green Works, for overhaul, by 23rd March 1976; noted at Pen Green Works, 1st May 1976; to Gretton Brook, by 10th July 1976; to BSC Steelworks Disposal Site, Heavy Mills, Corby, 29th December 1980; noted at Heavy Mills, 8th June 1982; scrapped on site by Shanks & McEwan Ltd of Corby, August 1982.

D9555   Swindon   1965   87E   4/69   P   D9555
new 22nd October 1965; the last locomotive built at Swindon Works; new fifteen months after the first of its class; withdrawn 26th April 1969; sold to NCB; seen at BR Canton Depot, Cardiff, 2nd March 1970; despatched from BR Canton Depot, 4th March 1970; noted in Tyne Yard, 5th March 1970; arrived at NCB Burradon Colliery, Dudley, 6th March 1970; seen at Burradon Colliery, 21st August 1970; given number 9107/57; to Gosforth Depot, for tyre turning, August 1974; noted at Gosforth Depot, 17th August 1974; to Ashington Central Workshops, for repairs, September 1974; to Ashington Colliery, 7th February 1975; to Burradon Colliery, Dudley, March 1975; to Ashington Colliery, by 4th June 1975; to Backworth Colliery, 3rd December 1975; noted at Backworth Colliery, 9th December 1975; to Ashington Colliery, 15th August 1980; to Lambton Engine Works, Philadelphia, for overhaul and repaint, 11th November 1981; to Ashington Colliery, 10th March 1982; noted

in dark blue livery with number 9107/57, Ashington Colliery, 14th April 1982; derailed/damaged in collision with 38 at Lynemouth Colliery, 16th October 1985; repaired at Lynemouth Colliery; noted at Lynemouth Colliery, 27th June 1986; to Ashington Colliery, 28th November 1986; noted at Ashington Colliery, 29th November 1986; seen with 'NCB North East Area, Plant No. 9107/57, Serial No. D9555' on cabside, Ashington Colliery, 20th April 1987; re-sold to Rutland Railway Museum, Cottesmore, and moved 24th September 1987; seen restored to BR green livery with number D9555, Rutland Railway, 31st May 1990; to Northampton & Lamport Railway, on loan, August 1998; seen with number D9555, Northampton & Lamport Railway, 16th August 1998; to Rutland Railway Museum, Cottesmore, 27th October 1998; to Old Oak Common Depot, London, open day, August 2000; returned to Cottesmore; re-sold to Dean Forest Railway, Lydney, about March 2002; to EWS Toton Depot, for tyre turning, 10th June 2003; returned to Dean Forest Railway, Lydney, 11th June 2003; to East Lancashire Railway, Bury, for gala, 9th July 2014; to Dean Forest Railway, Lydney, 30th July 2014; to Midland Railway, Butterley, for repairs, 1st December 2021; noted in green livery with number D9555, Butterley, 10th June 2023.

The vast majority of class 14 were sold into industrial service, depriving British locomotive builders of much-needed orders. 28 (D9547) poses at Stewarts & Lloyds Minerals Ltd, Corby Quarries during an IRS rail tour, 23rd June 1973. *(Adrian Booth)*

SECTION 20:

LMSR and British Railways built 0-6-0 diesel electric locomotives, numbered 12033-12138, and introduced 1945. Fitted with an English Electric 6KT engine developing 350bhp at 680rpm, and driving wheels of 4ft 0½in diameter. Later classified TOPS Class 11.

12049   Derby   1948   1E   10/71   F   12049
new 31st December 1948; withdrawn, 3rd October 1971; sold to Day & Sons; despatched from BR Bletchley, October 1972; to Day & Sons (Brentford) Ltd, Roadstone Terminal, Brentford Town Goods Depot, London, October 1972; to BR Old Oak Common Depot,

London, for repairs, October 1976; returned to Day & Sons, January 1977; noted in green livery with no number, with 'DAY Aggregates' on side, 28th August 1991; donated to Mid Hants Railway, Ropley, Hampshire, and moved by road on 28th July 1998; noted in green livery with number 12049, Mid Hants Railway, May 2009; severely damaged in an engine shed fire, 26th July 2010; to European Metal Recycling, Kingsbury, for scrap, 3rd November 2010; scrapped, July 2011; 12082 (which see) was later obtained and renumbered 12049 to replace this original locomotive.

12050   Derby   1949   9A   7/70   F   12050
new 19th February 1949; withdrawn, 11th July 1970; moved to Newton Heath Depot, for storage, and seen there on 18th March 1971; sold to NCB; despatched from BR Newton Heath Depot, April 1971; to NCB Philadelphia Locomotive Shed, County Durham; used for spares, June 1971; noted at Philadelphia, 24th August 1971 and 15th June 1972; remains scrapped on site, late June 1972.

12052   Derby   1949   5A   6/71   P   12052
new 16th April 1949; withdrawn, 13th June 1971; sold to NCBOE; despatched from BR Crewe Diesel Depot, early December 1971; arrived at Tyne Yard, 10th December 1971; to Derek Crouch (Contractors) Ltd, NCBOE Widdrington Disposal Point, 14th December 1971; noted in blue livery, Widdrington, 17th April 1972; noted freshly repainted red with white roof, with number MP228, Widdrington, 15th October 1972; seen at Widdrington, 21st May 1977; seen in pink livery with number MP228, being used as a source of spares for 12093, Widdrington, 4th November 1985; seen at Widdrington, 8th November 1987; re-sold to Scottish Industrial Railway Centre, Dalmellington, and moved 2nd October 1988; stored at Dunaskin shed; moved to the Scottish Industrial Railway Centre's site at the former Minnivey Colliery, 8th May 1994; to Scottish Industrial Railway Centre, Dunaskin shed, Dalmellington, 11th March 2002; re-sold to Caledonian Railway, Brechin, and moved about 8th April 2002; noted in red livery with no number, Brechin, July 2006; noted in black livery with number 12052, Brechin, 25th October 2020 and 13th August 2022.

12054   Derby   1949   6A   7/70   F   12054
new 4th June 1949; withdrawn, 25th July 1970; seen at 6A Chester Depot, 26th July 1970; sold to Adams; despatched from 6A Chester Depot, 15th September 1971; to A.R. Adams & Son, Newport; used as a hire locomotive (see Appendix A); scrapped, April 1984.

12060   Derby   1949   9A   2/71   F   512 / 2233/512
new 12th November 1949; withdrawn, 7th February 1971; sold to NCB; despatched from 9D Newton Heath Depot, initially in 17-40 freight working to Healey Mills, 30th March 1971; arrived at NCB Derwenthaugh Locomotive Shed, Blaydon, County Durham, early April 1971; to Philadelphia Locomotive Shed, 16th April 1971; renumbered 512, May 1971; noted with number 512, Philadelphia, 13th June 1973; noted freshly repainted in yellow and green, Philadelphia, 16th October 1976; noted with number 512, Philadelphia, 6th September 1980; to Philadelphia Locomotive Shed, 1983; noted at Philadelphia Locomotive Shed, 2nd December 1983; offered for sale, July 1985; noted at Philadelphia Locomotive Shed, 20th October 1985; scrapped on site by C.F. Booth Ltd of Rotherham, November 1985.

12061   Derby   1949   8J   10/71   F   4
new 21st November 1949; withdrawn, 10th October 1971; noted at Springs Branch Depot, 25th October 1971; sold to NCB; despatched from Springs Branch Depot, Wigan (ran 9Z10), 11th December 1972; delivered to NCB Nantgarw Coking Plant, Treforest, 11th

December 1972; interesting to note that the three diesels purchased by Nantgarw were older than the steam locomotives they were to replace; noted in BR green livery, Nantgarw, 1st January 1973; noted in blue livery with yellow cab and with number 4, Nantgarw, 16th February 1974; noted at BR Canton Depot, Cardiff, for tyre turning, 4th December 1974; noted at BR Canton Depot, Cardiff, 27th December 1974; to Nantgarw Coking Plant, Treforest, about January 1975; to BREL Swindon Works, for repairs, 10th December 1981; to Nantgarw Coking Plant, Treforest, 8th February 1982; last coke produced, Nantgarw Coking Plant, 20th November 1986; towed to south end of colliery sidings, Nantgarw, 21st December 1986; re-sold to Vale of Neath Railway Society, Cae Groes Terrace, Aberdulais, and moved by road on 23rd August 1987; noted at Aberdulais, 3rd April 1988; seen in green livery with number 4, Aberdulais, 2nd April 1991; to Gwili Railway, Bronwydd Arms, 13th September 1991; re-sold to Heritage Shunters Trust, Rowsley, and moved 11th June 2004; restoration began, but mechanical problems meant it was not viable to restore; to European Metal Recycling, Attercliffe, Sheffield, for scrap, 27th March 2013; scrapped, 28th March 2013.

12063   Derby   1949   8F   1/72   F   5
new 17th December 1949; withdrawn, 1st January 1972; sold to NCB; despatched from BR Springs Branch Depot, Wigan (ran 9Z10), 11th December 1972; delivered to NCB Nantgarw Coking Plant, Treforest, 11th December 1972; noted in BR blue livery, Nantgarw, 1st January 1973; noted in blue livery with yellow cab and with number 5, Nantgarw, 16th February 1974; to BR Canton Depot, Cardiff, for repairs, January 1977; noted at Canton Depot, Cardiff, 12th January 1977; returned to Nantgarw Coking Plant; last coke produced, Nantgarw Coking Plant, 20th November 1986; towed to south end of colliery sidings, Nantgarw, 21st December 1986; noted at Nantgarw, 6th October 1987; scrapped on site, November 1987.

12071   Derby   1950   8F   10/71   F   6
new 26th August 1950; withdrawn, 9th October 1971; sold to NCB; despatched from BR Springs Branch Depot, Wigan (ran 9Z10), 11th December 1972; to NCB Nantgarw Coking Plant, Treforest, 11th December 1972; noted in BR blue livery, Nantgarw, 1st January 1973; noted in blue livery with yellow cab and with number 6, Nantgarw, 16th February 1974; to BR Canton Depot, Cardiff, for tyre turning, 23rd December 1974; noted at Canton Depot, Cardiff, 27th December 1974; to Nantgarw Coking Plant, Treforest, about January 1975; to Canton Depot, Cardiff, for repairs, about November 1976; despatched from Canton, working 9E76 (per Train Notice 227), to Nantgarw Coking Plant, Treforest, 11:46 on Monday 15th November 1976; to BR Ebbw Junction Depot, Newport, for repairs, March 1977; noted at Ebbw Junction Depot, Newport, 21st March 1977; to BREL Swindon Works, for repairs, 6th July 1977; noted at Swindon Works, 24th July 1977; to Nantgarw Coking Plant, Treforest, 2nd October 1977; to BREL Swindon Works, for repairs, September 1980; noted in green livery with yellow cab and number 6, Swindon Works, 5th and 29th October 1980; to Canton Depot, Cardiff, for change of traction motor, 4th February 1981; to Nantgarw Coking Plant, Treforest, 13th February 1981; last coke produced, Nantgarw Coking Plant, 20th November 1986; towed to south end of colliery sidings, Nantgarw, 21st December 1986; moved to National Smokeless Fuels, Aberaman Phurnacite Plant, Abercwmboi, July 1987; noted at Abercwmboi, 10th November 1987 and 25th September 1989; re-sold to C.F. Booth Ltd, Rotherham, and moved by road on 18th July 1990; seen in blue livery with number 6, C.F. Booth Ltd, 13th October 1990 and 8th August 1992; re-sold to South Yorkshire Railway Preservation Society, Meadowhall, Sheffield, and moved 24th August 1992; used for spares, June 1995; remains to Coopers (Metals) Ltd, Attercliffe, Sheffield, for scrap, June 1995; scrapped.

12074　Derby　　　　　　1950　6A　　　1/72　F　12074

new 9th September 1950; withdrawn, 1st January 1972; moved to Crewe Depot, for storage, early 1972; sold to NCBOE; despatched from BR Crewe Depot, June 1972; to Johnsons (Chopwell) Ltd, NCBOE Swalwell Disposal Point, Whickham, County Durham; noted at Swalwell, 17th June 1972; noted in green livery, Swalwell DP, 18th December 1979; seen in green livery with number 12074, Swalwell DP, 2nd August 1983 and 16th April 1989; Swalwell DP closed on 4th June 1989; re-sold to Harry Needle Railroad Company; to South Yorkshire Railway Preservation Society, Meadowhall, Sheffield, 21st June 1989; noted in green livery with number 12074, Meadowhall, 27th August 1989; to European Metal Recycling, Kingsbury, about June 2001; scrapped, July 2002.

12077　Derby　　　　　　1950　8F　　　10/71　P　12077

new 14th October 1950; withdrawn, 9th October 1971; sold to Cashmore Ltd; despatched from BR Springs Branch Depot, September 1973; to Cashmore Ltd, Great Bridge, Staffordshire, September 1973; re-sold to Midland Railway, Butterley, Derbyshire, and moved 16th December 1978; seen at Butterley, 13th March 1979; noted repainted in green livery, Butterley, 13th April 1979; seen in green livery with number 12077, Butterley, 8th February 1987 and 20th August 2012.

12082　Derby　　　　　　1950　6G　　　10/71　P　12049

new 17th November 1950; withdrawn, 3rd October 1971; sold to UKF Fertilisers Ltd; despatched from BR Chester Depot, 27th March 1973; to UKF Fertilisers Ltd, Ince Marshes, Ellesmere Port; to Manchester Ship Canal Company, Ellesmere Port, on loan, 16th July 1974; noted at MSC Ellesmere Port, 16th September 1974; to UKF Fertilisers Ltd, Ince Marshes, 25th October 1974; to BREL Swindon Works, for repairs and re-paint in BR blue livery, by road, 21st December 1977; noted at Swindon Works, 15th February 1978; to UKF Fertilisers Ltd, Ince Marshes, by road, 6th April 1978; noted at Ince Marshes, 14th March 1980; seen in blue livery with number 12082, Ince Marshes, 11th June 1986; re-sold to Harry Needle Railroad Company; moved to South Yorkshire Railway Preservation Society, Meadowhall, Sheffield, December 1991; noted in blue livery, Meadowhall, 4th May 1992; to Cobra Railfreight, Wakefield, West Yorkshire, on hire, March 1993; returned to South Yorkshire Railway Preservation Society, 9th December 1994; to RFS (Engineering) Ltd, Doncaster, for repairs, 28th November 1995; to Cobra Railfreight, Wakefield, on hire, 5th April 1996; noted in blue livery with number 12082, Cobra, 8th May 1996; to RFS (Engineering) Ltd, Doncaster, for repairs, 15th October 1997; to South Yorkshire Railway Preservation Society, 18th December 1997; to Barrow Hill Engine Shed, Staveley, 24th June 1999; received Railtrack registration number 01553; to Wabtec, Doncaster, 21st December 2000; displayed at Doncaster Works open weekend, 26th and 27th July 2003; to Lafarge, Hope Cement Works, Derbyshire, on hire, 6th August 2003; to Barrow Hill Engine Shed, Staveley, by May 2004; to Whitemoor Yard, March, on hire, 29th September 2004; but unsuitable and returned to Barrow Hill Engine Shed, Staveley, 30th September 2004; to Midland Railway, Butterley, 2004; to Harry Needle Railroad Company, Long Marston, July 2005; noted in yellow and grey livery with numbers 12082 and 01553, Long Marston, June 2007; to Deanside Transit, Glasgow, on hire, January 2008; to Barrow Hill Engine Shed, Staveley, September 2010; re-sold to Mid Hants Railway to replace their original 12049 (which see) that had suffered fire damage; noted in green livery with number 12049, Barrow Hill Engine Shed, October 2010; moved to Mid Hants Railway, Ropley, 1st November 2010.

12083   **Derby**   **1950**   **12A**   **10/71**   **P**   **12083 / M413**

new 24th November 1950; withdrawn, 9th October 1971; sold to Tilcon Ltd; noted at Carlisle Kingmoor Depot, 9th June 1973; to Tilcon Ltd, Swinden Lime Works, Grassington, early July 1973; noted at Tilcon Ltd, 3rd July 1973; to BR Doncaster Depot, for repairs, September 1974; returned to Tilcon Ltd, October 1974; seen in blue livery with number 12083, Grassington, 8th June 1981 and 1st April 1983; noted in blue livery, with numbers 12083, 201276 and M413 on cabside, Tilcon, 10th April 1995; re-sold to Harry Needle Railroad Company; to South Yorkshire Railway Preservation Society, Meadowhall, Sheffield, 21st May 1998; to Battlefield Line, Shackerstone, for storage, 1st August 2001; re-sold to a preservation group at Shackerstone, November 2006; noted in faded unrestored grey livery with number 12083, Shackerstone, 13th April 2013 and 30th April 2023.

12084   **Derby**   **1950**   **5A**   **5/71**   **F**   **514 / 2233/514**

new 9th December 1950; withdrawn, 16th May 1971; sold to NCB; despatched from 5A Crewe Diesel Depot, October 1971; to NCB Burradon Colliery, Dudley, Northumberland, October 1971; noted at Burradon Colliery, 25th October 1971; to Philadelphia Locomotive Shed, County Durham, 25th November 1971; renumbered 514, December 1971; to Silksworth Colliery, Sunderland, 7th April 1972; to Hylton Colliery, Castletown, early July 1972; noted with number 514, Hylton Colliery, 6th July 1972 and 24th July 1973; to Philadelphia Locomotive Shed, 3rd March 1975; given plant number 2233/514; to Easington Colliery, 22nd December 1975; to Blackhall Colliery, 5th January 1976; to Bates Colliery, Blyth, 5th April 1976; seen at Bates Colliery, 21st May 1977; noted in shed, receiving new engine, Bates Colliery, 14th July 1978; to Lambton Engine Works, Philadelphia, 25th February 1982; noted at Lambton Engine Works, 4th August 1982 and 22nd July 1983; to Philadelphia Locomotive Shed, 21st October 1983; noted at Philadelphia Locomotive Shed, 2nd December 1983; used for spares; noted at Philadelphia Locomotive Shed, 20th October 1985; scrapped on site by C.F. Booth Ltd of Rotherham, November 1985.

12085   **Derby**   **1950**   **12A**   **5/71**   **F**   **12085**

new 23rd December 1950; withdrawn, 22nd May 1971; sold to Thos. W. Ward Ltd; noted at Carlisle Kingmoor Depot, 28th April 1973; despatched from BR Carlisle Kingmoor Depot; to Thos. W. Ward Ltd, Devonshire Dock, Barrow-in-Furness, May 1973; noted at Thos. W. Ward Ltd, 25th January 1974; scrapped, about June 1976.

12088   **Derby**   **1951**   **8J**   **5/71**   **P**   **12088**

new 2nd June 1951; withdrawn, 11th May 1971; sold to NCBOE; moved to Springs Branch Depot, June 1972; despatched from Springs Branch Depot, Wigan, July 1972; to Johnsons (Chopwell) Ltd, NCBOE Swalwell Disposal Point, Whickham, County Durham, late July 1972; noted in green livery at Swalwell DP, 24th August 1972; seen at Swalwell DP, 29th December 1973; seen in green livery with number 12088, Swalwell DP, 2nd August 1983; noted at Swalwell DP, 23rd March 1989; re-sold to Harry Needle Railroad Company, 31st May 1989; Swalwell DP closed on 4th June 1989; moved to South Yorkshire Railway Preservation Society, Meadowhall, Sheffield, June 1989; noted in green livery with number 12088, Meadowhall, 27th August 1989; to Johnsons (Chopwell) Ltd, Widdrington Disposal Point, Northumberland, on hire, 28th May 1996; noted at Widdrington, 4th September 1996; given registration 01564, date not known; to Steadsburn Opencast Site / Disposal Point, from 2007 (west of Widdrington and operative from mid-2007 to 2011); to Butterwell Disposal Point, near Linton, Northumberland, 8th November 2011; re-sold to Aln Valley

Railway, Lionheart Enterprise Park, Alnwick, and moved 6th December 2012; noted in blue livery with number 12088, Aln Valley Railway, July 2014.

12093　　Derby　　　　　1951　　5A　　　5/71　　P　12093

new 18th August 1951; withdrawn, 16th May 1971; sold to NCBOE; despatched from BR Crewe Diesel Depot, December 1971; arrived at Tyne Yard, 10th December 1971; to Derek Crouch (Contractors) Ltd, NCBOE Widdrington Disposal Point, Northumberland, 15th December 1971; noted in green livery, Widdrington, 17th April 1972; noted in red livery with white roof and number MP229, Widdrington, 24th July 1974; seen in pink livery with number MP229, with RE 1597 of 1953 registration plate, Widdrington, 21st May 1977 and 4th November 1985; put up for sale, February 1987; seen at Widdrington, 8th November 1987; re-sold to Scottish Industrial Railway Centre, Dalmellington; noted on a low-loader, Newcastle upon Tyne, 8th October 1988; delivered to Scottish Industrial Railway Centre, Dunaskin shed, Dalmellington, 9th October 1988; re-sold to Caledonian Railway, Brechin, and moved about 8th April 2002; noted with number MP229, Brechin, 5th April 2004; noted in black livery with number 12093, Brechin, July 2006; noted in green livery with number 12093, Brechin, June 2016.

12098　　Derby　　　　　1952　　9A　　　2/71　　F　12098

new 16th February 1952; withdrawn, 27th February 1971; sold to NCB; despatched from 9D Newton Heath Depot, initially in 06-30 freight working to Healey Mills, 31st March 1971; delivered to NCB Derwenthaugh Locomotive Shed, Blaydon, County Durham, early April 1971; to Philadelphia Locomotive Shed, 16th April 1971; renumbered 513 and 2100/513, June 1971; noted with number 513, Philadelphia, 13th June 1973 and 6th September 1980; to National Smokeless Fuels Ltd, Lambton Coking Plant, about September 1983; noted at Lambton Coking Plant, 26th October 1983 and 4th November 1985; works closed and locomotive disused, Lambton, 2nd May 1986; put out to tender (reference S358), 18th June 1986; re-sold to Stephenson Railway Museum, Middle Engine Lane, North Shields, and moved by road on 5th January 1987; re-sold to Harry Needle Railroad Company; to South Yorkshire Railway Preservation Society, Meadowhall, Sheffield, 9th December 1997; to European Metal Recycling, Kingsbury, by road, for scrap, July 2001; noted in green livery with number 12098, Kingsbury, 18th February 2004 and 18th January 2005; scrapped, June 2006.

12099　　Derby　　　　　1952　　1E　　　7/71　　P　12099

new 1st March 1952; withdrawn, 17th July 1971; sold to NCBOE; despatched from BR Bletchley, April 1972; to Murphy Bros Ltd, NCBOE Lion Disposal Point, Blaenavon, April 1972; noted at Lion DP, Blaenavon, 31st May 1972; noted in yellow livery with no number, Lion DP, Blaenavon, 26th September 1972; rail traffic ceased at Lion DP, June 1975; noted disused, Lion DP, 27th August 1975; to Taylor Woodrow Construction Ltd, NCBOE Cwm Bargoed Disposal Point, 23rd October 1975; noted at Cwm Bargoed, 7th April 1979; to Hargreaves Industrial Services Ltd, NCBOE British Oak Disposal Point, Crigglestone, August 1981; seen in yellow livery with number 12099, Crigglestone, 31st October 1981; to NCBOE Bowers Row Disposal Point, Astley, 11th February 1983; noted at Bowers Row, 26th February 1983; repainted in orange livery at Bowers Row, March 1983; seen in orange livery with no number, Bowers Row, 11th June 1983; disused by August 1988; re-sold to C.F. Booth Ltd, Rotherham, and moved 21st February 1989; seen in orange livery with no number, C.F. Booth Ltd, 26th February 1989 and 25th March 1990; re-sold to Severn Valley Railway, and moved by road to Kidderminster on 26th March 1990; seen in orange livery with no number, Severn Valley Railway, Kidderminster Station, 31st March 1991; seen in black livery with number 12099, Severn Valley Railway, Kidderminster, 6th May 1991;

towed to SVR Bridgnorth, 2nd August 1991; underwent lengthy repairs; to SVR Kidderminster, 15th December 1992; underwent lengthy overhaul including fitting with vacuum braking equipment; undertook first shunt, 1998; noted in black livery, 13th August 2006; appeared in SVR Wartime event, in black livery with temporary number WD40, June 2013; noted in black livery with number 12099, Severn Valley Railway, 30th September 2014; ownership transferred to charitable trust, January 2018; noted in black livery, 5th October 2019; normally used as Kidderminster pilot, 2024.

12119   Darlington   1952   50B   11/68   F   509
new 9th September 1952; withdrawn, 18th November 1968; sold to NCB; despatched from 50B Dairycoates Depot, Hull, 6th February 1969; arrived at NCB Philadelphia Locomotive Shed, County Durham, 7th February 1969; noted at Philadelphia, 14th March 1969; renumbered 509, August 1969; seen at Philadelphia, 21st August 1970; allocated plant number 2233/509; noted with number 509, Philadelphia, 25th October 1971 and 25th October 1978; to Lambton Engine Works, Philadelphia, for repairs, November 1980; to Philadelphia Locomotive Shed, January 1981; noted at Philadelphia Locomotive Shed, 2nd December 1983; put up for sale, July 1985; noted at Philadelphia Locomotive Shed, 20th October 1985; scrapped on site by C.F. Booth Ltd of Rotherham, November 1985.

12120   Darlington   1952   50B   12/68   F   510
new 11th September 1952; withdrawn, 28th December 1968; sold to NCB; despatched from 50B Dairycoates Depot, Hull, 6th February 1969; arrived at NCB Philadelphia Locomotive Shed, County Durham, 7th February 1969; noted at Philadelphia, 14th March 1969; renumbered 510, August 1969; seen at Philadelphia, 21st August 1970; noted with number 510, Philadelphia, 25th October 1971; to Whittle Colliery, Newton-on-the-Moor, June 1978; to Philadelphia Locomotive Shed, June 1979; seen in green livery with number 510, Philadelphia Locomotive Shed, 26th June 1979; to Lambton Engine Works, Philadelphia, early July 1979; used for spares; noted with engine removed, Lambton Engine Works, 5th July 1979; remains scrapped on site by L. Marley of Stanley, March 1980.

12122   Darlington   1952   40B   7/71   F   12122
new 24th September 1952; withdrawn by BR as 'surplus', 4th July 1971; sold to NCBOE; despatched from 40B Immingham Depot, 30th January 1972; to Murphy Bros Ltd, NCBOE Lion Disposal Point, Blaenavon; suffered front end collision damage when hit by runaway wagons, February 1972; noted at Lion DP, 18th June 1974; rail traffic ceased at Lion DP, June 1975; noted disused, Lion DP, 27th August 1975; to Taylor Woodrow Construction Ltd, NCBOE Cwm Bargoed Disposal Point, October 1975; noted disused, Cwm Bargoed, 27th May 1977, 7th April 1979 and 1st June 1980; to Hargreaves Industrial Services Ltd, NCBOE British Oak Disposal Point, Crigglestone, August 1981; used for spares; seen in green livery with number 12122, Crigglestone, 31st October 1981; noted derelict, Crigglestone, 3rd March 1985 and 17th September 1985; remains scrapped on site by Rawden of Barnsley, October 1985.

12131   Darlington   1952   30A   3/69   P   12131
new 18th November 1952; noted in ex-works condition, being hauled from Darlington to March by WD 90608, 19th November 1952; withdrawn, 2nd March 1969; sold to NCB; despatched from 30A Stratford Depot, by rail, March 1969; spent a couple of days at Chart Leacon Depot being serviced and receiving attention to its traction motors; delivered to NCB Betteshanger Colliery, Kent, 25th March 1969; noted at Betteshanger Colliery, 19th April 1969; noted in NCB Kent colours but retaining number 12131, Betteshanger Colliery,

14th June 1969; seen with plant number 1802/B3, Betteshanger Colliery, 15th October 1972; to Snowdown Colliery, Kent, 22nd June 1976; noted at Snowdown Colliery, 23rd June 1976; never worked at Snowdown Colliery, used for spares; seen in blue livery with numbers 12131 and 1802/B3, Snowdown Colliery, 13th September 1979; re-sold to North Norfolk Railway, Sheringham, and moved 25th April 1982; seen in black livery with number 12131, Sheringham, 27th August 1986, 23rd May 1993 and 11th April 2017.

12133   Darlington   1952   40B   2/69   F   511

new 2nd December 1952; withdrawn, 4th January 1969; sold to NCB; despatched from 40B Immingham Depot, 9th May 1969; to NCB Philadelphia Locomotive Shed, County Durham, May 1969; renumbered 511 and 2233/511, August 1969; noted at Philadelphia, 24th May 1970; seen at Philadelphia, 21st August 1970; noted with number 511, Philadelphia, 13th June 1973; to Lambton Engine Works, Philadelphia, 1979; seen in blue livery with number 511, and with a 40B shed plate on cab back, Philadelphia Locomotive Shed, 26th June 1979; noted at Philadelphia Locomotive Shed, 4th December 1980; to Whittle Colliery, Newton-on-the-Moor, about March 1981; to Lambton Engine Works, Philadelphia, 23rd April 1981; noted at Lambton Engine Works, 18th June 1981; to Philadelphia Locomotive Shed, 13th August 1981; noted at NCB Ashington, July 1983; to Lambton Coking Plant, about September 1983; noted at Lambton Coking Plant, 26th October 1983 and 2nd December 1983; to Lambton Engine Works, Philadelphia, 20th July 1985; put on sale, July 1985; noted at Philadelphia Locomotive Shed, 20th October 1985; scrapped on site by C.F. Booth Ltd of Rotherham, 24th to 26th November 1985.

SECTION 21:

British Railways built 0-6-0 diesel electric locomotives, numbered 15211-15236, and introduced 1949. Fitted with an English Electric 6KT engine developing 350bhp at 680rpm, and driving wheels of 4ft 6in diameter. Later classified TOPS Class 12.

15222   Ashford   1949   73C   10/71   F   15222

new 3rd September 1949; withdrawn, 31st October 1971; sold to Cashmore Ltd; despatched from BR Hither Green Depot, May 1972; noted at Acton Yard, 6th May 1972; left Acton Yard, 8th June 1972; arrived at Cashmore Ltd, scrap merchants, Newport, June 1972; re-sold to John Williams Ltd, Blaenyfan Quarry, Kidwelly, 1974; used briefly as a stationary generator; various parts removed and remains lay derelict for four years; remains scrapped on site, September 1978.

15224   Ashford   1949   75C   10/71   P   15224

new 22nd October 1949; withdrawn, 31st October 1971; sold to NCB; despatched from BR Selhurst, October 1972; to NCB Betteshanger Colliery, Kent, October 1972; seen with 'National Coal Board, Kent, Colliery No.1802/B5' on sides, Betteshanger Colliery, 15th October 1972; to Snowdown Colliery, Kent, 27th May 1976; noted at Snowdown Colliery, 21st June 1976; seen in blue livery with numbers 15224 and 1802/B5, Snowdown Colliery, 13th September 1979; withdrawn in 1981; re-sold to North Downs Railway; left Snowdown Colliery, 9th October 1982; arrived at BR Hove Goods Yard (the only member of its class to be preserved), 12th October 1982; noted at Hove Goods Yard, 8th April 1983 and 5th May 1983; to Brighton Works Locomotive Association, Preston Park Car Sheds, Brighton, May 1983; re-sold to Lavender Line, Isfield, East Sussex, and moved June 1985; noted in green livery with no number, Lavender Line, 26th September 1985; noted in green livery with number 15224, Lavender Line, 9th July 1994; to Spa Valley Railway, Tunbridge Wells,

21st January 1998; to Lavender Line, Isfield, 1st February 1998; re-sold to Spa Valley Railway, Tunbridge Wells, and moved about July 2008; noted in green livery with number 15224, Spa Valley Railway, 2nd August 2013 and 1st August 2015; noted at Spa Valley Railway, in store, 6th July 2024.

15231   Ashford   1951   73F   10/71   F   TILCON
new 20th October 1951; withdrawn, 31st October 1971; sold to Tilcon Ltd, Swinden Lime Works, Grassington, and moved June 1972; seen in blue livery with no number (but still fitted with Boxpok wheels), Grassington, 8th June 1981; seen disused, Grassington, 1st April 1983; scrapped on site, February 1984.

SECTION 22:

Ruston & Hornsby Ltd built, 3ft 0in gauge, 4-wheel diesel mechanical locomotive, number ED10, built in 1958 (Ruston's class 48DS). Fitted with a Ruston 4YC engine developing 48bhp at 1375rpm, three speed gearbox, and driving wheels of 2ft 6in diameter. No TOPS classification.

ED10   RH   411322 1958   BSD   2/65   P   ED10 / 11
ex-works, 28th February 1958; used from new in Departmental service by British Railways to push narrow gauge bolster wagons loaded with wooden sleepers into the creosote impregnation chambers, at Beeston Sleeper Works; system closed, early 1965; to Thos. W. Ward Ltd, Sheffield, February 1965; re-sold to Cleveland Bridge & Engineering Co Ltd, Darlington, May 1966; thereafter carried a small plate reading 'Cleveland Bridge & Engineering No. E9'; used by CB&E Co on the Tinsley Viaduct, Sheffield, contract; noted working at Tinsley Viaduct, 19th July 1966; re-sold to Shephard Hill & Co Ltd (contractors), February 1970; fitted with rubber tyres and stabilisers and used on a contract to construct three miles of hover-train track for Tracked Hovercraft Ltd, Earith, Cambridgeshire, from 1971; noted in yellow livery on hover-track, Earith, 17th September 1972; project cancelled in 1973 and site closed 7th September 1974; re-sold to E. Hampton, Church Farm, Fenstanton, St Ives, Huntingdonshire, for preservation, 1975; re-sold to Irchester Narrow Gauge Railway Trust, Irchester Goods Shed, Northamptonshire, and moved 28th September 1987; to Irchester Country Park, Northamptonshire, 8th June 1988; noted in yellow livery, Irchester, 17th June 1990; re-gauged to metre gauge, May 1991; noted in green livery with number ED10 and name EDWARD CHARLES HAMPTON, Irchester, 16th August 1998 and 26th August 2012; noted in green livery with number ED10, Irchester, 21st August 2016; later given number 11.

SECTION 23:

Ruston & Hornsby Ltd built, 1ft 6in gauge, 4-wheel diesel mechanical locomotive, number ZM32, built 1957 (Ruston's class LAT). Fitted with a Ruston 2VSH engine developing 20bhp at 1200rpm, two speed gearbox, and driving wheels of 1ft 4¼in diameter. No TOPS classification.

ZM32   RH   416214 1957   ZJ   3/64   P   ZM32 / HORWICH / 11
ex-works, 12th September 1957; used from new at Horwich Works, Lancashire; to S.E.E.C. Manchester, September 1965; sold to a buyer in British Honduras; the sale was later cancelled and the locomotive was placed in store at Liverpool Docks (a photograph exists

of it at Howitt Bros Ltd, Bootle, in December 1971); re-sold to R.P. Morris, Longfield, Kent, December 1971; to Alan Keef Ltd, Cote, Oxfordshire, 17th April 1973; rebuilt to 2ft 0in gauge; to Narrow Gauge Railway Centre of North Wales, Gloddfa Ganol, Blaenau Ffestiniog (still owned by R.P. Morris), 20th July 1976; re-sold to Michael Strange, 1998; to FMB Engineering, Oakhanger, Hampshire (for repairs and conversion back to 1ft 6in gauge), 6th October 1999; to Uppertown, near Ashover, Derbyshire, for storage, 30th May 2000; to Steeple Grange Light Railway, Wirksworth, Derbyshire, 9th June 2000; noted in BR blue livery with double-arrow logo, SGLR, July 2000; to Whaley Bridge, for overhaul including fitting with air brakes, 24th November 2002; to Steeple Grange Light Railway, Wirksworth, 15th January 2003; re-sold to the SGLR Company, 23rd March 2006; repainted in green livery; seen in green livery with plates ZM32, No.11 and HORWICH, SGLR Wirksworth, 29th June 2023.

SECTION 24:

English Electric Ltd built, 0-6-0 diesel electric locomotives, built 1956. Fitted with an English Electric 6RKT engine developing 500bhp at 750rpm, and driving wheels of 4ft 0in diameter. D0226 was diesel-electric transmission, and D0227 was diesel-hydraulic. These locomotives were tested by British Railways but were never incorporated into capital stock. No TOPS classification.

D0226	EE	2345	1956	- -	12/60	P	D0226 / VULCAN
	VF	D226					

given trials on British Railways from 1956 to December 1960; mainly based at Stratford Depot in London; originally given number D226 to match its works number but renumbered D0226 in August 1959 to avoid clash of number with new Type 4 (later Class 40) locomotive D226; noted at Bristol St Philip's Marsh Depot, 3rd January 1960; withdrawn, 31st December 1960; returned to English Electric Ltd, Vulcan Works, Newton-le-Willows, for storage, January 1961; to Keighley & Worth Valley Railway, Haworth (on permanent loan from English Electric), 18th March 1966; seen at Haworth, 8th January 1967; to BR Doncaster Depot, for tyre turning, 4th March 1979; returned to Keighley & Worth Valley Railway, March 1979; noted in green livery, Keighley & Worth Valley Railway, 30th August 1981; seen in maroon and orange livery with number D226 and VULCAN nameplate, Keighley & Worth Valley Railway, 20th June 1992; seen in blue livery with one horizontal yellow stripe, number D226 and VULCAN nameplate, Haworth, 13th May 2004; to Railfest, York, for display, 26th May 2004; returned to Keighley & Worth Valley Railway, June 2004; noted in green livery with number D0226 and VULCAN nameplate, Keighley & Worth Valley Railway, 30th May 2012 and 24th June 2017; noted in grey livery with number D0226 and VULCAN nameplate, Keighley & Worth Valley Railway, 12th June 2022.

D0227	EE	2346	1956	- -	9/59	F	D0227 / BLACK PIG
	VF	D227					

given trials on British Railways from 1956 to 1960; mainly based at Stratford Depot in London; originally given number D227 to match its works number but renumbered D0227 in August 1959 to avoid clash of number with new Type 4 (later Class 40) locomotive D227; returned to English Electric Ltd, 1960; to EES (ex-Robert Stephenson & Hawthorns Ltd works), Darlington, 1960; noted at Darlington works, 13th August 1964; scrapped soon after.

SECTION 25:

Ruston & Hornsby Ltd built, 4-wheel diesel mechanical locomotives (Ruston's class LB). Fitted with Ruston 3VSH engines developing 31bhp at 1800rpm, and 1ft 4¼in diameter wheels. Ruston & Hornsby built no less than 557 examples of this class, to 23 different narrow gauges, with the two examples recorded below being of 2ft 0in gauge. No TOPS classification.

85049   RH   393325 1956   CJ   c4/86 P 85049
ex-works, 15th February 1956; used from new by British Railways at the extensive Chesterton Junction Permanent Way Materials Depot, Cambridge, which was set up in the 1950s; the 2ft gauge system was used to convey reclaimed track fittings to various parts of the yard in flat wagons and skips; noted in faded green livery with no number, Chesterton, 17th March 1973; noted in yellow livery with number 85049, Chesterton, 11th May 1981; sold to Northamptonshire Ironstone Railway Trust, Hunsbury Hill, and moved there on 2nd August 1986; to Overland Railways, Chidham, near Chichester, for restoration, 1989; to Vobster Light Railway, Holwell Farm, Mells, Somerset, 13th January 1992; to Somerset & Avon Railway Company, Radstock, 25th June 1994; to Derbyshire Dales Narrow Gauge Railway, Rowsley, Derbyshire, February 1999; to Nunckley Narrow Gauge Railway, Rothley, Leicestershire, 15th March 2017; noted in bright yellow livery with number 85049, Rothley, 25th March 2018.

85051   RH   404967 1957   CJ   c4/86 P 85051
ex-works, 7th February 1957; used from new by British Railways at Chesterton Junction Permanent Way Materials Depot, Cambridge (as detailed above); noted in faded green livery with no number, Chesterton, 17th March 1973; noted in yellow livery, Chesterton, 11th May 1981; sold to Cadeby Rectory, Market Bosworth, Leicestershire, and moved there on 3rd July 1986; re-sold to Ashover Light Railway Society, Rowsley South, Derbyshire, and moved 6th May 2006; noted in yellow livery with number 85051, Rowsley South, 3rd April 2019.

SECTION 26:

Ruston & Hornsby Ltd built, 4-wheel diesel mechanical locomotives. Ruston's class 44/48hp (later restyled as class 48DL), fitted with a Ruston 4VRO engine developing 48bhp. Ruston & Hornsby built 147 of the 44/48hp class, and 1,127 of the 48DL class, to forty different gauges. Works numbers 187073, 198284 and 221615 were of 2ft 0in gauge and 224337 of 3ft 0in gauge. None of these locomotives was ever allocated a BR number. No TOPS classification.

-   RH   187073 1938   MQ   ?   F ?
ex-works, 23rd May 1938; delivered to Southern Railway, Meldon Quarry, Devon; spares were ordered for it, for delivery to Meldon Quarry, from 30th November 1938 to 5th May 1950, during which time it passed to ownership of British Railways; disposal not known.

-   RH   198284 1940   MQ   ?   F ?
ex-works, 18th March 1940; delivered to Southern Railway, Meldon Quarry, Devon; spares were ordered for it, for delivery to Meldon Quarry, from 21st May 1941 to 26th January 1950, during which time it passed to ownership of British Railways; disposal not known.

| - | RH | 221615 1943 | MQ | ? | F | ? |

ex-works, 2nd April 1943; delivered to Southern Railway, Meldon Quarry, Devon; spares were ordered for it, for delivery to Meldon Quarry, from 10th June 1944 to 5th April 1950, during which time it passed to ownership of British Railways; exported to Egypt (see Appendix C).

| - | RH | 224337 1944 | LSD | 1964 | P | 06/22/6/2 |

ex-works, 29th August 1944; used from new (named MONTY) by British Railways at the creosoting plant and workshops at Lowestoft Sleeper Depot, Suffolk; works closed, May 1964; sold to dealer A. King & Sons Ltd, Norwich, September 1964; re-sold to Lynite Concrete Co Ltd, Bury Road, Ramsey, date not known; used to transport concrete products from autoclaves to storage sheds; use of railway ceased in 1974; stored for fourteen years; re-sold to J. & K. Harris, scrapyard, Norwood Road Industrial Estate, March, Cambridgeshire, December 1988; seen in green livery with number 06/22/6/2 on side, and 'Chief Civil Engineer Eastern Region' on cab back, J. & K. Harris, 17th April 1993; re-sold to Andrew Wilson, Leeds, 1995; to Green's Industrial Services, Sibthorpe, Nottinghamshire, for storage, 14th June 1995; to Andrew Wilson, Leeds, 27th March 1997; re-sold to Apedale Valley Light Railway, Newcastle-under-Lyme, summer 2022.

SECTION 27:

Ruston & Hornsby Ltd built 0-6-0 diesel electric locomotives (Ruston's class 165DE). British Railways purchased five for Departmental work on the Western Region which were initially numbered PWM650 to PWM654. Fitted with a Ruston 6VPH engine developing 155hp, and driving wheels of 3ft 2½in diameter. Later renumbered 97650 to 97654. No TOPS classification.

| PWM650 | RH | 312990 1953 | 81D | 4/87 | P | PWM650 / 97650 |

97650   ex-works on a low-loader to Swindon Works, 5th January 1953; renumbered 97650, September 1979; it was allocated 'CCE Plant No.83650' but this number (and CCE yellow livery) was never carried; withdrawn, 22nd April 1987; stored at Reading Depot, April 1987; sold to Lincoln City Council, January 1990; moved to Holmes Yard, Lincoln, 28th February 1990; seen in BR blue livery with number PWM650 on cabside and 97650 on bonnet top, on a short length of track in Holmes Yard (outside former BR steam depot), Lincoln, 6th March 1992; to Appleby Frodingham Railway Preservation Society, Scunthorpe, for storage, 4th February 1994; to Lincolnshire Wolds Railway, Ludborough, June 1994; re-sold to Heritage Shunters Trust, Rowsley, and moved 11th January 2017; a major restoration started in 2021; seen in blue livery with number 97650, Rowsley, 24th July 2021.

| PWM651 | RH | 431758 1959 | 86A | 9/98 | P | PWM651 |

97651   ex-works on a low-loader to Swindon Works, 3rd July 1959; noted passing through Newark, 3rd July 1959; renumbered 97651, September 1979; seen in yellow livery with number 97651 and 'CCE Plant No.83651' on cabside, Gloucester Depot, 29th May 1990; seen inside shed, in same livery, Radyr, 20th April 1992; latterly stored at Canton Depot; withdrawn September 1998; sold for preservation; despatched from Canton Depot, Cardiff, to Northampton & Lamport Railway, Chapel Brampton, by road, 10th November 1998; noted in yellow livery with number 97651, 16th June 2002; re-sold to Strathspey Railway, Aviemore, and moved May 2008; noted in yellow livery with number 97651, Strathspey, June 2010; re-sold to Swindon & Cricklade Railway, and moved 14th August

2015; noted in BR green livery with number PWM651, Swindon & Cricklade Railway, 10th February 2019 and 17th September 2022.

PWM653 RH   **431760 1959   81D   8/92   F   97653**
97653   ex-works on a low-loader to Swindon Works, 11th September 1959; noted passing through Newark, 11th September 1959; officially renumbered 97653 in November 1979; seen in yellow livery with number 97653 and 'CCE Plant No.83653' on cabside, Radyr, 26th March 1982; withdrawn, 5th August 1992; later used as a source of spares for 97654 at Reading Depot; noted at Reading Depot, April 1998; sold for preservation; despatched from Reading Depot, 6th November 1998; delivered to Yorkshire Engine Company, Long Marston, November 1998; allocated Yorkshire Engine Company works number L163; acquired by John Payne from receivers of Yorkshire Engine Company, November 2001; locomotive remained at Long Marston; used for spares and stripped down to just the frames; remains to Brian Hirst Recycling, Bullington Cross, Andover, Hampshire, 1st August 2011; scrapped, 2011.

PWM654 RH   **431761 1959   81D   5/97   P   PWM654**
97654   ex-works on a low-loader to Swindon Works, 2nd October 1959; noted passing through Cheltenham, 6th October 1959; renumbered 97654 and 'CCE Plant No.83654' and painted yellow, June 1982; used on works trains during remodelling of the approach to Paddington Station, late 1992/early 1993; repainted (still in yellow, with white tyres), 1995; stored at Reading Depot; transferred for 'departmental service' to Scotland, by road, December 1996; based at Slateford Plant Depot, near Edinburgh; withdrawn, 7th May 1997; sold to Heritage Shunters Trust; despatched from Slateford, early April 2005; arrived at Heritage Shunters Trust, Rowsley (in final BR livery of engineer's yellow), 6th April 2005; used at Rowsley in April 2005; seen with number 97654, Rowsley, 9th March 2007; repainted in BR blue livery with number PWM654, 2009; seen in blue livery with number PWM654, Rowsley, 24th July 2021.

SECTION 28:

Ruston & Hornsby Ltd built, 4-wheel diesel mechanical locomotive (Ruston's class 20DL). Fitted with a Ruston 2VSO engine developing 20hp, and wheels of 1ft 4½in diameter. Ruston & Hornsby built 1,198 of this class, to 37 different gauges, with works number 202005 being of 2ft 3in gauge. This locomotive was never allocated a British Railways number. No TOPS classification.

---   RH   202005 1940   HHC   ?   F   ?
ex-works, 23rd July 1940; the Great Northern Railway established a sleeper depot at Hall Hills, close to Boston Docks in Lincolnshire. Sleepers were manufactured using timber brought in at the adjacent docks. 202005 was supplied to the LNER in July 1940 and was used on an internal 2ft 3in gauge tramway to move timber from the stock pile area to the manufacturing plant. A Ruston works photograph shows it with 'L.N.E.R.' lettering on a protective steel plate in front of the radiator, whilst the locomotive had a special coupling and a canvas sided canopy. It later passed to the ownership of British Railways; the depot closed about 1963 and the locomotive was placed in store; sold to John S. Allan & Son Ltd, Mardyke Works, Cranham, near Upminster, Essex (via Rundle & John Philips & Co Ltd), by May 1967; exported to Singapore (see Appendix C).

LOCOMOTIVE APPENDICES:

APPENDIX A: A.R. ADAMS & SON, NEWPORT

According to a Newport trade directory advertisement of the 1930s the Adams business was established in 1893. The directory describes the firm as 'Engineers, General Smiths and Boiler Makers' and its Pill Bank Ironworks stood at the junction of the west side of Robert Street and Courtybella Terrace, Newport (ST312868). The site was served by a branch of the Tredegar Estates Line running along Courtybella Terrace. The rail-connected locomotive repair shop and 'Adams Yard' were opposite the ironworks (ST 313867) between the east side of Robert Street and Price Street. Adams had acquired this site and its private sidings by 1925, and the previously mentioned directory notes that repairs to marine, locomotive and stationary boilers and fireboxes were major activities at that time. Almost certainly overhauls and sales of locomotives had been a feature of the firm's trade from its early days, although the earliest authenticated locomotive transaction dates from 1920. The company hired out steam locomotives of standard and narrow gauges, plus the practice of hiring, or loaning, a locomotive to a customer (whilst its own engine was overhauled by Adams) seems to have been established early on. The final phase, from about 1968 onwards, saw Adams acquire a small fleet of second-hand 0-6-0 diesel locomotives, mostly surplus ex-BR class 03 and 04 machines. In this 'diesel era' the company hired locomotives to various concerns, which were mainly involved with the coal industry in South Wales. From the late 1960s locomotives under overhaul or off-hire were stabled and repaired in premises (either rented or acquired) previously owned by, or adjacent to, the United Wagon Company on King's Parade (ST319869) in the Town Dock area. Locomotives would be located inside the Old Dock workshops (for example in the cases of long-stored D2276 and the overhaul of King George V) or stored nearby inside a wired-off compound (the first mention of which was in December 1972). One report describes this as: "an 8ft high corrugated iron fence surmounted by barbed wire and quite impregnable!" Certain published reports have referred to this compound as being at Rowecord Engineering Ltd, but it is now understood that the compound was merely adjacent to Rowecord – and that this firm had also rented or acquired part of the former United Wagon Company premises, and had no connection to Adams. In this era locomotives were also sometimes to be found between hires in sidings in the Old Town Dock area. The company's offices and non-railway activities were relocated in the early 1970s to Coomassie Street (ST 315864) near the western end of the famous transporter bridge. In 1977 Adams ceased to use the King's Parade workshops and thereafter relocated its storage compound to Bolt Street Sidings. At this time Adams had four remaining locomotives, D2186, D2193, D2244 and 12054, of which the first three were usually found located at Bolt Street compound, whilst 12054 was out on hire. The Bolt Street compound was abandoned after the first three locomotives were scrapped in January 1981. When 12054 came off-hire in March 1981 it was thereafter kept at the Coomassie Street premises, and marked the end of Adams' locomotive hire business. Most readers who are interested in ex-BR locomotives will be aware of Adams, but unfortunately very few enthusiasts are known to have visited the company's premises. Definitive dated sightings are extremely rare, but known hirings are given below. It will be noticed that there are various gaps in the locomotive histories, and these periods may be explained by locomotives being stored at Adams' premises in Newport, or involving hitherto unknown hirings. A thorn in the side of enthusiasts was Adams' practice of repainting its diesel locomotives in green or blue livery and not re-applying their original

BR numbers. D2186, D2244, D2276 and 12054 are known to have retained their BR numbers in at least the early days of Adams ownership, but blank cab-sides on some locomotives led to some enthusiasts assuming their identities, and in the course of time assumptions took on the status of facts. Much research has taken place by a small group of enthusiasts, to try to untangle the Adams story, but it is accepted that this appendix is a work in progress. Any enthusiasts who visited Adams' premises or saw locomotives out on hire are asked to submit dated sightings and photographs to the author.

D2139: sold to Adams, Newport; moved 8Z21, Gloucester to Newport Alexandra Dock Sidings, 2nd December 1968; delivered to NCB Marketing Department, Gwent Coal Concentration Depot, Newport, 10th December 1968; to Old Dock Works (United Wagon Company), Newport, by March 1969, and noted there (now re-painted light green), 23rd March 1969 and 20th July 1969; to NCB Coal Products Division, Nantgarw Coking Plant, Treforest, late July 1969; noted at Nantgarw, 5th October 1969; to Adams, Newport, for repairs, January 1970; to Nantgarw Coking Plant, Treforest, by 21st August 1970; noted in green livery, Nantgarw, 12th December 1970; re-sold to NCB Coed Ely Coking Plant, Tonyrefail; noted in green livery with number 1, Coed Ely, 30th March 1971; noted on various dates to 1983; now preserved – see main listing.

D2178: sold to Adams, Newport, and delivered direct to Aberaman Colliery Washery, January 1970; a distinctive aid to its identification was that it carried an 84A shed plate (it was allocated to Laira from July 1962 to May 1969); to Wiggins Teape Ltd, Ely Paper Works, Cardiff, 24th February 1970; noted at Wiggins Teape on 28th February 1970, 20th August 1970 (when noted re-painted in green livery), 28th April 1971 (when noted with 84A shed plate), and 23rd August 1971; to Powell Duffryn Fuels Ltd, NCBOE Gwaun-cae-Gurwen Disposal Point, from about March 1972; noted at Gwaun-cae-Gurwen, 18th July 1972; noted being towed by D1612 back to Adams, Newport, 21st July 1972; noted at Adams, Newport, 21st December 1972, 10th March 1973 and 20th May 1973; re-sold to Coed Ely Coking Plant, Tonyrefail (received number 2), May 1974; noted at Coed Ely, 18th June 1974; now preserved – see main listing.

D2181: sold to Adams, Newport; moved 8Z21, Gloucester to Newport Alexandra Dock Sidings, 2nd December 1968; delivered to NCB Marketing Department, Gwent Coal Concentration Depot, Newport, 10th December 1968; noted at Gwent CCD, 23rd March 1969 and 12th July 1969; re-sold to Gwent CCD by 23rd January 1970; later scrapped – see main listing.

D2182: sold to Adams, Newport, and delivered direct to NCB Coal Products Division, Caerphilly Tar Works, 29th November 1968; worked at Caerphilly to February 1969; to Glyn Neath Disposal Point, on hire (or possibly on trial) where noted on 15th March 1969 and 6th April 1969; shortly after re-sold to Sir Lindsay Parkinson & Co Ltd and used at Glyn Neath Disposal Point; noted at Glyn Neath Disposal Point, on various dates between 26th May 1969 and 29th May 1972; re-purchased by Adams, about July 1972, and believed to have moved to Adams, Newport, by 10th March 1973; re-painted in green livery; re-sold to Lindley Plant and moved to Gatewen Disposal Point, New Broughton, September 1973; noted at Gatewen, 11th October 1973; now preserved – see main listing.

D2186: sold to Adams, Newport, and delivered direct to Aberaman Colliery Washery, 8th February 1970; a distinctive aid to its identification was that it carried a Swindon 1962 builders plate; noted in BR blue livery at Aberaman, 9th May 1970 and 21st August 1970; to Adams, Newport, about September 1970; to NCB Tower Colliery, Hirwaun, about January 1971; noted at Tower Colliery, 10th January 1971, 19th August 1971 (still in BR blue livery with BR number) and 28th August 1971; to Monsanto Chemicals, Newport, by July 1972; noted at Monsanto, 17th and 23rd July 1972; returned to Adams, Newport (United Wagon Company), about September 1972; overhauled and re-painted in green livery with no BR number; locomotive in green livery believed to be D2186 was noted at Adams, between 10th March 1973 and 22nd July 1973; to NCB Taff Merthyr Colliery, 23rd August 1973; returned to Adams, 7th December 1973; to Monsanto Chemicals, Newport, 1974 (not seen at Adams, 10th April 1974); noted at Monsanto, 23rd April 1974; returned to Adams; noted at Adams (United Wagon Company), 20th April 1975; noted at Adams, Bolt Street compound, on various dates between 24th August 1977 and 1st June 1980 (in green livery, with no engine); scrapped, January 1981.

D2193: sold to Adams, Newport, and delivered direct to Powell Duffryn Fuels Ltd, NCBOE Coed Bach Disposal Point, Kidwelly, September 1969; a distinctive aid to its identification was that it carried a Swindon 1961 builders plate (on its right hand side only); noted at Coed Bach, 10th January 1970; stored at Mahoney's, Newport, on behalf of Adams, for about three weeks, April/May 1970; noted at Mahoney, 4th May 1970; to Adams, Newport, May to June 1970; painted dark green with no number; to NCB Coal Products Division, Coed Ely Coking Plant, Tonyrefail, June 1970; noted in green livery, Coed Ely, 14th July 1970 and 12th December 1970; to NCB Mountain Ash Colliery, February 1971; noted at Mountain Ash, 31st March 1971; to Adams, about April 1971; noted in green livery at Adams (United Wagon Company), 18th August 1971; to NCB Coal Products Division, Nantgarw Coking Plant, Treforest, about May 1972; noted at Nantgarw, 31st May 1972 and 4th November 1972; to Adams, Newport, by March 1973; noted at Adams, Newport, on various dates between 10th March 1973 and 7th December 1973; noted at Monsanto Chemicals Ltd, Newport, 23rd April 1974; to NCBOE Coed Bach Disposal Point, Kidwelly; noted at Coed Bach, 12th July 1974 and 31st August 1974; to Adams, Newport, April 1975; noted in transit at BR Alexandra Dock Junction sidings, Newport, 20th April 1975; noted at Adams (United Wagon Company), on various dates between 12th October 1975 and 19th October 1977; to NCB Garw Colliery, Blaengarw, late October 1977; noted at Garw Colliery, 29th November 1977 and 20th April 1978; returned to Adams, Newport, 27th September 1978; noted at Adams, Newport (Bolt Street compound), 7th October 1978, 14th October 1979 and 1st June 1980; noted with number 2, 1st June 1980; scrapped, January 1981.

D2244: sold to Adams, Newport, and delivered direct to Monsanto Chemicals Ltd, Newport, August 1970; noted in BR green livery, Monsanto, 17th August 1970; to Adams, Newport, early 1971; re-painted in blue livery; to NCB Coed Cae Colliery, Heol-y-Cyw, near Pencoed, about March 1971; noted in blue livery with no number, Coed Cae, 28th March 1971; noted at Coed Cae on various dates from 30th March 1971 to 2nd September 1971; to NCB Ogmore Central Washery, Ogmore Vale, 13th March 1972; hired for a contract to install a new tippler, Ogmore, 20th March 1972 to 12th May 1972; to Adams, Newport, under its own power, 19th May 1972; to NCB Gwent Coal Concentration Depot, Newport; noted at Gwent CCD, 15th July 1972 and 17th September 1972; to Adams, December 1972; noted (with steel plates over all its windows) in Bolt Street Sidings, Newport, 21st December 1972; noted at Bolt Street on various dates from 1st January 1973 to 1st June 1980 (with no engine and with number 5); scrapped, January 1981.

D2276: sold to Adams, Newport, July 1970; used for spares; noted stored (in BR green livery) inside the Old Dock workshops (United Wagon Company), Pillgwenlly, Newport, on numerous dates between 18th August 1971 and 26th May 1977; scrapped, late May 1977.

12054: sold to Adams, Newport, and delivered direct to NCB Mountain Ash Colliery, 15th September 1971; noted at Mountain Ash on 18th September 1971 and 6th November 1971; to NCB Tower Colliery, Hirwaun, May 1972; noted in green livery, Tower Colliery, 16th July 1972 and 27th February 1973; to BR Canton Depot, Cardiff, for repairs, and noted there on 19th August 1973 and 3rd November 1973; to Adams, Newport, 20th November 1973; re-painted in light green livery; to NCB Mardy Colliery, 9th April 1974; noted at Mardy Colliery on various dates between 9th April 1974 and 15th September 1975; to BR Ebbw Junction Depot, Newport, 6th October 1975; noted undergoing repairs, 11th October 1975; to NCB Mardy Colliery, 23rd October 1975; noted at Mardy Colliery on various dates to 6th July 1979; to BR Canton Depot, Cardiff, for repairs, September 1979; noted on a low-loader, awaiting entry to Canton Depot, 5th September 1979; noted at Canton Depot, 20th September 1979, 14th October 1979 and 17th November 1979; to NCB Mardy Colliery, late November 1979; to Adams, Newport, by road low-loader, March 1981; noted at Adams, Coomassie Street, Newport, on various dates between 12th August 1981 and 11th February 1984; scrapped on site, April 1984.

note: Adams also owned three industrial 0-6-0 diesels: HE 5673 of 1963, ROGIE (DC 2218/VF D47 of 1947), and GWENT (DC 2252/VF D78 of 1948). The last named is preserved at the Mangapps Railway Museum where it has been modified to recreate a Wisbech and Upwell Tramway class 04, using the number 11104, which was not used in BR's actual class 04 numbering sequence. Adams previously had a large number of industrial steam locomotives of standard and narrow gauges.

APPENDIX B: T.J. THOMSON & SON, STOCKTON-ON-TEES

Thomson's business appears to have been established as metal merchants in 1871 with an office in Middlesbrough. Edgar Gilkes joined Thomas J. Thomson & Co in 1881 and Thomson & Gilkes then operated from the Millfield Iron Works at a site or sites in Stockton-on-Tees. Gilkes retired in February 1883. In 1928 the business, now titled T.J. Thomson & Son Ltd, was purchased by H.E.I Turner although the company's name remained the same and, in 1932, it moved to occupy the site of the former Moor Steel and Iron Works which it renamed the Millfield Works. These premises were situated on the west side of the LNER line from Yarm to Norton, a short distance south of Stockton Station. After 1940, traffic was exchanged with the railway company at Phoenix Sidings situated between the main line and the scrapyard. Here a Thomson locomotive collected condemned wagons and locomotives and propelled them into the yard where several sidings spread across the site, including some passing under gantries. Over the years Thomson's scrapped numerous main-line locomotives, including a number of Class 31s and, from the 1970s, many industrial diesels. The disappearance of much heavy industry in the area led to the rundown of the scrapyard from 2014. Most of the internal wagons were scrapped and the remaining locomotives sold to Ed Murray & Sons Ltd in 2015. The company continued to trade in scrap but stopped buying materials for processing on site in December 2016 and the remaining track was auctioned on 1st March 2017. Back in May 1970, Thomson's purchased three ex-BR Departmental locomotives, which were despatched to its Millfield Works from BR Thornaby Depot, having been stored at the latter location from about January 1969. All three were seen at 51L Thornaby Depot on 29th August 1969. A letter in 1970 from the BR Eastern Region Public Relations Officer at York stated that 56, 82, 87 were all officially withdrawn period-ending 27th October 1969. These locomotives were never used at Thomson's works, but were stored for over eleven years until scrapped in October 1981. The trio (which were all 4-wheel locomotives of the maker's class 88DS) were fitted with 88hp engines and diesel-mechanical transmission. When seen at Thomson's works by Adrian Foster on 24th October 1978, the trio were all in green livery with yellow and black striped cab backs and had 'British Railways – North Eastern Region Civil Engineering Department' along the frame sides.

Departmental numbers 82 and 87 (with 56 just visible on extreme left) at the Millfield Works on 24th October 1978. *(photograph by Adrian Foster)*

Departmental number	Builder	Works number	Year built
56	**RH**	**338424**	**1955**

ex-works, Ruston & Hornsby, 3rd February 1955; to stock in March 1955 and initially allocated to Chalk Lane Permanent Way Yard at Hull; transferred to Darlington Depot in 1963 for use by their CCE staff when lifting the Barnard Castle to Stainmore line; moved to Etherley Tip, near Bishop Auckland, at unknown date thereafter; noted working at Etherley, March 1968; stored at Darlington Depot from April 1968; noted at Darlington Depot on 14th April 1968; moved to Thornaby Depot about January 1969; stored in the roundhouse; withdrawn period-ending 27th October 1969; sold to T.J. Thomson & Son Ltd in May 1970; moved to Thomson's Millfield Works, July 1970; scrapped in October 1981.

82	**RH**	**425485**	**1958**

ex-works, Ruston & Hornsby, 4th December 1958; to stock in January 1959 and initially allocated to Dinsdale Welded Rail Depot, near Darlington; noted at Etherley Tip, near Bishop Auckland, in February 1967; transferred to Croft Store Yard, near Darlington, January 1968; noted out of use at Darlington Depot, 14th September 1968; moved to Thornaby Depot, for storage, April 1969; withdrawn period-ending 27th October 1969; sold to T.J. Thomson & Son Ltd in May 1970; moved to Thomson's Millfield Works, July 1970; scrapped in October 1981.

87   RH   463152   1961

ex-works, Ruston & Hornsby, 31st May 1961; to stock on 2nd June 1961 and initially allocated to Geneva Engineer's Yard, Darlington; transferred to Central Reclamation Yard, Darlington, December 1967; noted at Etherley Tip, near Bishop Auckland, on 23rd April 1968; stored at Darlington Depot, from April 1968; noted at Darlington Depot, 18th August 1968; moved to Thornaby Depot, for storage, April 1969; withdrawn period-ending 27th October 1969; sold to T.J. Thomson & Son Ltd in May 1970; moved to Thomson's Millfield Works, July 1970; scrapped in October 1981.

As a matter of interest: fellow Ruston & Hornsby departmental locomotives 83 (432477), 84 (432478), 85 (432489) and 86 (463151) were all also officially withdrawn period-ending 27th October 1969. 83 was scrapped at Walter Heselwood, Attercliffe, Sheffield, whilst 84, 85 and 86 were scrapped during July 1970 at W.H. Arnott Young & Co Ltd, Parkgate, Rotherham.

APPENDIX C: LOCOMOTIVES SOLD ABROAD

The very first confirmed ex-BR locomotives to be exported (but see the mystery locomotive and postscript below) were D3639 and D3649 which were shipped to the West Coast of Africa in March 1970. Thereafter at least a further thirty-two followed up to June 1982. A major player in this trade was the firm of Shipbreaking (Queenborough) Ltd which had yards at Cairnryan Port (Scotland) and in Kent. During this approximately twelve-year period examples of classes 03, 04, 06, 07, 08, 10 and 14 moved abroad, whilst two narrow gauge Ruston & Hornsby locomotives also fall within the remit of this appendix. Once these thirty-four locomotives were abroad it was extremely rare for any to be noted by British enthusiasts and reports giving positive sightings have been few and far between. Several locomotives have *never* been reported during the author's fifty-one years as compiler of the BRD records. Whilst preparing this volume for publication, a concerted effort has been made to update each entry with all known information. This is often sparse and there are many gaps in our knowledge. Anyone who can provide additional information is asked to contact the author.

D2010   Swindon   1958   51L   11/74   F   ?
03010   (from main listing); exported to Trieste Docks, Italy, May 1976; subsequent history not known; believed scrapped.

D2019   Swindon   1958   32A   7/71   F   1
(from main listing); exported to Stabilimento ISA, Ospitaletto, Brescia, Italy, August 1972; noted in yellow with red stripe livery with number 1, in use, Ospitaletto, 5th September 1991; company later known as Acciaierie ISA; noted in yellow with red stripe livery with number 1, in use, Ospitaletto, 30th August 1996; noted in use, May 1997; noted out of use on 30th August 1999, 8th March 2000 and 26th July 2000; subsequent history not known; believed scrapped.

D2032   Swindon   1958   32A   7/71   F   2
(from main listing); exported to Stabilimento ISA, Ospitaletto, Brescia, Italy, August 1972; company later known as Acciaierie ISA; painted in yellow livery, prior to 1987; noted in green livery with number 2, in use, 5th September 1991; later used as a source of spares for D2019; noted out of use, 30th August 1996, 5th June 1998, 30th August 1999 and 1st November 2002; subsequent history not known; believed scrapped.

D2033   Swindon   1958   32A   12/71   F   PROFILATINAVE 2
(from main listing); exported to Siderurgica Montirone SPA, Montirone, Brescia, Italy, August 1972; noted at Montirone, 5th September 1991; noted in green with yellow stripe livery, Montirone, in use, 30th August 1996, 7th June 1998, 30th August 1999, 6th July 2004; subsequent history not known; believed scrapped.

D2036   Swindon   1958   32A   12/71   F   PROFILATINAVE 1
(from main listing); exported to Siderurgica Montirone SPA, Montirone, Brescia, August 1972; noted in green with yellow stripe livery, Montirone, out of use, 30th August 1996 and 7th June 1998; noted in green with yellow stripe livery, being used for spares, Montirone, 6th August 2003; subsequent history not known; believed scrapped.

D2081   Doncaster   1960   31B   12/80   P   03081
03081   (from main listing); exported to Sobermai NV (dealer), Maldagem, near Bruges, Belgium, November 1981; to Tiense Suikerraffinerij, Genappe, Belgium, by 20th July 1991; noted in BR blue livery with number 03081, Genappe, 11th February 1995 and 25th May 1996; noted semi-derelict, Genappe, 13th September 1999; returned to England (see main listing).

D2098   Doncaster   1960   51A   11/75   F   ?
03098   (from main listing); exported to Trieste, Italy, May 1976; a locomotive thought to be D2098 was noted at Acciaierie Rumi, Montello, Italy in December 1990; subsequent history not known; believed scrapped.

D2128   Swindon   1960   82A   7/76   P   D2128
03128   (from main listing); exported, December 1976; to Sobermai NV (dealer) where fitted with a V&M power unit and torque converter, making it a diesel-hydraulic; to Zeebouw-Zeezand, contractor for Zeebrugge outer harbour construction contract (1977 to 1985); carried running number 6G1 whilst on this contract; noted at Gent Depot (Belgium Railways) for tyre turning, 26th March 1989; to Stoomcentrum, Maldegem, May 1989; returned to England (see main listing).

D2134   Swindon   1960   82A   7/76   P   D2134
03134   (from main listing); exported January 1977; to Sobermai NV (dealer) where fitted with a V&M power unit and torque converter, making it a diesel-hydraulic; to Zeebouw-Zeezand, contractor for Zeebrugge outer harbour construction contract (1977 to 1985); carried running number 6G2 whilst on this contract; noted at Zeebrugge, 18th November 1983; noted at Gent Depot (Belgium Railways) for tyre turning, 26th March 1989; to Stoomcentrum, Maldegem, May 1989; noted in blue and white livery, Maldegem, 27th December 1992; returned to England (see main listing).

D2153   Swindon   1960   51L   11/75   F   ?
03153   (from main listing); exported to Trieste, Italy, May 1976; subsequent history not known; believed scrapped.

D2156   Swindon   1960   52A   11/75   P   NPT
03156   (from main listing); exported to Altiforni e Ferriere di Servola, Trieste, Italy, May 1976; noted (in disused condition) at Ferramento Pugliesse, Terlizzi, Bari, 17th May 2006; noted (preserved and plinthed) in light blue livery with red wheels and black buffer beams, no number, Terlizzi, 26th September 2015 and 1st July 2020.

D2157   Swindon   1960   50C   12/75   F   NFT
03157   (from main listing); exported to Trieste, Italy, March 1977; initially stored at, or adjacent to, the premises of Acciaieria Ferriera Adriatica, Trieste Docks; noted at Trieste Docks, June 1977 and August 1978; later rebuilt by IPE Locomotori, Pradelle di Nagarole, Verona; to Trafilierre Carlo Gnutti SPA; this firm operated rolling mills founded in 1947 on two sites in Chiari, Brescia; locomotive use was at the site west of the station on the south side of the line; works closed, 1994; noted in yellow/mustard livery with no number, but with 2157 stamped on crank pins, Chiari, 28th August 1996; noted in yellow/mustard livery with no number, Chiari, 10th May 1997; scrapped on site, 23rd May 1997; works reopened (but no longer rail connected) about 1998.

D2164   Swindon   1960   30A   1/76   F   NFT
03164   (from main listing); exported to Trieste, Italy, March 1977; initially stored at, or adjacent to, the premises of Acciaieria Ferriera Adriatica, Trieste Docks; noted in BR blue livery with number 03164, Trieste Docks, June 1977 and August 1978; later rebuilt at IPE, Pradelle, Nagarole Rocca, Verona; to Trafilierre Carlo Gnutti SPA; this firm operated rolling mills founded in 1947 on two sites in Chiari, Brescia; locomotive use was at the site west of the station on the south side of the line; works closed, 1994; noted in yellow/mustard livery with no number, with D2164 stamped on crank pins, Chiari, 28th August 1996; noted in yellow/mustard livery with no number, Chiari, 10th May 1997; scrapped on site, 23rd May 1997; works reopened (but no longer rail connected) about 1998.

D2216   DC   2539   1955   30A   5/71   F   3
**   VF   D265**
(from main listing); exported to Stabilimento ISA, Ospitaletto, Brescia, Italy, August 1972; company later known as Acciaierie ISA; noted in yellow livery with number 3, out of use, October 1987 and 5th September 1991; noted in yellow livery, being used for spares (no engine), 30th August 1996; noted disused on 5th June 1998, 30th August 1999 and 8th March 2000; subsequent history not known; believed scrapped.

D2231   DC   2555   1956   16C   3/68   F   No.8001
**   VF   D281**
(from main listing); exported to Attilio Rossi of Rome, Italy, unknown date after June 1970; noted on the contract for relaying Milan to Como line, Lissone, 1973; noted at Lissone (positively identified per its worksplate), 29th June 1973; noted in Venice, 11th September 1981; re-sold by Rossi to IPE Locomotori, Pradelle di Nagarole, Verona, 1983; noted in yellow livery with number No.8001 and with 'Ditta Antonini, Verona' on cabside, Verona Depot, 31st May 1986; subsequent history unknown; believed scrapped by May 1997.

D2242   DC   2572   1956   55H   10/69   F   ?
**   RSH   7858**
(from main listing); exported, May 1972; docked at La Spezia Docks, aboard ship 'Anna Luhmann', May 1972; to Feralpi, Acciaierie di Lonato, Brescia, Italy, May 1972; noted (with 2242 stamped on rods) at Feralpi, Acciaierie di Lonato, Brescia, 1976; subsequent history not known; believed scrapped; by September 2015 the works had closed and a new company, Lonato SpA, was operating as a sugar distributor.

D2289   DC   2669   1960   70D   9/71   P   NPT
**   RSHD   8122**
(from main listing); exported, May 1972; docked at La Spezia Docks, aboard ship 'Anna Luhmann', May 1972; to Feralpi, Acciaierie di Lonato, Brescia, Italy, May 1972; per FS

documentation of 1974, it was allowed to work into FS exchange sidings from Acciaierie di Lonato; noted at Feralpi, Acciaierie di Lonato in 1976 and on various dates from 1991 to 2003; noted freshly re-painted in red livery, Lonato, 14th April 2005; noted at Lonato, 31st October 2009 (working), 4th November 2012 and 26th September 2015 (out of use); by September 2015 the works had closed and a new company, Lonato SpA, was operating as a sugar distributor; locomotive re-sold for preservation; loaded onto a low-loader, 6th June 2018; left Lonato, 8th June 2018; returned to England (see main listing).

D2295 DC 2675 1960 70D 4/71 F ?
RSHD 8128
(from main listing); exported, May 1972; docked at La Spezia Docks, aboard ship 'Anna Luhmann', May 1972; destination believed to be Siderurgica Meridionale Stefana Antonio S.P.A. Termoli, Italy; noted under overhaul, Termoli, 1974; Termoli closed and works demolished, 1982; subsequent history not known; believed scrapped.

D2432 AB 459 1960 65A 12/68 F ?
(from main listing); exported to Trieste, Italy, March 1977; initially stored at, or adjacent to, the premises of Acciaieria Ferriera Adriatica, Trieste Docks; noted at Trieste Docks, June 1977 and August 1978; subsequent history not known; believed scrapped.

D2993 RH 480694 1962 70D 10/76 F ?
07009 (from main listing); exported to Trieste, Italy, March 1977; initially stored at, or adjacent to, the premises of Acciaieria Ferriera Adriatica, Trieste Docks; noted at Trieste Docks, June 1977 and August 1978; to Attilio Rossi, Rome, date not known; because it was fitted with air brakes 07009 was suitable for this PW firm to use it on construction trains; subsequent history not known; believed scrapped by May 1997.

D3047 Derby 1954 70D 7/73 F 105
(from main listing); exported, February 1975; to Liberian American Swedish Mineral Co [LAMCO], Tokadeh Mine, Nimba, Liberia, February 1975; noted in orange livery with one horizontal white stripe, 1981; last used 29th November 1981; Nimba mine was worked out by late 1989 and LAMCO was unwilling to invest in further development at Tokadeh, so sold out to Liberian Mining Company [LIMCO] who worked the mine until 1992, when it was closed. Arcelor-Mittal obtained agreement to reopen the Tokadeh mine in 2006, but it took until 2011 before mining recommenced. The locomotives were gathered at the Yekepa shed and workshops for storage after withdrawal. Observed derelict, December 2011; subsequent history not known; believed scrapped.

D3092 Derby 1954 73C 10/72 F 101
(from main listing); exported, early May 1974; to Liberian American Swedish Mineral Co [LAMCO], Tokadeh Mine, Nimba, Liberia; noted in orange livery with one horizontal white stripe, 1981; spare locomotive by 1986; last used 23rd February 1987; extant, but observed derelict, December 2011; noted to carry plate "Overhauled & Modified, Derby 1974"; subsequent history not known; believed scrapped; see further history of the works in D3047 entry.

D3094 Derby 1954 73F 10/72 F 102
(from main listing); exported, early May 1974; to Liberian American Swedish Mineral Co [LAMCO], Tokadeh Mine, Nimba, Liberia, early May 1974; last used 14th June 1980; noted in orange livery with one horizontal white stripe, 1981; used for spares, 1986; observed derelict, December 2011; subsequent history not known; believed scrapped; see further history of the works in D3047 entry.

D3098   Derby   1955   73F   10/72   F   103
(from main listing); exported, early May 1974; to Liberian American Swedish Mineral Co [LAMCO], Tokadeh Mine, Nimba, Liberia, early May 1974; last used pre-1980; used for spares, 1986; observed derelict, December 2011; subsequent history not known; believed scrapped; see further history of the works in D3047 entry.

D3100   Derby   1955   75C   10/72   F   104
(from main listing); exported, early May 1974; to Liberian American Swedish Mineral Co [LAMCO], Tokadeh Mine, Nimba, Liberia, early May 1974; last used 14th March 1982; used for spares, 1986; observed derelict, December 2011; subsequent history not known; believed scrapped; see further history of the works in D3047 entry.

Mystery locomotives: an unidentified D31XX Class 10 locomotive (with no running number but with a Darlington-built worksplate) was photographed at Lissone, Lombardy, Italy, on 29th June 1973 (being used on the relaying of the Milano to Como line) and at Albate Camerlata on 23rd March 1974. It was owned by Attilio Rossi, a company involved with railway infrastructure works, with a headquarters in Rome. It was later observed at Mori, Trento, on 7th September 1981; then (in yellow livery with no number) at San Martino della Battaglia, Lombardy, on 1st January 1990; and at Vicenza, Veneto, on 29th August 1999. Another sighting (date unknown) was at Torricola, Rome. Its identity is not confirmed, but the prime suspect is D3137. In addition, during July 1969 an ex-BR diesel shunter was noted working at Imperia, northern Italy – and it has been suggested that this might be D3193, which was last reported at BR Derby Works in August 1967. Anyone with information is asked to contact the author.

D3639   Darlington   1958   36A   7/69   F   ?
(from main listing); exported, March 1970; to Conakry, Guinea, West Coast of Africa; noted at Conakry, June 1970; used on the construction of the 84-mile long Chemin de fer de Boke railway line, to transport bauxite ore from the mining town of Sangaredi to the port city of Kamsar; dismantled by 1976; scrapped, early 1980s.

D3649   Darlington   1959   36A   7/69   F   ?
(from main listing); exported, March 1970; to Conakry, Guinea, West Coast of Africa; noted at Conakry, June 1970; used on the construction of the 84-mile long Chemin de fer de Boke railway line, to transport bauxite ore from the mining town of Sangaredi to the port city of Kamsar; dismantled by 1976; scrapped, early 1980s.

D9505   Swindon   1964   50B   4/68   F   98 88 202 2202-4
(from main listing): exported about July 1975; noted at Zeebrugge, 9th August 1975; to Sobermai NV (dealer), Maldagem, near Bruges, Belgium; overhauled by Sobermai NV (dealer); noted at Sobermai, 1st May 1976 and 27th June 1976; later re-sold to Suikergroep N.V. Opperstraat, Moerbeke-Waas sugar factory, near Gent; noted at Moerbeke-Waas in 1984, 7th July 1987, 28th November 1992, 7th December 1994, 25th May 1996 and 22nd June 1997; scrapped on site, 1999.

D9515   Swindon   1964   50B   4/68   F   ?
(from main listing): exported to Bilbao, Spain, having been purchased by Cubiertas y MZOV S.A. (CMZ), on or shortly after 16th June 1982, via the agent Aiken Espanoila SA of Madrid; re-numbered P603-03911-001-CMZ; used on the construction of the new RENFE Northern line from Madrid Atocha to Madrid Chamartin; noted at Chamartin, 17th June 1983; following completion of this contract, stored at Chamartin Yard, by May 1984; noted at

Chamartin, 7th February 1986, 29th January 1988 and 2nd April 1988; re-sold to NECSO Entrecanales Cubiertas S.A. by 28th April 1997; scrapped, October 2002.

D9534   Swindon   1965   50B   4/68   F   ?
(from main listing): exported about July 1975; noted at Zeebrugge, 9th August 1975; to Sobermai NV (dealer), Maldagem, near Bruges, Belgium; overhauled by Sobermai NV (dealer); noted at Sobermai, 1st May 1976 and 27th June 1976; a source at Sobermai suggested it was re-sold to an industrial user named Ambrogio, Italy, later in 1976; another source reported it working at a steelworks near Brescia, by May 1997, and extant in June 2003; however, there are no confirmed sightings of this locomotive in Italy and its use there is doubtful. An alternative view exists that this locomotive was used for spares in the overhaul of D9505 and was scrapped at Sobermai sometime after 27th June 1976.

D9548   Swindon   1965   50B   4/68   F   937113106028
(from main listing): exported to Bilbao, Spain, having been purchased by Cubiertas y MZOV S.A. (CMZ), 16th June 1982, via the agent Aiken Espanoila SA of Madrid; re-numbered P-602-03911-002-CMZ; used on the construction of the new RENFE Northern line from Madrid Atocha to Madrid Chamartin; noted at Chamartin, 17th June 1983; following completion of this contract, stored at Chamartin Yard, by May 1984; noted at Chamartin, 7th February 1986 and January 1988; serviced and repainted for further use with CMZ; given new number P-602-03911002-CMZ; noted at Curtis, Galicia, near Coruna, 6th December 1989; the owner became NECSO Entrecanales Cubiertas S.A. on 28th April 1997; noted at Sagrera goods yard, Barcelona, 19th June 1998, with NECSO EC name painted on its side and numbered 937113106028; scrapped, October 2002.

D9549   Swindon   1965   50B   4/68   F   P-601-03911-003-CMZ
(from main listing): exported to Bilbao, Spain, having been purchased by Cubiertas y MZOV S.A. (CMZ), 16th June 1982, via the agent Aiken Espanoila SA of Madrid; used on the construction of the new RENFE Northern line from Madrid Atocha to Madrid Chamartin; noted at Chamartin, 17th June 1983; following completion of this contract, stored at Chamartin Yard, by May 1984; noted at Chamartin, 7th April 1988; serviced and repainted yellow for further use with CMZ; given new number P-601-0-3911-003-CMZ; noted at Santiago di Compostella, 1st June 1990 and 16th July 1991; noted at L'Aldea Amposta Tortosa, near Tarragona, 27th May 1996; the owner became NECSO Entrecanales Cubiertas S.A. on 28th April 1997; to Industrias Lopez Soriano SA, Calle de Miguel Servet, Zaragosa, by 2002; noted at Industrias Lopez Soriano SA, August 2002, 18th January 2003 and 7th May 2003; to Constanti, Tarragona, date not known; scrapped, March 2007.

-   RH   221615 1943   MQ   ?   F   ?
(from main listing – see section 26); exported to Egypt, date not known, but after April 1950; noted derelict at Nag Hammadi Sugar Factory, Egypt, 1982; the Nag Hammadi Sugar factory was built in 1895-1897 by French contractors Cail and Fives and was still in operation in 2019; the locomotive's subsequent history not known; believed scrapped.

-   RH   202005 1940   HHC   ?   F   ?
(from main listing – see section 28); exported to Singapore, date not known but after May 1967; subsequent history not known; believed scrapped.

note: at the time of publishing 9BRD there are some ex-BR locomotives whose disposal was unknown or not proven, but which remain suspects as having possibly been exported. These are: D2002, D2003, D2038, D2039, D2042, D2212, D2264, D2277 and D2278. If any reader has positive information as to the disposals of these nine, particularly if noted in the Queenborough area circa 1970 to 1972, the author will be pleased to receive details.

Shipbreaking (Queenborough) Ltd. exported a number of ex-BR locomotives but D2294 was not one of them. Seen there in the company of the BR version of the Drewry design (class 03) on 13th September 1979, it was scrapped in 1985. *(Adrian Booth)*

APPENDIX D: EX-LMS LOCOMOTIVES

In addition to the ex-BR locomotives in this book, there have also been a number of ex-LMS shunters which fall into the categories covered herein. Although these locomotives are not ex-BR shunters, they are closely related forerunners, and the five known *extant* examples are included here for the sake of interest and historical record. Other examples saw industrial service before being scrapped, whilst others went abroad in World War 2, some surviving to have post-war careers before being scrapped. Acknowledgement is due to E.V. Richards and his book *LMS Diesel Locomotives and Railcars* (RCTS 1996) for information below.

LMS 7050 : built for the LMS at the Dick, Kerr works at Preston, per an order from The Drewry Car Co Ltd of London EC2; allocated works number 2047 / EE 874 of 1934; worksplate shows 160hp; delivered as LMS 7050 to Preston, November 1934; to Agecroft, February 1935; regularly used for dock shunting at Salford; to Air Ministry, Stafford, on loan, August 1940; to Air Ministry, Leuchars, Fife, on loan, October 1940; to LMS Agecroft, August 1941; withdrawn from LMS stock, March 1943; sold to War Department and renumbered 224 and later 70224; to Feltham Central Ordnance Depot, Middlesex, February 1945; renumbered 846 in 1952; to West Hallam WD Depot, October 1953; to

Central Ordnance Depot, Stirling, August 1956; to Bicester Workshops, May 1957; to Central Ordnance Depot, Hilsea, June 1957; to Bicester, for repairs, April 1961; to Elstow, November 1961; to West Moors Army Supply Depot, Poole, April 1964; renumbered 240 in 1968; to RNVR Botley Depot, May 1974; to Sessay, North Yorkshire, mid-1976; to Warminster, 1977; re-sold for preservation and moved to National Railway Museum, November 1979; to Museum of Army Transport, Beverley, about 1984; seen with plate Reconditioned by Baguley S-435 of 1950 and brass nameplates RORKE'S DRIFT, Beverley Museum, 1st June 1985; to National Railway Museum, York, 2003; still at National Railway Museum in 2022.

LMS 7051 : built by the Hunslet Engine Company of Leeds (works number 1697 of 1932) as a demonstrator; underwent trials at Waterloo Main Colliery and on the LMS at Hunslet Lane Goods Yard; returned to Hunslet Engine Company, 30th October 1932; exhibited at British Industries Fair, February 1933; sold to the LMS, May 1933; delivered as 7401 but renumbered 7051 in June 1934; worked from Chester, Agecroft and Crewe South depots; requisitioned by the War Department and given number 27, August 1940; used at Capenhurst; returned to LMS (and renumbered 7051), 7th June 1941; returned to War Department and renumbered 70027, 5th August 1944; returned to the LMS, 2nd June 1945; withdrawn and sold to Hunslet Engine Company, December 1945; repaired at Derby Works, January to May 1946; used by Hunslet Engine Company as a works shunter; to London & Thames Haven Oil Wharves Ltd, Stanford-le-Hope, on hire, October 1949; returned to Hunslet Engine Company in 1951; resumed duties as works shunter; re-sold to Middleton Railway, Leeds, for preservation, September 1960; named JOHN ALCOCK, January 1961; to National Railway Museum, York, 12th January 1979; to Middleton Railway, Leeds, 5th October 1989; seen in black livery with number 7051, Middleton Railway, 11th June 2023.

LMS 7069 : built for the LMS by Hawthorn/Leslie (works number 3841 of 1935); initially worked from Crewe South depot; to Swansea East Dock depot, September 1939; to War Department 1940; dispatched to France as WD18 with British Expeditionary Force, 1st May 1940; abandoned at Nantes in 1940; captured, taken over, and used by German forces; recaptured near Le Mans in June 1944; re-used by the French at a General Reserve Munitions Depot until about 1957; to Chemin de Fer Mamers a St Calais, a privately run public railway in the Sarthe Region, France; renumbered No.7; disused and stored from 1973; to Louis Patry, Paris (equipment dealer) in mid-1974; disused until December 1987; shipped to Poole Docks, November 1987; to Swanage Railway, Dorset, 27th November 1987; to private site at Hamworthy, for overhaul, about 5th April 1991; to East Lancashire Railway, Bury, about December 1994; to Gloucestershire Warwickshire Railway, Toddington, 18th April 1998; to Vale of Berkeley Railway, Sharpness, 1st September 2015.

LMS 7103 : built at Derby Works in 1941 but delivered on-loan to War Department as number WD52, November 1941; later renumbered WD70052; withdrawn by the LMS in December 1942 and purchased by the War Department; shipped to Egypt, about December 1942; used in Egypt as MEF13; used in Tunisia in 1943; to Italy, April 1945; purchased by Italian Railways (FS) in 1946; renumbered 700.001; noted at Savona, Northern Italy, October 1978; withdrawn by FS about 1984; stored out of use at Arquata di Scrivia, near Genova, by April 1985; sold to Cariboni SPA, Colico (track maintenance contractors); noted at Vercelli, Italy, in 1994; to Museo Ferroviario Piemontese Store, Torino, Ponte Mosca Station, Italy, 1998; noted in yellow livery, 3rd August 2019.

LMS 7106 : built at Derby Works in 1941 but delivered on-loan to War Department as number WD55, December 1941; later renumbered WD70055; withdrawn by the LMS in

December 1942 and purchased by the War Department; shipped to North Africa, February 1943; to Tunisian Railways, Tunis, June 1943; to Algeria from November 1943 to March 1944; to Italy, March 1944; purchased by Italian Railways (FS) in 1946; renumbered 700.003; noted at Savona, Northern Italy, October 1978; withdrawn by FS about 1984; stored out of use at Arquata di Scrivia, near Genova, by April 1985; purchased by Transporto Ferroviario Toscana, Arezzo Pescaiola, Italy (who operate the branch line from Arezzo, west of Florence), 1991; overhauled 1991/92; noted in pristine black livery, in use, 16th September 2015; noted in black livery with number 98 83 2700 003-6 I-LFI, 12th October 2017 and 4th August 2019.

APPENDIX E: BOGUS 'EX-BR' LOCOMOTIVES

Over the years various preserved industrial diesel shunters have been given running numbers which purport to be BR numbers. Sometimes these are genuine BR numbers which were once carried by now-scrapped locomotives. Sometimes they are similar to BR numbers. Sometimes non-BR locomotives carry an authentic BR livery. Such locomotives are theoretically not relevant to this book because they are not genuine ex-BR machines. The existence of such bogus locomotives at preservation sites, however, can prove confusing and/or misleading to visiting enthusiasts (as witness many erroneous, although in good faith, captions to photographs posted on the internet) and could prove to be a minefield for future historians. It has been decided, therefore, to provide a basic list of known examples of bogus locomotives, as a matter of historical record, and thus explaining why what appear to be ex-BR locomotives at preservation sites are not in this book's main listings. The locomotives' true identities are also provided. No attempt has (or will) be made to provide detailed histories of these bogus locomotives and a simple list of known relevant numbers is considered to be all that is necessary. The list is supplemented by a few genuine ex-BR locomotives which have carried bogus TOPS-style numbers. Thanks to Alex Betteney, Brian Cuttell, and Robert Pritchard for their input to this list.

01003 (D2956, qv), 02101 (YE 2779), 02641 (HE 2641), 04110 (D2310, qv), 05101 (D2578, qv), 10119 (D4067, qv), 11103 (VF D297), 11104 (VF D78 and VF D297), 11230 (RSHN 7860), 11509 (RH 414304), 11510 (RH 281269), 11517 (RH 458641), 11520 (RH 319824), second 12049 (12082, qv), 12139 (EEDK 1553), 12589 (RSHN 7922), 14021 (D9521, qv), 14029 (D9529, qv), 14901 (D9524, qv), 15097 (MR 1930), 15099 (MR 2026), 97088 (RH 466630), 97649 (RH 327974), D2447 (AB 388), D2650 (HE 9045), D2700 (NB 27426), D2870 (YE 2677), D2875 (YE 2779), D2892 (YE 2782), D2899 (YE 2854), D2911 (NB 27876), D2957 (RH 319290 and RH 512572), D2959 (RH 382824 and RH 384139), D2960 (RH 281269), D2961 (RH 418596), D2962 (RSH 6980), D2971 (RH 313394), D2999 (BT 91), D3058 (RH 458961), DS48 (RH 305306), DS1169 (RH 305302), DS1174 (RH 458959), LMS 7120 (Derby) and Departmental 20 (RH 432479).

In addition 10077 (S 10077), 27414 (NB 27414), D1120 (EEV D1120) and DL26 (HE 5238) have received fictitious BR liveries and/or BR emblems, etc.

1.   D2090 at National Railway Museum, Shildon, on 11th November 2023.

(Adrian Booth)

2.   D2207 at North Yorkshire Moors Railway, Pickering C&W Depot, on 16th May 2009.

(Adrian Booth)

3. D2248 (incorrectly numbered 2243) at National Coal Board, Maltby Colliery, on 18th January 1976. (Adrian Booth)

4. D2332 at National Coal Board, Manvers Main Coal Preparation Plant, on 2nd May 1970. (Adrian Booth)

5.   D2420/06003 at Heritage Shunters Trust, Rowsley, on 1st October 2023.

(Brian Cuttell)

6.   D2511 at Keighley & Worth Valley Railway, Oxenhope, on 22nd July 2000.

(Mark Jones)

7.   D2595 at Ribble Steam Railway, Preston, on 23rd April 2022. (Robert Pritchard)

8.   D2613 at National Coal Board, Brodsworth Colliery, on 2nd May 1971.
(Adrian Booth)

9.   D2738 at National Coal Board, Killoch Colliery, on 29th May 1978.

(Adrian Booth)

10.   D2868 at Barrow Hill Engine Shed, on 21st September 2019.   (Adrian Booth)

11.   D2953 at Heritage Shunters Trust, Rowsley, on 24th July 2021.   (Adrian Booth)

12.   D2996/07012 at Barrow Hill Engine Shed, on 6th April 2019.   (Adrian Booth)

13. D3059/13059 at Bridge of Dun, Brechin, on 16th June 2008. (Robert Pritchard)

14. D3308/08238 at Dean Forest Railway, on 14th July 2021 .(Adrian Booth)

15. D3429/08359 at Chasewater Railway, on 4th September 2021.

(Robert Pritchard)

16. D3551/08436 at Swanage Railway, Dorset, on 9th June 2024. (Robert Pritchard)

17. D3558/08443 at Scottish Railway Preservation Society, Bo'ness, on 5th July 2024.
(Robert Pritchard)

18. D3689/08527 at Flixborough Wharf, on 15th March 2011.
(Adrian Booth)

19. D3819/08652 at Merehead Stone Terminal, Somerset, on 7th September 1994.

(Adrian Booth)

20. D3910/08742 at Barrow Hill Engine Shed, on 21st September 2019.

(Adrian Booth)

21.   D4092 at Barrow Hill Engine Shed, on 16th June 2016.   (Adrian Booth)

22.   D4137 at Great Central Railway, Loughborough, on 3rd October 2021.

(Robert Pritchard)

23.   D3666/09002 at Barrow Hill Engine Shed, on 5th April 2017.   (Adrian Booth)

24.   D4106/09018 at Hope Cement Works, on 15th February 2013.   (Adrian Booth)

25. D8568 at Severn Valley Railway, Kidderminster, on 16th May 2021.

(Robert Pritchard)

26. D9531 at Arnott Young, Parkgate, on 24th November 1968.   (Adrian Booth)

27.   D9537 at Corby Quarries, on 8th November 1975.   (Adrian Booth)

28.   12077 at Midland Railway, Butterley, on 27th August 2011.   (Adrian Booth)

29.　15224 at National Coal Board, Snowdown Colliery, on 8th September 1980.
(Robert Pritchard)

30.　ED10 at Tracked Hovercraft Ltd, Earith, on 17th March 1973.　(Robert Pritchard)

31. ZM32 at Steeple Grange Light Railway, Wirksworth, on 8th September 2022.
(Adrian Booth)

32. D226 at Keighley & Worth Valley Railway, Haworth, on 13th May 2004.
(Adrian Booth)

33.   85049/85051 at BR Chesterton Junction Permanent Way Materials Depot, Cambridge, on 17th March 1973.   (Robert Pritchard)

34.   06/22/6/2 at J. & K. Harris, March, Cambridgeshire, on 21st April 1989.

(Robert Pritchard)

35.   PWM650/97650 at BR Lincoln shed, on 6th March 1992.   (Adrian Booth)

36.   03164 and 03157 (behind) at the closed foundry of Trafilierie Carlo Gnutti, Chiari,
      (Brescia), on 10th May 1997.   (Mark Jones)

37.   D2289 at Heritage Shunters Trust, Rowsley, on 24th July 2021. (Adrian Booth)

38.   12054 at A.R. Adams & Son, Coomassie Street yard, Newport, on 29th August
        1983.   (John Scholes)

39. Adrian Booth (BRD Officer for 51 years, left) and John Wade (BRD locomotives enthusiast and preservationist, right) standing in front of John's D2953 at the Heritage Shunters Trust site at Rowsley, on 24th July 2021.

(collection Adrian Booth)